D1544518

THE
ELUSIVE
MALARIA VACCINE
MIRACLE OR MIRAGE?

THE
ELUSIVE
MALARIA VACCINE
MIRACLE OR MIRAGE?

IRWIN W. SHERMAN

Department of Biology
University of California
Riverside, California

Department of Cell Biology
Institute for Childhood and Neglected Diseases
The Scripps Research Institute
La Jolla, California

ASM PRESS

Washington, DC

Address editorial correspondence to ASM Press, 1752 N St. NW, Washington, DC 20036-2904, USA

Send orders to ASM Press, P.O. Box 605, Herndon, VA 20172, USA
Phone: 800-546-2416; 703-661-1593
Fax: 703-661-1501
E-mail: books@asmusa.org
Online: estore.asm.org

Library of Congress Cataloging-in-Publication Data

Sherman, Irwin W.
 The elusive malaria vaccine : miracle or mirage? / Irwin W. Sherman.
  p. ; cm.
 Includes bibliographical references and index.
 ISBN 978-1-55581-515-8 (hardcover)
 1. Malaria vaccine—History. I. Title.
 [DNLM: 1. Malaria Vaccines—history. 2. Malaria—prevention & control. WC 765
S553e 2009]

 QR189.5.M34S54 2009
 614.5'32—dc22

                       2009003425

Current printing (last digit)
10 9 8 7 6 5 4 3 2 1

In the hope that the children of the world will no longer face the scourge of malaria, I dedicate this book to my grandchildren, Zachary, Jack, and Nathaniel.

Contents

Preface

Cats may be killed by curiosity, but science dies for the lack of it.

G. G. Simpson

Malaria is one of humankind's greatest killers. In sub-Saharan Africa, every 30 s a child, usually under 5 years of age, dies from malaria. The disease afflicts 300 million to 500 million people, and 36% of the world's population, in more than 90 countries and territories, is still exposed to the risk of malaria. If the global population continues to grow as expected, almost 3.5 billion people will be living in areas affected by malaria by 2010. Although most people in Asia and the Americas now live in areas where the risk of malaria is low, serious problems remain in economically underdeveloped areas and countries affected by social disruption. Of great concern is that a surge of malaria infections is occurring in places previously free of the disease. Over the years, the malaria parasite, *Plasmodium*, has become resistant to a variety of medicines, and often the best medicines are either not available or too expensive for millions of people in developing nations. Further exacerbating the problem is the combination of insecticide resistance in the mosquitoes that transmit the parasite and economic constraints. Some scientists believe that a vaccine is our only hope to eradicate the disease.

The hunt for a malaria vaccine has been unfolding for a century. It has been punctuated by periods of intense activity and excitement followed by inaction and boredom. The quest has been enlivened by controversy; there have been wars of words, and errors have been made, sometimes honest but sometimes not. Ideas have been stolen, credit for discovery has gone unacknowledged, and there have been and continue to be intense rivalries with clashes of ego and even scandals. This book chronicles all these aspects of the hunt for a malaria vaccine. It tells of the slow and erratic research, the promises of success and the disappointing failures, and the fierce competition between the microbe hunters who have had as their single goal a protective malaria vaccine able to reduce morbidity and mortality by one of the world's most notorious assassins.

1

Hunting Microbes

A plague descended on the people. "They say that the disease first began south of Egypt, in Ethiopia . . . and then descended into . . . most of the King of Persia's domains. It attacked Athens suddenly, getting its grip, at first on the people of the Piraeus . . . later it advanced up into the city of Athens proper, and the deaths became much more numerous as a result. Those who had escaped the disease, had the most pity on those who were suffering and dying, because they knew what the pain was like and because they were now full of confidence. You see, the disease did not attack the same person twice, at least not so to kill him. So these people . . . held the vain belief that they would never die from any other disease in the future . . ." (507).

The nature of this plague, described by the Greek historian Thucydides in 431 BC, remains a mystery; the consequences of being infected and recovering, however, are clear: those surviving the plague were immune. The word "immune" derives from a time in malaria-ridden Rome when citizens were given freedom from some onerous duty owed to the Empire, but today the dictionary defines it as "protected from disease." For adults, "acute bouts of (malaria) fever are most unpleasant and sometimes fatal . . . A person chronically infected . . . is chronically tired, enfeebled, debilitated, hobbled, shackled. Little gets done. An attack of malaria weakens man, an attack of malaria strengthens man, because he is better able to resist future infections" (30). This fact alone has prompted legions of scientists to hunt for the means to immunize humans artificially—that is, to vaccinate them—against this disease. Before a malaria vaccine was ever dreamed of, however, it was necessary for hunters to trap and identify the disease-causing agent.

The Microbe Hunters

They were bold, persistent, and curious. They peered into a mysterious microscopic universe—worlds never seen before—inhabited with "a thousand different kinds of beings, some ferocious and deadly, others friendly and useful, many of them more important to mankind than any continent . . ." or planet (124). These hunters would grope, fumble, make mistakes, and, at times, rouse vain hopes. When they were able to identify the immensely small assassins, they even tamed some of them so they could be used as wonderful protective weapons: vaccines.

Antonie van Leeuwenhoek (1632 to 1723), an obscure linen merchant from Delft, The Netherlands, was the first of the microbe hunters. During Leeuwenhoek's lifetime, and for a thousand years before, poisonous exhalations from the swamps—for this reason they were called miasmatic fevers—were believed to cause ill health. In his own country "intermittent fever" was associated with the coastal marshes; in neighboring France it was called "paludisme" (literally swamp fever) and in England it was known as "ague" (from the Latin "acuta" used with "febris," meaning a sharp or acute fever). An English contemporary of Leeuwenhoek, Samuel Jeake, described "intermittent fever" in his diary: "About noon exactly or very neer, I was taken with a Tertian ague; cold, bot not Shaking; after it lay in my head from 6 PM till next morning & I sweat very much at night." Two days later he wrote: "a second fit, cold at first and vomiting . . . after an hot fit and sweat much in the night," and 2 days later he wrote again: "a 3rd fit, cold and vomiting, but not so bad as the second . . . this was the last fit" (131).

Leeuwenhoek first encountered magnifying glasses when he was 16 and working in Amsterdam as an apprentice and bookkeeper for a textile merchant, where these lenses were used to count thread densities for quality control purposes. At the age of 20 he returned to Delft, opened a linen-draper shop, and began to use tiny glass lenses ground with great care by his own hands and mounted in a special metal holder to magnify specimens other than textiles. With these microscopes, magnifying objects 300 times, Leeuwenhoek stumbled upon a hitherto unseen world. He turned his lenses on everything he could get ahold of. He described what he saw when he examined the white stuff between his teeth (tartar): tiny fishlike creatures that swam forward then whirled and tumbled. Some moved sluggishly and looked like bent sticks, and others resembled corkscrews. There was a living zoo in his mouth! In 1674 he gave the first accurate description of red blood cells, noting that these were oval in fish and frogs and round in humans. Examining the webbed foot of a frog, he saw the circulation of blood in the capillaries. And, during a bout of diarrhea—a "looseness"—Leeuwenhoek described (in 1681) what he saw with his magnifying lenses in a droplet of his own excrement: "I have sometimes seen animalcules a—moving very prettily. Their bodies were somewhat longer than broad, and their belly, which was flattened was furnished with sundry little paws, wherewith they made such a stir in the medium, and among the globules . . ." (130).

Leeuwenhoek had discovered *Giardia*, a parasite, but exceptional lens grinder and careful observer that he was, he did not make the connection between the swimming "wee beasties" and his own sickness. Leeuwenhoek, in examining his own blood or the blood of those suffering with "intermittent fevers" or ague, might have seen a clear body—an animalcule—within the "red globule," but the connection between it and illness would not have been made. Indeed, at the time of his observations of excrement and blood, no animalcules invisible to the naked eye were known to exist inside the human body. The prevailing thought about disease then, and one that would persist for another 200 years, was that illness resulted from an invasion of the body by miasmas—noxious, toxic vapors. By the middle of the 19th century, however, a time when there were better microscopes than those used by Leeuwenhoek, a profound change in thought concerning disease had occurred. This revolution was largely due to the work of two exceptional scientists, the Frenchman Louis Pasteur (1822 to 1895) and the German Robert Koch (1843 to 1910).

Pasteur, the son of a tanner from the village of Arbois, was a diligent but unexceptional student, who abandoned art to devote himself to science (123, 187). His high school work in chemistry was mediocre, but at the Ecole Normal Supérieure in Paris he achieved high marks in chemistry and physics; there also he learned the importance of applying the experimental approaches of chemistry to solving problems in biology and medicine. During his time as a doctoral student in chemistry, Pasteur used his microscope and discovered that the tiny crystals of tartaric acid existed as mirror images—right and left—much like a pair of hands. Later, when he was appointed Professor and Dean at the Faculty of Sciences in Lille, he was called upon to help beet sugar distillers who were having problems with their fermentations. Applying the microscopic technique used in studying crystals, Pasteur observed that in the healthy vats where sugar was fermented into alcohol there were tiny globules—yeasts—

whereas in the sick vats there were no yeasts but a tangled mass of rods—Leeuwenhoek would have called them sticks—each smaller than a yeast globule. These rod-like bacteria, called bacilli from the Latin "baculum," meaning rod, always produced the acid characteristic of sour milk. Pasteur concluded that fermentation to alcohol was due to live yeast and that the bacilli fermented the sugar to acid. He wrote: "It is those yeasts that my microscope showed me in the healthy beet vats, it is those yeasts that turn sugar into alcohol—it is undoubtedly yeasts that make beer from barley and it is certainly yeasts that ferment grapes into wine—I haven't proved it yet, but I know it" (124). Later, Pasteur would again use his microscope (as well as growth in a nutritious broth) to determine the kinds of bacteria that caused butter to become rancid, milk to go sour, and wine to turn ropy. Pasteur's conclusions flew in the face of the organic chemists who contended that fermentation was a chemical process and that "living ferments" such as yeast were its product, not its cause. In addition, and of equal importance for us even today, Pasteur devised a method to control the offending microbe: pasteurization.

But where did the microbes come from in the first place? Did they arise spontaneously from lifeless matter? Were they the result of noxious vapors (miasmas)? Pasteur placed some meat broth in a flask and boiled it until it was sterile; then he drew out the neck of the flask so that it was formed into the shape of an S. He did not seal the neck, so that air could pass freely into the broth by moving through the twisting neck. The long curving swan-like neck of the flask, however, trapped airborne microbes and prevented them from reaching the broth. In spite of the ease of access of the "vital force"—whatever it was—the flasks remained sterile. If Pasteur tipped the flask, however, so that some of the broth ran into the bend of the S and then back into the flask, the broth became contaminated with microbes. Obviously, more than broth plus air was needed to produce life. His swan-neck flask experiment was convincing proof that a vital force did not exist and that microbes come from other microbes. In this elegant but simple experiment, Pasteur had disproved the centuries-old theory of spontaneous generation.

In 1865, at the request of his former teacher the chemist Jean Baptiste Dumas, now Minister of Agriculture, Pasteur became involved in studies of a disease of silkworms (called pébrine because the sick worms were covered with black spots resembling pepper) which was devastating France's silkworm industry. For 6 years he struggled with pébrine disease; using the microscope he discovered that the spots contained a microbe, and experimentally he was able to pass the disease to healthy silkworms. He triumphantly announced: "The little globules are alive—they are parasites! The little corpuscles are not only the sign of the disease, they are its cause" (124). Pasteur now traveled the countryside showing the farmers how to keep their healthy silkworms away from all contact with mulberry leaves that had been soiled by the sick worms. He had no sooner settled the problem of pébrine than another sickness of silkworms would appear. Now prepared for the problem, he was able to find the microbe responsible for this disease much more quickly. At the age of 45, Pasteur basked in the glory of having saved the silkworm industry of his beloved country, and he boldly imagined a day when it would be in the power of humans to make parasitic diseases disappear from the face of the Earth. Pasteur's research on silkworms provided a transition between studies of fermentation and the diseases of animals, silkworms first and humans later. He would over the next decade carry out studies of chicken cholera, anthrax, childbed fever, and rabies. Some of these studies would lead to the means for control of a contagious disease through preventive measures and vaccines, but not all were easily accepted at the time.

Puerperal (from the Latin "puer," meaning child) fever or childbed fever, a serious form of septicemia contracted by some women during or shortly after childbirth or abortion, serves as a good example of how resistant to change the medical community can be. In 1843, the American physician Oliver Wendell Holmes published *The Contagiousness of Puerperal Fever*, concluding that the disease was frequently carried from patient to patient by physicians and nurses (235). He suggested hand washing, wearing of clean clothing, and avoidance of autopsies by those assisting in birthing to prevent the spread of puerperal fever. His medical contemporaries ridiculed Holmes' conclusions.

In 1844, Ignaz Semmelweis, a physician in the First Obstetric Division of the Vienna General Hospital in Austria who was unaware of Holmes' essay, noted a 16% mortality rate in his Division from childbed fever, a rate very substantially higher than the 2% mortality rate in the Second Division, where midwifery students were trained (348). He also noticed that the fever was rare in women who gave birth before arriving at the hospital. Semmelweis found that doctors in the First Division performed autopsies each morning on women who had died the previous day, but that the midwives were not required or allowed to perform such autopsies. A very clear association between the autopsies and puerperal fever was made when his colleague, Jakob Kolletschka, died of septicemia after accidentally cutting his hand while performing an autopsy. Although Semmelweis did not identify the contagious agent, he ordered all doctors and students working in the First Division to wash their hands in chlorinated lime solution before starting ward work, and again before each vaginal examination. The mortality rate from puerperal fever in his Division fell from 18% in May 1847 to less than 3% in June to November of the same year. Despite these results, he too was treated with skepticism and ridicule. And in 1879, even after Pasteur had identified the spherical bacterium (*Streptococcus pyogenes*) in the blood of women with puerperal fever, concluding that it was the cause of the disease, and warned hospital personnel that they carried the microbe from infected to healthy women, his advice was rejected by the medical community largely because preventive medicine was viewed as a threat to their profession.

While Pasteur was examining spoiled wines and sick silkworms, Robert Koch was studying medicine at the University of Göttingen (54). Koch served during the Franco-Prussian War and then became a physician in the small town of Wollstein. Although isolated from the world of science and the investigations being carried out by Pasteur, the small, serious, and nearsighted Koch had a gift for attention to detail and for simple and ingenious techniques to isolate and characterize microbes. Unlike Pasteur, Koch was not prone to jumping to conclusions, and his pursuit of disease-causing microbes came from a curiosity about medical questions. His initial entry into microbe hunting came during an outbreak of anthrax.

Anthrax is primarily a disease of sheep and cattle, but in humans it can cause skin ulcers or a virulent pneumonia known as woolsorter's disease. As early as 1849, Franz Pollender had observed large rod-shaped bacteria in the blood of anthrax-infected sheep, but he did not publish his findings until 1855. In 1863 Casimir-Joseph Devaine published similar observations and found identical bodies in human skin ulcers. However, none of this evidence was compelling proof that these specific microbes were the cause of anthrax. Indeed, it could have just as easily been concluded that the bacteria were the result of the disease or innocent bystanders. For his 28th birthday (1871) Koch was given a microscope by his wife. He started using the microscope with the same aimlessness of Leeuwenhoek, examining everything he could get his hands on until he examined a drop of blackened blood from anthrax-infected sheep. There he discovered what others had already seen—tangled threads that were alive. Like Devaine, he transferred the blood into rabbits and from the infected rabbits to mice. He was able to grow the bacteria outside the body in the fluid taken from the eyeballs of cattle, and this too could be introduced into clean mice, which then became infected. Koch infected mice in a series of mouse-to-mouse inoculations, and even after 20 passages the bacteria remained true to form and their capacity to cause disease was undiminished. These experiments ruled out the possibility that a poison from the original animal caused the disease in mice, since only an agent capable of multiplication within the body could produce such a result after being diluted 20-fold.

Koch devised a simple but elegant method, a hanging drop. On a glass microscope slide Koch placed a drop of fluid obtained from the center of the eye of a freshly butchered calf, and into it he put a small piece of spleen from a mouse that had died from anthrax. Over this he carefully placed a thicker glass slide with its center scooped out, and then he ringed the surface with petroleum jelly. When the two slides were placed together, they stuck together to form an airtight "sandwich." By quickly turning the glass "sandwich" upside down, Koch was able to suspend the drop over the space of the scooped-out thicker slide. This preparation allowed him to sit for hours with his microscope, examining the contents of the drop and not have it dry out. He saw the anthrax microbes divide and observed the transformation of the rods into beadlike spores. When fresh medium was added to the spores, they transformed into active motile rods. The extreme hardiness of the spores, Koch explained, was the reason for the persistence of the anthrax germ in the soil in contaminated pastures and on the wool itself to cause woolsorter's disease. Understanding the natural history of the disease, Koch suggested, was the means for control of anthrax: proper disposal of the contaminated carcasses.

Convinced that he had solved the riddle of anthrax, Koch visited Germany's leading microbe hunter, Professor Ferdinand Cohn at the University of Breslau, and demonstrated his experiments. Finding the results convincing, Cohn sponsored Koch's paper, "The Etiology of Anthrax. Based on the Life Cycle of *Bacillus anthracis*." With this work, Koch had placed Pasteur's germ theory of disease on a firm experimental footing. Furthermore, he enunciated the essential conditions for assigning a specific microbe unambiguously to a particular sickness, conditions now called Koch's postulates:

1. The parasite must occur in every case of the morbid disease.

2. The parasite does not occur in other diseases.

3. After isolation and repeated growth in pure culture, the parasite is able to produce the same morbid disease when introduced into a healthy animal.

Koch advocated the cultivation of pure strains of microbes, urged abandonment of speculative work, and showed how studies of microbes could aid in the prevention and treatment of a disease.

Pasteur and Koch first met in the summer of 1881 at the 7th International Congress of Medicine held in London. Pasteur, at 59, was at the height of his fame, having published on fermentation and spontaneous generation and having solved the problems of pébrine disease, soured milk, and spoiled wine. At the time, he had carried out preliminary studies on anthrax transmission and was in the process of producing vaccines for chicken cholera and anthrax through attenuation. He could, however, be irritating. He had a flamboyant style in speeches, and he was haughty and often contemptuous of the work of others. Koch, aged 38, demonstrated the plate culture technique that had led him to identify the agent of anthrax, *Bacillus anthracis*. He could also be irritating, being both arrogant and insensitive. For the most part Pasteur ignored Koch's work, but Koch attacked Pasteur's work on vaccines, accusing him of having impure cultures as well as making errors in inoculations. This attack may have resulted partly from an upstart's feeling resentful at being ignored by the grand master and may have been partly a legitimate reflection of differences in techniques, but nationalism also played an important role in their bitter rivalry. France had been humiliated in losing the Franco-Prussian War (1870 to 1871), a war in which Koch was exempt because of nearsightedness; he had volunteered anyway, becoming a physician in a field hospital where he witnessed severe battle wounds and the devastating effects of typhoid fever. In 1868 Pasteur had suffered a massive cerebral hemorrhage. Initially he lost the ability to speak or to use the limbs on the left side of his body, but gradually his speech and powers of thought returned. He never regained the use of his left hand, however. The construction of his Paris laboratory was interrupted by the war. His only son, Jean-Baptiste, had enlisted and was gravely ill with typhoid. During the siege of Paris, the Pasteur family was forced to flee, first to Geneva and then to Lyon. The war affected everything Pasteur held dear—family, country, and science—yet he was physically unable to fight back. He became obsessed with a hatred of Germans and their nation. He swore: "each of my studies will bear the epigraph: Hatred to Prussia, vengeance!" (123, 187).

This trio of microbe hunters—Leeuwenhoek, Pasteur, and Koch—differed in significant ways: Leeuwenhoek was what we would call today an amateur, a curious observer of "animalcules"; Pasteur, trained as a chemist and more interested in broad philosophical questions and larger scientific issues than specific medical problems, established Leeuwenhoek's animalcules as agents of disease (the germ theory); and Koch, educated as a physician, unequivocally assigned a specific animalcule to a particular medical malady. Despite their bitter antagonism, the studies that Koch and Pasteur carried out were complementary and would have far-reaching consequences for malaria. Koch would make his major contributions to hygiene and public health measures for malaria control and, through his associates, to protection through chemotherapy (Paul Ehrlich) and serum therapy (Emil von Behring), while Pasteur would make primary medical contributions to immunization through microbe attenuation.

Laveran's Animalcule

The antiquity of human malaria is reflected by the records in the Ebers Papyrus (ca. 1570 BC), in clay tablets from the library of Ashurbanipal (2000 BC), and in the classic Chinese medical text the Nei Ching (AD 100). These describe the typically enlarged spleen, periodic fevers, headache, chills, and fever. Malaria probably came to Europe from Africa via the Nile Valley or because of closer contact between Europeans and the people of Asia Minor. The Greek physician Hippocrates (ca. 460–ca. 370 BC) discussed in his *Book of Epidemics* the two kinds of malaria: one with recurrent fevers every third day (tertian) and another with fevers every fourth day (quartan). He also noted that those living near marshes had enlarged spleens. Although Hippocrates did not describe a malignant kind of tertian malaria in Greece, there is clear evidence of the presence of this malaria in the Roman Republic by 200 BC. Indeed, the disease was so prevalent in the marshland of the Roman Campagna that the condition was called the Roman fever. Since it was believed that the fever recurred during the sickly summer season due to vapors emanating from the marshes, it was called by the Medieval Italian name mal' aria, literally, "bad air." Over the centuries malaria spread across Europe, reaching Spain and Russia by the 12th century; by the 14th century it was in England. Malaria was brought to the New World by European explorers, conquistadors, colonists, and African slaves (442).

Henry the Navigator of Portugal (1394 to 1460) inspired the passion for exploration of Africa and Asia. In 1442 two of Prince Henry's ships brought back with them a dozen Africans whom they had captured on the west coast of Africa during an unprovoked attack on an African village. From this time onward, contact with the remainder of the world increased. Initially, the Portuguese had little inclination to venture further into the interior because there seemed to be little of commercial value, but after 1482 Portugal sent an expedition to obtain gold that was available in the villages that lined the Bight of Benin. By the 16th century 10% of the world's gold supply was being sent to Lisbon. In the early 17th century the English attempted to obtain the riches of Africa's interior, but their efforts failed largely as a result of malignant fevers. With the settling of the Americas, however, a demand was created for another African commodity, one more valuable than gold—slaves.

Slaves became an important part of the American economies, which were heavily dependent on crops such as cotton, sugar cane, and tobacco. Slaves were also needed because the local labor force, the Amerindians, had died from European diseases, and the Europeans themselves were highly susceptible to the tropical diseases endemic to the Caribbean. Africans, on the other hand, were relatively immune to tropical diseases such as yellow fever and malaria, and so they could work in the Americas. The Spanish kings initially employed the Portuguese as intermediaries for obtaining their slaves, but as their conquests in the West Indies and mainland America increased, so too did the demand for slaves, and they inaugurated a system of special contracts under which they bestowed a monopoly on chosen foreign nations, corporations, and individuals to supply slaves for their American possessions. Soon other European countries including Britain, Holland, and France were in competition with one another for these sources of human flesh. Though the French were also one of the players in the slave trade, when the Napoleonic Wars began (in 1792) and Britain blockaded France and other trading nations, it was primarily the British ships that moved as many as 50,000 slaves per year. The Atlantic slave trade reached its peak between 1740 and 1790, and it is estimated that from 1515 to 1870 Europeans brought 11 million people from Africa to the Americas on slave ships.

Many of the early expeditions to Africa failed due to deadly fevers, especially malaria, and those who came down with the disease and did not die suffered greatly: "I wanted to sit up, but felt that I didn't have the strength to, that I was paralyzed. The first signal of an imminent attack is a feeling of anxiety, which comes on suddenly and for no clear reason. Something has happened to you, something bad. If you believe in spirits, you know what it is: someone has pronounced a curse, and an evil spirit has entered you, disabling you and rooting you to the ground. Hence the dullness, the weakness, and the heaviness that comes over you. Everything is irritating. First and foremost, the light; you hate the light. And others are irritating—their loud voices, their revolting smell, their rough touch. But you don't have a lot of time for these repugnances and loathings. For the attack arrives quickly, sometimes quite abruptly, with few preliminaries. It is a sudden, violent onset of cold. A polar, arctic cold. Someone has taken you, naked, toasted in the hellish heat of the Sahel and the Sahara and has thrown you straight into the icy highlands of Greenland or Spitsbergen, amid the snows, winds, and blizzards. What a shock! You feel the cold in a split second, a terrifying, piercing, ghastly cold. You begin to tremble, to quake to thrash about. You immediately recognize, however, that this is not a trembling you are familiar with from earlier experiences—when you caught cold one winter in a frost; these tremors and convulsions tossing you around are of a kind that any moment now will tear you to shreds. Trying to save yourself, you begin to beg for help. What can bring relief? The only thing that really helps is if someone covers you. But not simply throws a blanket or quilt over you. The thing you are being covered with must crush you with its weight, squeeze you, flatten you. You dream of being pulverized. You desperately long for a steamroller to pass over you. A man right after a strong attack . . . is a human rag. He lies in a puddle of sweat, he is still feverish, and he can move neither hand nor foot. Everything hurts; he is dizzy and nauseous. He is exhausted, weak, and limp. Carried by someone else, he gives the impression of having no bones and muscles. And many days must pass before he can get up on his feet again" (265).

The human losses and severity of the symptoms of malaria provoked horror as well as an intense interest on the part of the Europeans. The Europeans were quite aware that the native Africans were comparatively free of this disease. The nature of their immunity, however, was unknown, and some of the European whites feared that if it was survival of the fittest they would eventually be replaced by the better-protected blacks. Such a thought was not very comforting to the whites, and so they concocted fanciful explanations. One was that the Africans had a greater facility for sweating and that this threw off the noxious vapors. Another was that black Africans were spared because of the "offensive odors from their persons" or because of their thick skins.

The Europeans eventually overcame the "immunity problem" by means of drug therapy (and by force). Although the disease agent responsible for malaria would remain unknown for many years, a break-through in treatment—a triumph of empiricism—occurred in 1640 with the discovery of the fever bark, called by the Indians of Peru "kin-kina." The Jesuits brought the drug to Europe. Despite its successes in curing malaria, however, there were failures because samples of bark varied greatly in potency. In 1820, though, when the cinchona alkaloids (which were the active principle) were isolated and crystallized by two French chemists, Pierre Pelletier (1788 to 1842) and Joseph Caventou (1795 to 1887), and named quinine, it could be prescribed in a powder form of known strength, and its reliability increased. By 1854 quinine powder was generally available. With quinine it was possible for Europeans to enter Africa and live without fear of death. Quinine therapy combined with the introduction of the breech-loaded rifle gave the Europeans increased confidence, and exploration moved into high gear with a move toward the Rift Valley. Not only was East Africa strategically placed, but it also had great plains, immense game, and snow-capped mountains. The British of India were encouraged to spend their leaves there, and a new brand of intrepid explorer was born. These explorers were of the safari type, and their adventures could be recorded by sketches and camera. Big-game hunting and ivory trading became the rage. Missionaries in increasing numbers came to Africa, and they opened the continent even more.

The period of European colonial expansion, termed by some writers "the scramble for Africa," began in the 1880s. European powers were able to pick up huge parcels of land cheaply and with little risk from armed resistance since they now possessed two critical resources: quinine and firearms (repeating rifles and machine guns). In 1884 Germany laid claim to Togo, Cameroon, Tanganyika, and Namibia. The British, who already had a presence in West Africa (Gambia and Nigeria), secured land around Mombassa and Berbera to ensure a safe sea route to their outposts in India, and then Germany proclaimed sovereignty over Dar-es-Salaam (present-day Tanzania). Great Britain formalized occupation of the Guinea Coast (an old geographic designation for the West Coast of Africa that now stretches from Senegal to Angola), and the French, who were already in Algeria, established themselves on the right bank of the Congo. The Italians, who entered the competition late with, as Bismarck put it, "such a large appetite and such poor teeth," had taken out a small bite of East Africa, and both Portugal and Belgium occupied land in Africa. In this way the delicate balance of power in Europe was preserved, but power could be expressed by European nations in the less sensitive area of colonial lands.

The most common experience with malaria was that it was a sickness of wetlands, hence the French name "paludisme" (from the Latin "palus" meaning swamp) (212, 469). The disease was generally believed to result from human exposure to the poisonous vapors (miasma) that had seeped into the swamp water from the soil or to be acquired by drinking that water. Indeed, this seemed quite plausible since, in 1834, almost all of the 120 French soldiers who had embarked on the ship *Argo* in Bone, Algeria, bound for Marseille suffered from malaria and 13 died; this was attributed to the ship having taken on drinking water drawn from a swamp. But from 1866 onward—a time when Pasteur's germ theory was in full flower—there were various reports of a sighting of the causative agent of malaria. Of particular interest was one in 1879 by two investigators in Italy, Edwin Klebs and Corrado Tommasi-Crudeli, who found a bacillus in the mud and waters of the Pontine marshes in the malarious Roman Campagna. When cultured on a fish gelatin medium, the bacteria developed into long threads, and when lymph was added, spores developed. When the bacteria were injected into rabbits, fever resulted, the spleen was enlarged, and the same bacteria could be reisolated from the sick rabbits. They named the bacillus *Bacillus malariae*. The finding, quite naturally, attracted worldwide attention. Lending credence to their claim were the considerations that the studies had satisfied Koch's postulates and that both Klebs and Tommasi-Crudeli were well-respected scientists. Klebs held the Professorship of Pathology in Prague and had conducted research on the relationship of bacteria to disease with Koch; Tommasi-Crudeli, who had studied with the eminent pathologist Rudolph Virchow in Berlin, was Director of the Pathological Institute in Rome with a specialty in malarial fevers. Initially, supporting evidence for *B. malariae* as the agent of malaria was forthcoming. In 1880 Ettore Marchiafava, a loyal first assistant to Tommasi-Crudeli, and Giuseppe Cuboni reported isolation of the same bacillus from the bodies of patients who had died from malaria, and when studies in Klebs' laboratory found that quinine killed *B. malariae*, there appeared to be little doubt that malaria was a bacterial disease.

In the United States, the U.S. Board of Health commissioned Major George Sternberg, a well-trained bacteriologist, to try to repeat the experiments of Klebs and Tommasi-Crudeli in the area around New Orleans that was affected by malaria. He found bacteria similar to *B. malariae* in the mud from the Mississippi delta; however, after the bacteria were injected into rabbits, the fevers produced were atypical of malaria in that they were not periodic. Sternberg was also able to cause the disease in rabbits by injecting them with his own saliva. He concluded that the disease was septicemia, not malaria, and suggested that the bacteria were contaminants. Further questions were raised when Patrick Manson, while working as a medical missionary in China, showed that mosquitoes carry filariae—the roundworms that cause the disease elephantiasis—but was unable to cultivate *B. malariae* from the blood of patients with tertian malaria by using sterilized water from malaria marshes. Others suggested that the germicidal effect of quinine was not specific. This lack of reproducibility certainly acted to undermine the belief that malaria was a bacterial infection; more telling, however, was the discovery made by an obscure French military physician, Charles-Louis Alphonse Laveran, working in malaria-ridden North Africa.

Laveran (1845 to 1922) began hunting the malaria parasite after he was posted to Algeria in 1878 (433). He was then 33 years of age and was following in his father's footsteps as an Army doctor and Professor at the Ecole du Val-de-Grace in Paris. Earlier Laveran had studied at the Public Health School in Strasbourg, where he received his medical degree (in 1867). Strasbourg provided the best training in microscopy of all the French medical schools, and Laveran became an accomplished microscopist. Following residency in the Strasbourg civil hospitals, he served as an ambulance officer during the Franco-Prussian war and was involved in the battles at Gravelotte and St. Privat and in the siege of Metz. At Metz he experienced the horrors of septic and gangrenous wounds, dysentery, and typhoid. After the capitulation of Metz and his detention by the Germans, he was allowed to return to France, where he was attached to hospitals in Lille and Paris. In 1874 after a competitive examination, he was appointed Chair of Military Diseases and Epidemics at the Ecole du Val-de-Grace, a position previously established for and occupied by his father. At Val-de-Grace Laveran became well acquainted with several physicians who had worked in the French territory of Algeria studying the epidemiology, quinine therapy, and pathology of malaria. One in particular, L. F. Achille Kelsch (1841 to 1911), a pathologist who had studied malaria pigment, exercised a profound influence on the young Laveran.

One of the signal characteristics of malaria in cadavers is the enlarged and blackened spleen and liver. This discoloration is the result of the accumulation of a brownish-black pigment first described in 1716 in the spleen by the Italian Giovanni Lancisi. Later (1831), in *Reports of Medical Cases*, the English physician Richard Bright noted that in cadavers of patients with periodic fever the brain was the color of black lead. In 1847 the German pathologist-psychiatrist Heinrich Meckel, after observing the accumulation of pigment in the blood of a malaria patient with an enlarged spleen, proposed that the pigment itself, called hemozoin, was the cause of the disease. In the 1870s, Kelsch's attention as a pathologist had been drawn to the pigment, and like Meckel he observed that hemozoin was almost always contained in the white blood cells but occasionally was enclosed in a clear body. He provided no further description; however, in hindsight it is evident that he had seen the pigment-bearing malaria parasite. In the main, during his time in Algeria Kelsch studied malaria in autopsy material but on occasion was able to examine blood from patients living with malaria, and in almost all cases there was the telltale pigment. Although he recognized the presence of pigment as diagnostic for malaria and even found that the pigment appeared in the blood at the time of fever, Kelsch never discovered the causal agent itself. Why? As a pathologist Kelsch looked at only dead material, and even freshly drawn blood would routinely be fixed before undergoing microscopic examination, hence his failure. Kelsch lacked what his fellow Frenchman Pasteur described as being essential to discovery, i.e., "in the fields of observation, chance favors only the mind which is prepared." Two years after arriving in Algeria, Laveran satisfied Pasteur's aphorism.

Laveran has been described as "bespectacled with sharp features and a small trim beard" (311). He was reputed to be extraordinarily precise, meticulous, singularly sharp-minded, incisive, and self-opinionated. In short, he did not suffer fools gladly. Initially, Laveran spent much of his time looking at autopsy material but also examined fresh specimens. His microscope was not a good one and he used no stains, but he was patient and determined. On 6 November 1880 while examining a drop of unstained fresh blood from a feverish artilleryman, he saw several transparent mobile filaments—flagella—emerging from a clear spherical body (433). He recognized that these bodies were alive and that he was looking at an animal, not a bacterium or a fungus. Subsequently he examined blood samples from 192 malaria patients; in 148 of these he found the telltale crescents. Where there were no crescents, there were no symptoms of malaria. Laveran also found spherical bodies in or on the blood cells of those suffering with malaria. Remarkably, and in testimony to his expertise in microscopy, Laveran's discovery was made using a microscope with a dry lens giving a magnification of only 400 diameters. He named the parasite *Oscillaria malariae* and communicated his findings to the Société Médicale des Hôpitaux on

24 December 1880. The drawings in his paper provide convincing evidence that without use of stains or a microscope fitted with an oil immersion lens, Laveran had seen the development of the malaria "animalcule." Laveran was anxious to confirm his observations on malaria parasites in other parts of the world, and so he traveled to the Santo Spirito Hospital in Rome, where he met with two Italian malariologists (one of whom was Ettore Marchiafava, Tommasi-Crudeli's assistant, and the other was Angelo Celli, Professor of Hygiene) and showed them his slides. The Italians, whose chief interest was *B. malariae,* were unconvinced and told him that the spherical bodies he had seen were nothing more than red blood cells that were degenerating as a result of *B. malariae* or some other agent.

In 1882 Eugène Richard, a French military surgeon working in Algeria, described seeing the parasites described by Laveran, although he believed that they were within the red blood cell and not on the surface, as had been claimed by his colleague Laveran. Despite this confirmation, the vast majority of the medical world remained skeptical of Laveran's findings. In 1883 Marchiafava and Celli claimed to have seen the same bodies as described by Laveran but without any pigment granules. They also denied the visit by Laveran 2 years earlier. The Italians were unsuccessful in growing the parasites outside the body of the malaria patient. The lack of consensus on the causative agent was not simply due to differences in interpretation of what was seen with the microscope, it was a matter of focus: the Italians emphasized the smallest forms in the blood, whereas Laveran concentrated on the whiplike filaments. While the search for the parasite itself occupied most research during this period, there was a serendipitous finding of great significance: malaria was an infectious disease, but not one that could be contracted by simply being exposed to a patient with a fever, as would be the case in persons with influenza. In 1884, Carl Gerhardt deliberately induced malaria for therapeutic purposes in two patients with tertiary syphilis by injection of blood from another patient suffering with intermittent fever, and then cured them all with quinine. A year later, Marchiafava and Celli, working on the wards of Rome's Santo Spirito Hospital, gave multiple injections intravenously and subcutaneously to five healthy subjects. Parasites were recovered from three of the five who came down with malaria; all recovered after quinine treatment. Clearly, it was the blood of a malaria patient that was infectious, not his breath.

Staining of microbes and tissues became popular in the 1880s, largely due to the studies of Paul Ehrlich in Koch's Berlin laboratory. Ehrlich introduced at least 12 aniline dyes as stains, and especially useful was methylene blue for coloring bacteria and live nerve fibers. Ten years later, after returning from Egypt where he had gone to recuperate from tuberculosis and knowing that methylene blue stained malaria parasites in the blood, he administered the dye to two patients who had been admitted to the Moabite Hospital in Berlin suffering from mild malaria. Both recovered, and although later methylene blue was found to be ineffective against the more severe forms of malaria, this was the first instance of a synthetic drug being used with success against a specific disease. The advances in staining techniques allowed malaria parasites to be more easily seen under the microscope; it was not until late 1884 or early 1885, however, that Marchiafava and Celli abandoned their use of fixed stained smears and began to study fresh blood as had Laveran. Examining drops of liquid blood, they observed the ameba-like movements of the parasite within the red blood cell, and hence they called it *Plasmodium* (from the Latin "plasmo" meaning "mold"). They also witnessed emerging whiplike filaments (called flagella) from the clear spherical bodies within the red blood cell, although they questioned the significance of the flagella in the disease. The differences between the interpretations of Laveran and of Marchiafava and Celli are now evident: for several years the Italians examined only dried stained specimens and so did not see any movement of the parasite that had caused Laveran to give it the name *Oscillaria* (212). Using both fixed and liquid preparations, Marchiafava and Celli were able to trace the development of the small nonpigmented bodies within the red cell; they also described the hemozoin as the by-product of the destruction of the red cell's hemoglobin by the growing parasite. In 1886 during a visit to Europe, Major George Sternberg visited Celli at the Santo Spirito Hospital. Celli drew a drop of blood from the finger of a malaria patient and was able to show Sternberg the ameba-like movement of the parasite and the emergence of flagella. Sternberg returned to the United States and, working with blood taken from a malaria patient in the Bayview Hospital in Baltimore, was able to find Laveran's parasite in William Welch's laboratory at Johns Hopkins University. A year later Welch separated the two kinds of malarias with 48-h fever peaks; one would be named *P. vivax* and the other he named *P. falciparum* because it had sickle-shaped crescents (and "falcip" in Latin means "sickle or scythe").

Beginning in 1885, Camillo Golgi (1843 to 1926) of the University of Pavia, convinced by the Italian confirmation of Laveran's observations, felt that malaria parasites deserved further study (269). In 22 patients with a 72-h fever cycle (whose agent was later named *P. malariae*), he traced the tiny, unpigmented bodies of Marchiafava over 3 days until they grew to fill the red cell, and on the day of the fever paroxysm he found the pigment to concentrate in the center of the parasite as it divided. Golgi discovered that the parasite reproduced asexually by fission and correlated the clinical course of fever with destruction of the red blood cell to release the parasite. In 1886, when he noted that in both the 48-h and the 72-h fevers there were no crescents, he effectively had distinguished the three kinds (species) of malaria based on fever symptoms. Golgi received the Nobel Prize in 1906, though not for his malaria work but rather for his descriptions of the microscopic anatomy of the nervous system with the silver impregnation staining technique. Golgi's studies of malaria were taken a step further by Marchiafava and his student Amico Bignami. Cases of 48-h and 72-h fever cycle malarias occurred throughout the year in Italy, whereas in the autumn and summer they were outnumbered by a much more severe 48-h type, called aestivo-autumnal or malignant malaria. Later, it would be recognized that the malignant disease was caused by the malaria parasite with crescents, *P. falciparum*.

Clinicians in the United States, however, continued to be skeptical of the significance of Laveran's discovery to malaria as a disease. William Osler, the premier blood specialist of his day, on hearing a paper presented at the inaugural meeting of the American Association of Physicians (June 1886) by W. T. Councilman, working in Welch's Johns Hopkins laboratory, in which he found flagellated parasites in 80 attempts, challenged his findings. Osler also questioned the role of the flagellated bodies discovered by Laveran, finding them improbable and contrary to all past experience of flagellated organisms occurring in the blood. By late 1886, however, after verifying the existence of the parasites with his own eyes—and postponing his Canadian vacation to examine the blood of every malaria patient he could find—Osler became a convert to the doctrine of "Laveranity." Osler published his findings on 28 October 1886 and in his 1889 treatise *Hematozoa of Malaria*. Osler's paper was read by Henry VanDyke Carter, a pathologist working at the Grant Medical College in India. Previously, VanDyke Carter had been unable to find the malaria parasite in the blood, but with Osler's guidance he succeeded. Carter published his findings of three kinds of malaria parasites in India, but his report received little notice for a decade among his colleagues in the Indian Medical Service. Despite this, malaria parasites were now being identified elsewhere: by Metchnikoff in Russia, Morado and Coronado in Cuba, Anderson in Mauritius, and Atkinson in Hong Kong. By 1890 almost all the world believed in both the existence of Laveran's animalcules and their being the cause of the disease malaria. The significance of Laveran's observation of the release of motile filaments (flagella)—exflagellation—would remain unappreciated, however, until William MacCallum and Eugene Opie of Johns Hopkins University made some seminal observations (212).

In 1873 the railroad magnate Johns Hopkins died, leaving $3.5 million to found a university and hospital in Baltimore, MD. It was to be modeled after the great German universities where both research and teaching were emphasized. In 1876 Hopkins opened with a laboratory, and a medical school was added later (1893). One of the earliest members of the Hopkins laboratory was William Henry Welch. Welch had trained in Europe, where he studied with some of the leading scientists including Ferdinand Cohn, and learned bacteriology from several of Koch's protégés. However, Welch's genius was not restricted to the laboratory bench but was also evident in his exceptional knowledge and an instinct for judging those with scientific talent. Welch recruited to Hopkins, among others, the clinician William Osler and the malariologist William S. Thayer. Thayer had done clinical work on malaria (prevalent at that time in the environs of Baltimore) and was a master of the literature; he persuaded two of his medical interns, William MacCallum and Eugene Opie, to follow up Laveran's observations with the malaria-like parasites found in the blood of birds. In 1897, Opie described some of these parasites in wild-caught birds. During that same summer, MacCallum, on vacation outside of Toronto, Canada, studied one of these "malarias" named *Haemoproteus*, where the male and female sex cells (gametocytes) in the blood are clearly different from one another even in unstained preparations. (This is unlike human malarias, where the gametocytes are very similar in appearance.) One type, the hyaline or clear form, put out Laveran's flagella, whereas the granular forms freed themselves from the red cell and remained quiescent. Observing the two forms in the same field under the microscope, he found the released flagella to invade and unite with the hyaline form to produce a wormlike gliding form. MacCallum immediately recognized that the flagella were sperm-like, that the granular forms were egg-like, and that he had witnessed fertilization to form a vermicule (later called the ookinete). On his return to Baltimore, MacCallum confirmed his discovery in a woman suffering from subtertian (falciparum) malaria. In his 1898 publication, MacCallum described the ookinete: "The movement is slow and even . . . with the pointed end forward. It can move in any direction readily . . . Often it is seen to rotate continually along its long axis. The forward progression . . . occurs with considerable force . . . pushing directly through the obstacle. The ultimate fate and true significance of these forms is difficult to determine." But then he incorrectly concluded: "it is reasonable to suppose . . . it is the much sought resistant stage" (302).

Miasma to Mosquito

Although Ronald Ross, a Surgeon-Major in the Indian Medical Service, was probably the most unlikely person to solve the puzzle of how humans "catch" malaria, he did so (311). Ross was born on Friday 13 May 1857 in the foothills of the Himalayas, where his father was an officer in the British Army stationed in India, which, at that time, was a part of the British Empire. As a boy of 8, his parents shipped him to England to receive a proper British education. He was a dreamer, and although he liked mathematics, he preferred wandering around the countryside, observing and collecting plants and animals. At Springhill Boarding School, which he attended from the age of 12, he began to write poetry, painted watercolors, and thought of becoming an artist. But, his father insisted that he study medicine in preparation for entry into the Indian Medical Service. Therefore, at age 17, young Ronald began his medical studies. He was not a good student, not because of laziness but because he had so many other interests and could not concentrate on medicine. He preferred composing music to learning anatomy, and he wrote epic dramas rather than writing prescriptions. Publishers rejected these "great works," and so he had them printed at his own expense. He eventually did pass his medical examination (after failing the first time), worked as a ship's doctor, and then entered the Indian Medical Service. Although India was rife with disease—malaria, plague, and cholera—Ross busied himself writing mathematical equations, took long walks, wrote poetry, played the violin, and studied languages. Occasionally he used his microscope to look at the blood of soldiers ill with malaria, but he found nothing. He shouted to all who could hear: "Laveran is wrong. There is no germ of malaria" (124). In 1894 Ross returned to England on leave. By that time, he had spent 13 years in India and had few scientific accomplishments: he wrote a few papers on malaria for the Indian Medical Gazette and claimed (without any real evidence) that malaria was primarily an intestinal infection. His hunt for the way malaria was transmitted from person to person began on 9 April, when the 37-year-old Ross visited with the 50-year-old Patrick Manson at his home at 21 Queen Street in London.

Manson (1844 to 1922) received his medical training at the University of Aberdeen (1866) and then served as Medical Officer (1871 to 1873) to the British-run Chinese Imperial Maritime Customs Office in Amoy, a subtropical port in China (311). There he studied the transmission via mosquito of the worm that causes elephantiasis. Returning to England in 1889, he developed a lucrative consulting practice and was also appointed Medical Advisor to the Colonial Office in London. At the time of Ross' visit, Manson was physician to the Seaman's Hospital Society at Greenwich and a lecturer on Tropical Diseases at St. George's Hospital and Charing Cross Hospital Medical Schools, where he had access to malaria contracted by sailors and others in the tropical regions of West Africa and India. Manson was a dedicated and experienced clinician as well as an expert microscopist, and he had been shown Laveran's animalcule by H. G. Plimmer (1856–1918) of the University of London, who in turn had been able to see the parasite under the guidance of Marchiafava during a visit to Rome. In his primitive laboratory on the top story of his house, night after night Manson watched the release of Laveran's flagella from the crescent forms. At the Seaman's Hospital, Manson took a drop of fresh blood from a sailor ill with malaria and showed Ross both Laveran's parasite peppered with the black-brown malaria pigment and the release of flagella. One day, as the two were walking along Oxford Street, Manson said: "Do you know Ross, I have formed the theory that mosquitoes carry malaria . . . the mosquitoes suck the blood of people sick with malaria . . . the blood has those crescents in it . . . they get into the mosquito stomach, shoot out those whips . . . the whips shake themselves free and get into the mosquito's carcass . . . where they turn into a tough form like the spore of an anthrax bacillus . . . the mosquitoes die . . . they fall into water . . . people drink a soup of dead mosquitoes and they become infected" (311).

Manson based this idea on his own work with elephantiasis. Working in China, he had demonstrated that after the microscopic blood stages of the worm, i.e., the microfilariae, are taken up by the mosquito (called the gnat of Amoy) they continue to mature within the mosquito's body. Then, believing in the current notion that mosquitoes feed only once in their lifetime, he postulated that the filariae escape into the water in which the mosquito dies; the worms are subsequently swallowed, and in this way the human contracts the disease. Manson's idea was based on several false assumptions and analogies. One was the earlier report (1870) of Alexei Fedchenko working in central Asia, who had shown that the guinea worm, *Dracunculus*, when released into water, entered into a tiny one-eyed crustacean, *Cyclops*, which, when swallowed in drinking water, caused an infection. Moreover, Manson did not realize that mosquitoes could feed more than once in a lifetime and did not die in the water after their eggs were laid after a single feeding. Thus, a string of seemingly "logical" deductions led to Manson's presumption that malaria parasites must be transmitted by ingestion. Manson had concocted a romantic story that was mostly a guess on his part, but the younger and inexperienced Ross took it as fact.

Manson was a fatherly type whose presence and bearing attracted attention and commanded respect. A medical student described Manson thus: "An afternoon with him was a memorable experience. His great knowledge, his familiarity with the literature and his sympathetic observations were brought to illuminate a problem and give point to a conclusion. He was kindly and at the patient's bedside was surrounded by a clutch of enthusiastic students. He patiently listened to the theories and queries of others and nothing gave him greater pleasure than suggesting a line of research to others" (310). For these reasons it was quite natural for Manson to assume the role of Ross' mentor, and he encouraged him to study mosquito transmission of malaria in order to validate Manson's theory.

Ross left England in March 1895 and reached Bombay a month later. In June 1895, encouraged by Manson's passionate plea, he captured various kinds of mosquitoes, although he had no idea what kind they were. He set up an experiment using the water in which an infected mosquito had laid her eggs, her young had been observed swimming, and the offspring had been allowed to die. The water with its dead and decaying mosquitoes was then given to a volunteer to drink. The man came down with a fever, but after a few days no crescents could be found in his blood. The same experiment was repeated with two other men who were paid for their services. Failure again. It thus appeared that drinking water with mosquito-infected material did not produce the disease. Ross began to think that perhaps the mosquitoes did have the disease but that they probably gave it to human beings by biting them and not by being eaten. He began to work with patients whose blood contained crescent-shaped malaria parasites and with mosquitoes bred from larvae. The first task was to get the mosquitoes to bite the patients. It was like looking for a needle in a haystack. There are more than 2,500 different kinds of mosquitoes, and at that time there were no good means of identifying most of them. Initially, Ross worked mostly with the gray and striped-wing kind. When these mosquitoes were dissected, the whiplike flagella were found in the mosquito stomach; however, no further development occurred. This result was no more informative than what Laveran had seen in a drop of blood on a microscope slide nearly 20 years earlier. Today we understand (as Ross did not) why this was so: the gray mosquitoes are *Culex* and those with striped wings are *Aedes,* and these do not carry human malaria. Instead, the one that he should have used was the brown, spotted-wing mosquito, *Anopheles*, but Ross did not recognize this for the entire year he dissected mosquitoes. Each mosquito dissection had required hours of effort at the microscope, and yet his prodigious labors had yielded nothing more than thousands of mosquito carcasses.

Then, at the age of 40 and having spent 17 years in the Indian Medical Service, Ross turned from the insusceptible kind of mosquito to the susceptible, brown, spotted-wing one. On 16 August 1897, his assistant brought him a bottle in which mosquitoes were being hatched from larvae. It contained "about a dozen big brown fellows, with fine tapered bodies hungrily trying to escape through the gauze covering of the flask which the angel of fate had given my humble retainer!" (403). He wrote: "My mind was blank with the August heat; the screws of the microscope were rusted with sweat from my forehead and hands, while the last remaining eyepiece was cracked. I fed them on Husein Khan, a patient who had crescents in his blood. There had been some casualties among the mosquitoes, and only three were left on the morning of 20 August 1897. One of these had died and swelled up with decay." At 7 a.m., Ross went to the hospital, examined patients, attended to correspondence and dissected the dead mosquito, without result. Then—a significant reminder of one of those unknown quantities in the equation to be solved—he examined another. No result again. He wrote in his notebook: "At about 1 p.m., I determined to sacrifice the last mosquito. Was it worth bothering about the last one, I asked myself? And, I answered myself, better finish off the batch. A job worth doing at all is worth doing well. The dissection was excellent and I went carefully through the tissues, now so familiar to me, searching every micron with the same passion and care as one would have in searching some vast ruined palace for a little hidden treasure.

Nothing. No, these new mosquitoes also were going to be a failure: there was something wrong with the theory. But the stomach tissues still remained to be examined—lying there, empty and flaccid, before me on the glass slide, a great white expanse of cells like a large courtyard of flagstones, each one of which must be scrutinized—half an hour's labor at least. I was tired and what was the use? I must have examined the stomachs of a thousand mosquitoes by this time. But the angel of fate fortunately laid his hand on my head, and I had scarcely commenced the search again when I saw a clear and almost perfectly circular outline before me of about 12 microns in diameter. The outline was too sharp, the cell too small to be an ordinary stomach cell of a mosquito. I looked a little further. Here was another, and another exactly similar cell.

The afternoon was very hot and overcast; and I remember opening the diaphragm of the substage condenser of the microscope to admit more light and then changing the focus. In each of these, there was a cluster of small granules, black as jet, and exactly like the black pigment granules of the . . . crescents. I made little pen-and-ink drawings of the cells with black dots of malaria pigment in them. The next day, I wrote the following verses and sent these to my dear wife:

> Seeking his secret deeds
> With tears and toiling breath,
> I find thy cunning seeds,
> O million-murdering death."

Here was the critical clue to the manner of transmission. Ross had shown that 4 or 5 days after feeding on infected blood, the mosquito had wartlike oocysts on its stomach. He did not know if these kept on growing, however, nor how the mosquitoes became infective. He planned to answer these questions shortly thereafter, but before that work could begin, he reported his findings to the *British Medical Journal* in a paper entitled "On some peculiar pigmented cells found in two mosquitoes fed on malarial blood." It appeared on 18 December 1897.

Ross knew that he could wrap up the unfinished work in a matter of a few weeks, but then he was struck by a blow from the Indian Medical Service by being ordered to proceed to Calcutta immediately. As soon as he arrived in Calcutta, he set his hospital assistants the task of hunting for the larvae and pupae of the brown, spotted-wing mosquitoes. Soon he had a stock of these and set about getting them to bite patients who were suffering from malaria. By flooding the ground outside the laboratory, he hoped to simulate rain puddles, and with this he began to learn about mosquito breeding. "If I am not on the pigmented cells again in a week or two," he wrote to Manson, "my language will be dreadful" (68).

In Calcutta, Ross was given a small laboratory. There were two Indian assistants who had already been working there when he arrived, but they were old men and not very intelligent, so he engaged two younger men, Purboona and Mahomed Bux. Both of these he paid out of his own monies. Since there were not many malaria cases in the Calcutta hospitals, Ross turned to something that Manson had suggested earlier: the study of mosquitoes and malaria, as seen in birds. Pigeons, crows, larks, and sparrows were caught and placed in cages on two old hospital beds. Mosquito nets were put over the beds, and then at night infected mosquitoes were put under the nets. Before much time had passed the crows and pigeons were found to harbor malaria parasites in their blood; also, he found the pigmented cells which he previously had spotted in the stomachs of mosquitoes that had been fed on infected larks. Ross became certain of the whole life history of the parasite in the mosquito, except that he had not actually seen the ookinetes, described by MacCallum, turning into oocysts. This was the last stage in the study. He found that the size of the oocysts depended exactly on the length of time since the mosquitoes had been fed on infected blood. They grew to their maximum size about 6 days after the mosquito had fed on infected blood. They left the stomach after this time, but he did not know what happened to them then.

One day while studying some sparrows, he found that one was quite healthy, another contained a few of the malaria parasites, and the third had a large number of parasites in its blood. Each bird was put under a separate mosquito net and exposed to a group of mosquitoes from a batch that had been hatched out from grubs in the same bottle. Fifteen mosquitoes were fed on the healthy sparrow; not one parasite was found in their stomachs. Nineteen mosquitoes were fed on the second sparrow; every one of these contained some parasites, although in some cases not very many. Twenty insects were fed on the third, badly infected, sparrow; every one of these contained some parasites in their stomachs and some contained huge numbers.

Ross wrote in his *Memoirs* (403): "This delighted me! I asked the medical service for assistance and a leave, but was denied this. I wanted to provide the final story of the malaria parasite for this meeting; but, I knew that time was very short. I still did not have the full details of . . . the change from the oocysts in the mosquito's stomach into the stages that could infect human beings and birds. Then, I found that some of the oocysts seemed to have stripes or ridges in them; this happened on the 7th or 8th day after the mosquito had been fed on infected blood." He continued, "I spent hours every day peering into the microscope. The constant strain on mind and eye at this temperature is making me thoroughly ill." He had no doubt these oocysts with the stripes or rods burst—but he did not know what happened to them. He asked himself: if they burst, did they produce the same stages that infected human blood?

Then, on 4 July 1898, Ross got something of value. Near a mosquito's head, he found a large branch-looking gland. It led into the head of the mosquito. Ross mused, "It is a thousand to one that it is a salivary gland. Did this gland infect healthy creatures? Did it mean that if an infected mosquito fed off the blood of an uninfected human being or bird, then this gland would pour some of the parasites . . . into the blood of the healthy creature?"

On 21 and 22 July 1898, Ross took some uninfected sparrows, allowed mosquitoes (which had been fed on malaria-infected sparrows) to bite them, and then, within a few days, was able to show that the healthy sparrows had become infected. This was the proof—this showed that malaria was not conveyed by bad air. After all, men and birds don't go around eating dead mosquitoes! On 25 July, now sure, he sent off a triumphant telegram to Patrick Manson, reporting the complete solution; 3 days later, Manson spoke at the British Medical Association meeting in Edinburgh, describing the long and painstaking research Ross had been carrying out for years past. Ross' findings were communicated on 28 July 1898 and published in the 20 August issue of *The Lancet* and the 24 September issue of the *British Medical Journal*.

Ross' discovery of infectious stages in the mosquito salivary glands in a bird malaria appeared to be the critical element in understanding transmission of the disease in humans. However, Manson rightly cautioned: "One can object that the fact determined for birds do not hold, necessarily, for humans." Ross and Manson wanted to grab the glory of discovery for themselves and for England but in such a quest they were not alone.

A Matter of Priority: Ross versus Grassi

Toward the end of the 19th century, important discoveries were being made about tropical diseases. Most of these were made by physicians with an insatiable curiosity and the simplest of tools. There were microscopes that could be used to examine microbes stained with aniline dyes, and it was possible to grow these outside the body in a nutritious broth. Their goal was to identify the agent of disease and determine how one caught that particular disease. During this period, as now, being first with an original theory or discovery usually led not to wealth but to an individual's fame and prestige as well as glory for his nation. For those with little or no training in science (or even for microbe hunters who were better prepared), the path from concept to discovery was central to scientific legitimacy. It remains so today.

Hunting the microbe responsible for periodic fevers (as well as other infectious diseases) involved more than jealousy; the rivalries could be bitter and pernicious and could border on slander. The work done by others was frequently overlooked, and oftentimes little credit was given for an original theory or a novel finding. In this respect the Italian and German workers were never slow to claim their due with regard to malaria. The Italians had many advantages for discoveries concerning malaria: the Roman Campagna was one of the most malarious areas in the world, and therefore many patients who could contribute to clinical, epidemiological, and biological studies. In spite of this, the malaria "animalcule" itself had not been discovered by the enthusiastic and able Italian students of malaria equipped with their better microscopes fitted with oil immersion lenses and facilities, but by an obscure French military physician working in a colonial outpost equipped with a microscope with a dry lens having little more magnification than the one used by Leeuwenhoek 200 years earlier. In addition, the claim by the Italians that a bacterium was the causative agent of malaria not only led to confusion but also retarded progress in understanding the disease. On the positive side, Italian physicians did make significant contributions to the subject: there were descriptions of the complete life cycle of the parasites in the blood, the kinds of malarias were differentiated from one another, and it was recognized that the malaria paroxysm occurred when the parasites destroyed the red cells and were set free in the blood. They missed the significance of Laveran's flagella, however, and the fact that the malaria parasite had another life stage in the mosquito. It was in this latter arena of malaria research that the battle for priority and originality was fought (212, 311).

Despite the fact that during his time in India Ross suffered from considerable handicaps at the outset of his research, including an ignorance of the literature on the work carried out by the Italians, little experience in bacteriology save for a short course taken in London, limited training in

microscopy, a lack of familiarity with staining methods, and a total ignorance of the different kinds of mosquitoes and their physiology and behavior, he succeeded in discovering the life cycle of the malaria parasite in
the mosquito and its role in transmission. Ross was a doer, indefatigable,
tenacious, single-minded, and impatient with those he regarded as intel-
lectually inferior and as obstructionists. Some who knew him considered
him "chronically maladjusted" and a "tortured man." As mentioned in
the previous section, when Ross was on leave from India, and following a
visit to London where he heard Manson's theory that mosquitoes carry
malaria, the 37-year-old Surgeon Major of the Indian Medical Service returned to Secunderbad, India, to "follow the flagellum." From that time
forward, Manson would be Ross' source of encouragement and information, as well as advice. On more than one occasion Manson was able to
provide an answer to the difficulties Ross encountered. One time he
wrote: "Look upon it as a Holy Grail and yourself as Sir Galahad and
never give up the research, for to be assured you are on the right track.
The malaria germ does not go into the mosquito for fun . . . it is there
for a purpose and that purpose depend on it—is in its own interest.
Germs are selfish brutes" (311).

During the summer of 1896, Manson learned that Amico Bignami at
the Santo Spirito Hospital in Rome had attacked his mosquito theory with
one of his own, namely, that the mosquito passed on the parasite in feeding, not dying. In reaching this conclusion, Bignami placed special emphasis on his inoculation experiments: transfusion of blood from a patient
with periodic fever into healthy persons resulted in a disease that could
be cured by quinine. Although Bignami did not isolate the parasite, it was
clear to him that the disease with which he was working was malaria. He
reasoned that the mosquito acted in a fashion similar to a hypodermic
needle, giving the parasite direct entry into the blood. On hearing of Bignami's claim, Ross became angry and began to suspect that the Italians
were starting to steal Manson's ideas. That same year, after Ross had carried out studies that convinced him that Manson's hypothesis of infection
by mosquito ingestion was incorrect, he wrote: "The belief is growing on
me that the disease is communicated by the bite of the mosquito. What do
you think?" Manson encouraged caution: "The parasite might not develop in any old mosquito. Possibly different species of mosquito may
modify the malaria germ" (68, 311). Indeed, Manson was correct. Ross
had done most of his work with mosquitoes (the grey *Culex* and the
brindled *Aedes*) that do not transmit human malaria. By October 1896,

Ross agreed with Manson that the mosquitoes with which he was working were not the right sort. In the summer of 1897, after he began to use the brown, dapple-winged mosquito, he found that when the mosquitoes had fed on a patient with crescents in the blood, these crescents changed into pigment-bearing globules in the mosquito stomach and became larger over time. The brown-black pigment in the globules was indistinguishable in color and shape from that found in finger blood from a malaria patient. Ross thought he was "on it" and believed (but did not prove) that the pigmented cells had transformed from Laveran's flagellum. This discovery occurred on 20 August 1897, a day Ross called "Mosquito Day." In September he wrote a short note on his observations and sent them to Manson for communication to the *British Medical Journal* under the title "On some peculiar pigmented cells found in two mosquitoes fed on malarial blood." Ross wrote Manson: "I really believe the problem is solved, though I don't like to say so . . . I have hardly restrained myself from writing 'pigment' to you . . ." Manson replied to the letter on 29 October 1898: "And you have to thank the plasmodium that it is fool enough to carry its pigment along with it into the mosquito's tissues. Otherwise I suppose you would not have spotted him" (68).

Ross' work on human malaria was interrupted by his transfer to Calcutta, where there were very few clinical cases. At Manson's suggestion, Ross began to work on the bird malaria *P. relictum* transmitted by *Culex* mosquitoes. Later that summer Ross recognized that "one single experiment with crescents will enable me to bring human malaria into line with (bird malaria)—they are sure to be the same" (68). But he never did the experiment. He left India for good on 16 February 1899.

During the period from 1894 to 1898—a time when Ross was working alone in India—the Italians published virtually nothing except for the 1896 paper by Bignami. But by the middle of July of 1898—a time when Ross' proof was complete and partly published—the Italians, led by Giovanni Battista Grassi, began to work in earnest on the transmission of malaria (212). Grassi received a medical degree from the University of Pavia (1875); however, he never practiced medicine, and by the time the work on transmission began he held the Chair in Comparative Anatomy in Rome. He was world renowned for his studies in zoology, including unraveling the complex and mysterious life history of the eel (1887) as well as the roundworm *Strongyloides*; he was able to diagnose hookworm disease by finding eggs in the feces, he identified fleas as vectors of dog and mouse tapeworms, and he wrote a monograph on the marine chaetognaths. As early as 1890, he had worked with avian malaria, and this led quite naturally to studies of human malaria; together with Amico Bignami and Antonio Dionisi at the Santo Spirito Hospital, he attempted (in 1894) to determine whether mosquitoes from the malarious areas were transmitters of the disease. In this, they were unsuccessful.

Grassi was small in stature and delicate but with a serious and forceful will. He had long, black, curly hair, beetling eyebrows, and an unkempt shaggy beard. He usually wore a battered felt hat, tattered clothes, and dark, smoked glasses. He seemed to enjoy a good quarrel and was pleased when such an opportunity presented itself. He was also very proud of the research he did and was resentful of those who questioned his authority. Grassi recognized that insofar as malaria was concerned there remained two main tasks: to demonstrate the developmental cycle of the human parasite in the mosquito and to identify the kind of mosquito that transmitted human malaria. To this end, he assembled a team of colleagues to make an all-out push. The team consisted of Dionisi, who had worked with avian malaria and was to test Ross' findings; Bignami, who would test his mosquito "bite theory"; and Grassi, who knew the different kinds of mosquitoes and who would survey the malarious and nonmalarious areas and, by comparing the mosquito populations, try to deduce which species were possible transmitters. Bignami, together with Giuseppe Bastianelli, a careful microscopist who knew his malaria parasites well, would follow Ross' trail to determine the parasite's development in the mosquito.

Beginning on 15 July 1898, Grassi began to examine the marshes and swamps of Italy. Where Ross was patient and perseverant and willing to carry out a seemingly endless series of trial-and-error experiments, Grassi was methodical and analytical; he was also able to distinguish the different kinds of mosquitoes. Grassi observed that "there was not a single place where there is malaria . . . where there aren't mosquitoes too, and either malaria is carried by one particular blood sucking mosquito out of the forty different kinds of mosquitoes in Italy . . . or it isn't carried by mosquitoes at all" (124). Working in the Roman Campagna and the surrounding area, Grassi collected mosquitoes and at the same time recorded information on the incidence of malaria among the people. (In effect, he was carrying out an epidemiologic study.) Soon it became apparent that most of the mosquitoes could be eliminated as carriers of the disease because they occurred where there was no malaria. There was, however, an exception. Where there were "zanzarone," as the Italians called the large brown spotted-wing mosquitoes, there was always malaria. Grassi recognized that the zanzarone were *Anopheles*, and he wrote: "It is the anopheles mosquito that carries malaria . . ." With this work Grassi was able to prove that "It is not the mosquito's children, but only the mosquito who herself bites a malaria sufferer . . . it is only that mosquito who can give malaria to healthy people." In September 1898 Grassi read a paper before the prestigious Accademia Nazionale dei Lincei, founded in 1603 and counting Galileo Galilei as one of its most famous members, in which he stated: "If mosquitoes do carry malaria, they are most certainly the zanzarone—*Anopheles*—not any of the 30–40 other species" (124).

Meanwhile, the German government had dispatched Robert Koch, who previously had shown little interest in malaria, to German East Africa to solve the malaria problem. During his 1897 to 1898 sojourn in Tanganyika, he claimed that the natives of the Usambra Mountains called malaria "Mbu" because they thought the fever was carried by the Mbu, or the mosquito, which used to bite them as they moved into the lowlands. Upon his return to Germany in June 1898, Koch delivered a lecture at Berlin's Kaiserhof Hotel under imperial patronage with a band playing the national anthem, in which he suggested that infected mosquitoes laid eggs on the human body, and afterward malaria parasites emerged to enter the bloodstream. Koch was determined to solve the malaria puzzle by carrying out a research program in Italy, in particular in the Roman Campagna, where Grassi's own research was being carried out. Koch arrived triumphantly in August, visited Rome and its hospitals, and departed Italy in October after failing to solve the mystery of malaria transmission. Grassi was irritated by the way he and his colleagues had been virtually ignored when Koch "invaded" Italy, presumably to grab the prize. Indeed, Koch meant to show that "only in the cloudy north was it possible for the star of science to shine for the illumination of the sleepy brains of the degenerate Italian race" (212). Grassi felt threatened by Koch and rushed to publish his findings and establish priority, lest Koch claim to have solved the riddle for himself and Germany. In October 1898 Grassi announced that he and his colleagues were experimenting with *Culex* and *Anopheles* mosquitoes and expected a definitive solution to the question of transmission shortly. A month later, *Culex* had been eliminated. In papers published in November and December 1898, Grassi and his colleagues reported on the development of the parasite in *Anopheles*. He infuriated Koch by sending him copies as a "Christmas present" and later declaring that "the victory was completely ours" (212).

Grassi found that the development of human malaria in *Anopheles* was as had been predicted from Ross' studies of bird malaria. He acknowledged the assistance of Bignami and Bastianelli but stated that the credit for the final completion belonged to himself. He also angered Ross with his faint praise, stating, "It [the life cycle of *Plasmodium*] finds confirmation in that observed by Ross with the malaria of birds in the grey (*Culex*) mosquito" (212). Although the Italians attempted to place their work on an equal footing with and parallel to Ross', the fact is that they had a copy of his official report and had followed his published procedures step by step. Ross accused them of piracy. Grassi, as egotistical and stubborn as Ross, fought back, stating he had worked independently of Ross and had made his discoveries without any prior knowledge of what Ross had done. He stigmatized Ross' bird malaria research saying that the figures and descriptions were incomprehensible and that he doubted whether they were of any value as a guide to what happens in human malaria since similar parasites may have different life cycles. In 1902, when the Nobel committee was considering splitting the prize for Physiology or Medicine between Ross and Grassi, Koch stood opposed. Indeed, Grassi was considered by Koch to be an enemy, and he called him a charlatan with neither brains nor ethics. Ross alone received the Nobel Prize, but despite this award for "work on malaria, by which he has shown how it enters the organism and thereby has laid the foundation for successful research on this disease and methods of combating it," for the remainder of his life he was embittered. This was largely due to his easily taking offense and magnifying a petty slight out of all proportion. Even apologies from Bignami and Bastianelli and their dissociation from Grassi's increasingly arrogant contentions did not mollify Ross' contempt for the Italians.

Ross claimed that it was only after Grassi had read his work on the transmission of malaria in birds that he recognized that human malaria occurred only in areas where *Anopheles* existed and that *Culex* was not involved, since Grassi did not publish this or the development of the parasite in these mosquitoes until late 1898. Ross wrote in his *Memoirs*: "They . . . had this paper of mine before them when they wrote their note. Their statement was . . . a deliberate and intentional lie, told in order to discredit my work and so to obtain priority . . . Many of the items . . . are directly pirated from my . . . results . . . stolen straight from me . . ." (403). For decades the arguments over priority and originality raged between Ross and Grassi. In truth, the contributions of Grassi and Ross could be considered to be of separate but of equal importance, and as the distinguished protozoologist C. M. Wenyon wrote in 1926, "Grassi may have been and probably was influenced and guided to some extent by what he had heard of Ross' discoveries, but nevertheless he and his colleagues were the first to obtain the absolutely scientific proof of the specific relationship of anophelines to human malaria" (212).

The importance of the master-apprentice relationship in establishing priority and originality is documented extensively in the correspondence between Patrick Manson and Ronald Ross (68). From 1895 to 1899, Ross wrote 100 letters to Manson, and Manson replied in 89 letters. Manson was not only a dedicated and experienced clinician, he was also a medical statesman advising the Colonial Office on matters of training in tropical diseases for medical candidates who would be posted to the colonies. In 1895 during a 6-month period when Ross was on leave from India, he was a constant visitor to Manson's home, and Manson became a sponsor of his. On Ross' return to India and in support of his investigations, Manson wrote to the Secretary of the India Office in London: "To our national shame be it said that few, very few of the wonderful advances in the science of the healing art . . . have been made by our countrymen . . . French, Germans, Italians, Americans and even Japanese are shockingly ahead of us . . . this is very humiliating . . . but in malaria there is a chance for an Englishman to rehabilitate our national character. I have no hesitation in saying that at the present moment Ross is the best man in India to carry on malaria investigation . . ." (311).

In October 1895, Manson warned Ross that Laveran was inclined to take up the mosquito hypothesis: "by next summer the French and Italians will be working on it. So for goodness sake hurry up and save the laurels for good old England." In response Ross wrote Manson in November, "We won't let Laveran assume the honor of the mosquito theory. Until you came up with this explanation no one had the slightest idea as to what the flagellated bodies were. Laveran . . . thought them to be organs of locomotion!" It was two years later, on 9 July 1898, that Ross wrote Manson, "I think I now may say Q.E.D. and congratulate you on the mosquito theory indeed." And in October he wrote, "These observations prove the mosquito theory of malaria by . . . Manson. His brilliant induction . . . indicated the true line of research it has been my part to merely follow his direction" (68).

Shortly after Ross' appointment to the Liverpool School of Tropical Medicine in 1899, however, his cordial relationship with Manson began to deteriorate (203). It became worse over the ensuing decade. In 1912, when Ross left Liverpool and Manson retired from the London School of Hygiene and Tropical Medicine and his post at the Colonial Office, Manson recommended his former student W. T. Prout for the vacated post at the Liverpool School and wrote that were Prout to be appointed, "teaching (in Liverpool) would improve." Ross took this as a personal affront to his scientific standing and sued for libel. Manson was dismayed, apologized, and paid the legal fees. Things worsened considerably after Manson's death in 1922. Ross, who had been a Surgeon-Major in the Indian Medical Service, now dismissed Manson as "nothing more than a medical man" and reflected on Manson's contributions: "Manson's brilliant induction regarding the appearance of flagellated bodies . . . indicated some suctorial insect as a possible second host, but it was far from proving the proposition. My discovery showing sporozoites enter the salivary gland . . . had not been suggested in any way by Manson. But, I wished to give as much credit as possible to him; and I therefore gave him publicly the whole credit for my work . . . But one must be a very dull person not to be able to distinguish here the exaggeration of gratitude. My praise was exaggerated . . . he could scarcely be credited with all my findings, many of which had been made independently of him. I doubt . . . he would ever have made my observation of August 1897 and July 1899. The work was done by me, alone . . . not by his instructions as is frequently pretended. (Hence) when I received the Nobel Prize I did not feel called upon to divide it with Manson."

The deterioration of their close relationship probably was due to their competing on the same turf and to the fact that Manson did not support Ross' appointment to the more prestigious London School. Ross increasingly resented Manson's having acted as his spokesman and alter ego. Manson was an attractive speaker. His presentations of Ross' findings were simple and lucid, and he had great rapport with his audience, but it was in Manson's voice that the announcements were made. There is a bitter irony to this: Ross was not a particularly impressive speaker—he had a slightly hesitant and rambling style—his writing was better than his lecturing. Nevertheless, in Ross' mind he had been denied the pleasure of hearing for himself the applause for his discoveries. Further, his dissatisfaction also had to do with his perceived lack of respect for Manson, who more and more had become regarded as more eccentric than venerable and who was ridiculed in some circles as "Mosquito Manson." A close association between the two of them, Ross must have felt, was no longer an asset.

A further insult Ross perceived was Manson's collaboration with the Italian workers on human malaria experiments. In 1900, when Manson sensed that the Colonial Office and the British public were unconvinced of the idea of transmission of malaria by mosquitoes, he devised two simple experiments. The first involved subjects living in the Roman Campagna in mosquito-proof huts where the windows and doors were protected by wire screening and the beds were covered with mosquito netting. The men lived under these conditions from early July to October, and none became ill with malaria despite not taking quinine. In the second experiment, conducted at the Santo Spirito Hospital, Bignami and Bastianelli allowed *Anopheles* mosquitoes to feed on patients with the benign tertian malaria, *P. vivax*. Then the mosquitoes bearing oocysts were shipped in special cages to London, where they were applied to the back of the hand so they could bite Patrick Manson's 23-year-old son, Thurburn, a medical student. The first experiment failed, but the other two succeeded, and parasites were found in Thurburn's blood 15 days after the feeding. He was successfully treated with quinine but suffered two relapses in the following year. Perhaps the greatest insult to Ross, though, was Grassi's dedication of his monumental and magnificently illustrated monograph *Studi di uno Zoologo sulla Malaria* (*A Zoological Study of Malaria*) to Manson.

Rejecting Manson as a mentor, Ross sought the favor of others. He began courting Laveran. He accorded Laveran priority for the mosquito hypothesis, writing, "Will you permit me to conclude with an expression of satisfaction . . . the announcement to you, who, not only originated our correct knowledge of these subjects, but from the first divined that the mosquito is connected with the propagation of these parasites." In July 1898, Ross sent specimens to Laveran that showed the development of the parasite in the mosquito. Laveran responding by promising to communicate the findings to the French Academy of Medicine; he did so in January 1899. Ross' paper gave Laveran sole credit for the mosquito hypothesis. In his Nobel Prize acceptance speech, Ross identified Laveran as his primary mentor. In doing so, he reinforced his claim to originality against any possible threat of Manson's influence. In addition, he was now able to share in Laveran's prestige. Laveran was an asset to Ross since he was neither a competitor nor a threat to Ross' claim for originality and research priority. In addition to his courting of Laveran, Ross,' fearing that the Italians might take full credit for discovering the transmission of malaria by mosquitoes, sought the favors of Koch. In 1901 he wrote to Koch for support, and with noblesse oblige Koch gave credit to Ross for the proof of the mosquito as the vector for malaria, since he had published first. Koch was a figure of great stature, and thus Ross obtained status without fear that Koch would be a competitive threat.

In Ross—a contentious, touchy, and combative individual—we witness how primary intellectual kinship can be asserted with a prestigious scientist with whom one never trained. Although less common today, it nevertheless does exist. Although such claims of intellectual descent or kinship are less common today, one may still come across them. Repudiation of an active, recorded scientific collaboration with a mentor, however, can be difficult if not impossible even within the present context of scientific research. As the stories of the hunt for a malaria vaccine unfold, the goal of originality in scientific theory and discovery will be seen to generate ambivalence in master-apprentice relationships much as it did with Ross and Manson a century ago.

Ross flagged the "dapple-wing" mosquito as involved in malaria transmission, and Grassi specifically identified that mosquito as *Anopheles*. Once *Anopheles* had been identified as the vector of malaria, methods for controlling it became possible. Of the 450 species of *Anopheles*, only 50 are capable of transmitting the disease, and of these only 30 are considered efficient vectors. Generally speaking, the measures for prevention of malaria are to keep infected mosquitoes from feeding on humans, to eliminate the breeding sites of mosquitoes, to kill mosquito larvae, and to reduce the life span of the blood-feeding adult. Contact with adult mosquitoes can be prevented by using insect repellents, wearing protective clothing, using impregnated mosquito netting, and installing window screens in houses. Breeding sites can be controlled by draining water, changing its salinity, flushing, altering water levels, and clearing vegetation. Adult mosquitoes can be destroyed by using sprays, and larvae can be destroyed by larvicides. Prevention depends on education and treatment of the human population. Employing all these strategies has helped to eradicate malaria from many temperate parts of the world, but in the tropics and in developing countries, especially those with limited budgets for mounting public health campaigns and where there is parasite and insecticide resistance, malaria is on the rise.

In Africa the most efficient vector is *Anopheles gambiae*, which can breed in small temporary pools of water such as those formed by footprints, hoofprints, or tire tracks. In other areas, *A. stephensi*, which can breed in wells or cisterns, is the vector. In the 1900s larvicides in the form of oil and Paris green (bright green powdered copper acetoarsenite, which is extremely poisonous and is sometimes used as an insecticide or fungicide), as well as drainage, were introduced to limit mosquito-breeding sites in water. These measures had outstanding success in reducing transmission in some parts of the world. Later DDT was introduced as a component of an eradication campaign, but by the early 1960s it had become clear that eradication could not be accomplished by DDT because of the emergence of resistant mosquitoes, because DDT had negative ecological side effects, and also because the parasites were becoming resistant to the antimalarial drugs used for treatment. By 1969 the World Health Organization (WHO) formally abandoned its eradication campaign and recommended that countries employ control strategies. Today, attempts at control involve using insecticide-impregnated bed nets, spraying with ecologically less disruptive but more expensive insecticides, and treating the human population with antimalarials. Some health practitioners and philanthropic agencies contend that a protective vaccine not only would assist in control but might even lead to eradication.

A Hidden Life Revealed

Ronald Ross, shortly after discovering that the mosquito was a vector for *Plasmodium*, presumed that after entry into the bloodstream, the inoculated sporozoites burrowed into red blood cells immediately. His rival Battista Grassi, however, suggested that the nucleus of the sporozoite was so different from that found in the blood stages that a considerable degree of transformation would be necessary to convert one directly into the other. Pursuing this line of thought, Grassi hypothesized in 1901 that an intermediate stage occurred somewhere in the body—a preblood (exoerythrocytic) form—and that this would carry out the necessary transformation. Grassi's hypothesis quickly fell apart when 2 years later Fritz Schaudinn (1871 to 1906), a preeminent microbe hunter, claimed to have observed the direct penetration of the red cell by the sporozoite. Schaudinn, it is said, took ripe and ruptured oocysts from a mosquito infected with *P. vivax*, placed them in a warmed, dilute drop of his own blood obtained from a blood blister he got from rowing, and then peered through his microscope for six uninterrupted hours (212). So persuasive was Schaudinn's description—sporozoites pushed into the red cell, first making a dent, then penetrating with their pointed tail, and lastly pulling themselves inside by peristaltic jerks—that even Grassi did not pursue the matter further. "Schaudinn's curious delusion lay like a spell over subsequent investigators" for decades (212). However, "science is a study of errors slowly corrected," and soon indirect evidence questioned both the observations and conclusions reported by Schaudinn. First, nobody had managed to confirm Schaudinn's microscopic findings, and second, in the treatment of patients with tertiary syphilis the effects of quinine were found to be markedly different in blood-induced infections from those that were sporozoite induced. In this so-called malariatherapy, the common practice was to treat tertiary syphilis by inducing a high fever via malaria (mostly *P. vivax*) by direct inoculation of blood or by inoculating sporozoites by mosquito bite or in isolated salivary glands or in entire ground-up mosquitoes. The blood-inoculated patients were cured with quinine, whereas those with sporozoite-induced infections relapsed after the same quinine therapy. Even more telling were the observations of the American Clay Huff, the British Sydney Price James, and the Australian Neil Fairley. During World War II, with the help of Australian Army volunteers, Fairley measured the incubation period, i.e., the time it took for parasites to appear in the blood after a mosquito-induced infection; in *P. vivax* it was 8 days and in *P. falciparum* it was 5 days (212). He also found that during the incubation period the blood was not infectious by transfusion. (Another overlooked clue was the case of Manson infecting his son with *P. vivax*-infected mosquitoes obtained from the Roman Campagna and the parasites not appearing in the blood for 15 days.) Malaria parasites must have been lurking somewhere in the body, but it was not known where.

Beginning in the mid-1930s, Huff and Bloom (in 1935) and James and Tate (in 1937) and then others observed malaria parasites developing in endothelial cells and macrophages before they appeared in the red blood cells in bird as well as in lizard malarias (245). These preblood stages were called exoerythrocytic (EE) forms. Following Huff and Coulston's (249) landmark description of the development of *P. gallinaceum*, from the entrance of sporozoites into the skin of chickens to the appearance of parasites in the blood, Huff (246) boldly suggested: "Since indirect evidence for . . . exo-erythrocytic stages in mammalian malarias is good it would appear advisable to adopt their presence in sporozoite-induced infections as a working hypothesis . . ."

In 1945 Colonel Sydney Price James told P. C. C. Garnham, then a young Medical Officer in Kenya, not to return from East Africa until he had found preblood forms in a mammalian malaria. James' gentle insistence proved stimulating to Garnham, and 2 years later, after James' death, Garnham found EE stages in the liver of a *P. kochi*-infected African monkey in the Medical Research Laboratory in Nairobi (182). Shortly thereafter, Garnham joined Henry Shortt at the London School of Tropical Medicine and Hygiene, where work began with *P. cynomolgi* in rhesus monkeys, with the expectation that the findings would relate to *P. vivax*. There were many attempts and many failures; success was finally achieved, however, when 500 infected mosquitoes were allowed to bite a single rhesus monkey and then for good measure the infected mosquitoes were macerated in monkey serum, and this brew was also injected into the monkey. Seven days later the monkey was sacrificed and its organs were removed for microscopic examination. Shortt expected that the EE stages would be found in locations similar to those described for bird malarias. This turned out not to be the case. Instead, the site of preblood stages of *P. cynomolgi* was the liver, as had been the case with *P. kochi*. Shortt and Garnham promptly reported their findings in a paper published in *Nature* (449). From that time forward, preblood stages have been described for the nonhuman primate malarias (91), the human malarias (107, 109, 448), and many of the rodent malarias (285).

In *P. falciparum* infections, the disappearance of parasitized red cells from the peripheral circulation (as evidenced by simple microscopic examination of a stained blood film) may be followed by a reappearance of parasites in the blood. This type of relapse, called recrudescence, results from an increase in the number of preexisting blood parasites. *P. vivax* and *P. ovale* also relapse, although the reappearance of parasites in the blood is not from a preexisting population of blood stages and occurs after cure of the primary attack. The source of these blood stages remained controversial for many years, but in 1980 the origin of such relapses was identified. During studies of relapsing malarias induced by sporozoites, Krotoski and coworkers (277, 278) found dormant parasites, called hypnozoites, within liver cells. The hypnozoites, by an unknown mechanism, are able to initiate full EE development and then go on to establish a blood infection. Undoubtedly, hypnozoites were responsible for the two relapses suffered by Manson's son Thuburn in 1900—a year after he was bitten by *Anopheles* carrying *P. vivax*—although at the time this was not recognized.

The EE stages of parasites in monkeys and humans are difficult to study because of ethical considerations, the scarcity of suitable species of monkeys, the narrow range of human parasites adapted to primates, the severity of disease in humans and primates, and the expense. Humans are more plentiful, but the necessary numbers of volunteers willing to undergo liver biopsy are difficult to find. Moreover, even when a mosquito inoculates dozens of sporozoites into a human and these successfully invade liver cells and develop into an EE form, only a few dozen may be present in a 3-lb organ. It has been estimated that if a mosquito inoculated 100 sporozoites, there would be a single EE form found among a billion liver cells. Truly, the EE form was a needle in a haystack.

Today it is recognized that *Plasmodium* has three developmental stages. Humans are infected through the bite of a female anopheline mosquito when she injects sporozoites from her salivary glands during blood feeding (Fig. 1). The inoculated sporozoites travel via the bloodstream to the

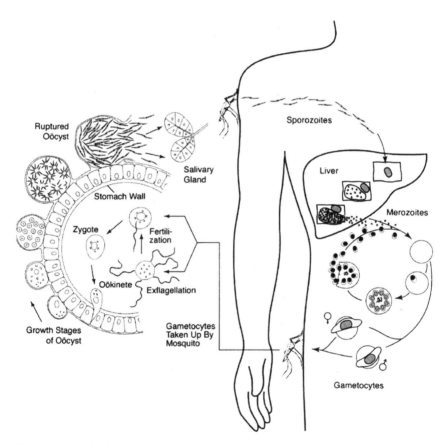

Figure 1   The life cycle of the human malaria parasite *Plasmodium falciparum*.

liver, where they enter liver cells. Within the liver cell, the nonpigmented parasite, the EE stage, multiplies asexually to produce 10,000 or more infective offspring. These do not return to their spawning ground, the liver, but instead invade erythrocytes. By asexual reproduction of parasites in red blood cells (called erythrocytic schizogony or merogony), infectious offspring (merozoites) are released from erythrocytes. These can invade other red cells to continue the cycle of parasite multiplication, with extensive red blood cell destruction and deposition of hemozoin. In some cases the merozoites enter red cells but do not divide; instead, they differentiate into male or female gametocytes. When ingested by the female mosquito during a blood meal, the male gametocyte divides into flagellated microgametes that escape from the enclosing red cell; these microgametes swim to the macrogamete, one fertilizes it, and the resultant motile zygote, the ookinete, moves across the stomach wall. This encysted zygote, now on the outer surface of the mosquito stomach, is an oocyst. Through asexual multiplication, threadlike sporozoites are produced in the oocyst, which bursts to release sporozoites into the body cavity of the mosquito. The sporozoites find their way to the mosquito salivary glands, where they mature, and when this female mosquito feeds again, sporozoites are introduced into the body and the transmission cycle has been completed.

2

Malaria, the Sickness

Although some 170 species of *Plasmodium* have been described, only 4 are specific for humans. The human malarias, *P. falciparum, P. vivax, P. ovale,* and *P. malariae*, are transmitted through the bite of an infected female anopheline mosquito when she injects sporozoites from her salivary glands during blood feeding. It is the asexual reproduction of malaria parasites in red blood cells and their ultimate destruction with release of infectious offspring (merozoites) that is responsible for the pathogenesis of this disease. Anemia is the immediate pathologic consequence of parasite multiplication and destruction of erythrocytes, and suppression of red blood cell production in the bone marrow can also occur. During the first few weeks of infection, the spleen is palpable because it is swollen from the accumulation of parasitized red blood cells and the proliferation of white blood cells. At this time it is soft and easily ruptured. If the infection is treated, the spleen returns to normal size; in chronic infections, however, the spleen continues to enlarge, becoming hard and blackened due to the accumulation of malaria pigment. The long-term consequences of malaria infections are an enlarged spleen and liver and organ dysfunction (197).

The primary attack of malaria begins with headache, fever, anorexia, malaise, and aching muscles (551). This is followed by paroxysms of chills, fever, and profuse sweating. The fever spike may reach 41°C and corresponds to the rupture of the red cell as merozoites are released from the infected cell. There may be nausea, vomiting, and diarrhea. Such symptoms are not unusual for an infectious disease, and it is for this reason that malaria is frequently called "the great imitator." Then, depending on the species, the paroxysms tend to assume a characteristic periodicity. In *P. vivax, P. ovale,* and *P. falciparum* the periodicity is 48 h; because the fevers recur every third day, they are called "tertian malarias," while *P. malariae* infection, with its periodicity of 72 h and fevers recurring every fourth day, is called "quartan malaria." *P. malariae* may persist in the body for up to four decades without causing signs of pathology. If the infection is not synchronous and there are several broods of parasites, the periodicity may occur at 24-h intervals.

P. vivax, which accounts for 45% of all malaria cases, results in severe and debilitating attacks but is rarely fatal; hence, it is called benign tertian malaria. It is able to remain in the body for a long period without reinfection as a result of the persistence of "sleeping stages" (hypnozoites) in the liver. Indeed, *P. vivax* and *P. ovale* may remain in this state of suspended animation for 3 to 5 years before there is any sign of a blood infection. *P. malariae, P. ovale*, and *P. vivax* attack red blood cells of a particular age and hence do not cause severe anemia that may lead to death; *P. falciparum*, however, unlike the other three kinds, is indiscriminate in its preference, invading red blood cells of all ages; as a result it causes infections that are more severe and which can result in a mortality rate of 25% in untreated adults. Falciparum malaria is referred to as malignant tertian malaria to distinguish it from the other tertian malarias (vivax and ovale malarias).

Complications of malaria include kidney insufficiency, kidney failure, fluid-filled lungs, neurologic disturbances, and severe anemia. In pregnant females, falciparum malaria may result in stillbirth, lower than normal birth weight, or abortion. Nonimmune persons and children may develop cerebral malaria, a consequence of the mechanical blockage of small blood vessels and capillaries in the brain because of the sequestration of infected red blood cells. *P. falciparum* may survive for a relatively short time in the body after causing an acute infection or may kill, but most often it remains as a chronic infection. If relapse occurs in falciparum malaria, it is as a result of an increase in the numbers of preexisting forms in the blood, previously too small to be detected microscopically; this type of relapse is termed recrudescence. Falciparum malaria accounts for 50% of all clinical malaria cases and is responsible for 95% of deaths. It was probably unable to gain a foothold in African populations until there were large sedentary populations, and it is thought to have emerged coincident with the development of agriculture some 4,000 to 10,000 years ago.

Finding Parasites

Although it is possible to diagnose malaria in a drop of liquid blood (as did Laveran), the task is made much easier by staining a blood film. In 1891 the Russian pathologist Dimitri Romanowsky used the available stains of the day, particularly methylene blue and eosin, to stain heat-fixed blood films.

The story is told that he forgot to replace the stopper on the bottle of methylene blue and that when he came to use it again it was mucky and full of mold. He used it anyway and counterstained the film with eosin. To his amazement, the cytoplasm stained pink and the DNA of the nucleus stained purple. The method was difficult to reproduce, since to obtain this differential staining the methylene blue solution had to be "aged" or "got moldy." As a result, J. H. Wright and W. B. Leishman made modifications of the Romanowsky method, but none were easy to handle or reproduce.

In 1899 Bernard Nocht, the chief medical officer of Hamburg Harbor in Germany, developed an interest in using stained blood samples to diagnose malaria among returning sailors who had acquired the disease in the tropics. He was stymied, however, by the lack of reproducibility of the Romanowsky method. When he became Director of the Institute for Maritime and Tropical Disease in Hamburg, he enlisted the help of Gustav Giemsa (1867 to 1948), a pharmacist with extensive training in chemistry and bacteriology, to solve the problem (169). Giemsa succeeded in identifying the compound in "aged" methylene blue as azure B (trimethylthionine) and found that a stable stock solution with reproducible staining properties could be obtained by mixing eosin with azure B in a glycerol-methanol mixture (29).

Diagnosis of a malaria infection today is not much different from in the days of Alphonse Laveran. It involves using a finger prick, taking a drop of blood from the fingertip, smearing this out into a thin film on a microscopic slide, fixing and staining it, and then examining the slide under a light microscope. On the basis of clinical patterns supplemented by the stained blood film, the human malaria parasites (*P. malariae*, *P. falciparum*, *P. vivax*, and *P. ovale*) can clearly be distinguished from one another. By using a light microscope, as few as 5 to 10 parasites in a small drop of blood can be seen. Ordinarily, *P. vivax* infects no more than 2% of the red cells whereas in *P. falciparum* infections the percentage can reach 10 to 20 times this value.

Control before the Vaccine

Since all of the pathology of malaria is due to parasites multiplying in the blood, most antimalarial drugs are directed to these rapidly dividing stages (42, 197, 402, 442). The earliest of these drugs was quinine, derived from the bark of the cinchona tree and isolated in 1820 by two French chemists,

Pierre Pelletier and Joseph Caventou. Quinine continues to be used, but completion of the 5- to 7-day regimen for cure is poor because of unpleasant side effects such as bitter taste, tinnitus, nausea, and vomiting, and so it remains limited to use by injection. Chloroquine and amodiaquine are synthetic antimalarial drugs that can be taken by mouth; they were developed in the 1940s. They were the mainstay of the unsuccessful malaria eradication program of the 1950s. Both act rapidly and can be taken prophylactically once a week. The problem with chloroquine is drug-resistant *P. falciparum*. Mefloquine (trade name Lariam) and halofantrine (trade name Halfan) were synthesized in the 1960s to counter parasite resistance to chloroquine. Mefloquine, which acts similarly to quinine, can be taken orally once a week before, during, and after exposure. In the late 1980s, Burroughs Wellcome began a program for the rational design of antimalarials. The result was atovaquone, a mimic of the vitamin coenzyme Q or ubiquinone. It blocks DNA synthesis by the parasites, which parasites die as a result. Atovaquone is too costly to treat those living in areas where malaria is endemic, but it is used to treat the pneumonia caused by *Pneumocystis* in AIDS patients in developed countries and has been used prophylactically for travelers when used in combination with proguanil, under the trade name Malarone. Primaquine is a drug used against the stages in the liver in the vivax malarias; if treatment is successful, relapse is prevented. A newer analog of primaquine called tafenoquine was developed through collaboration between the U.S. Army and GlaxoSmithKline, and it is the first new replacement drug since primaquine was introduced more than 60 years ago. At the time when Patrick Manson experimentally infected his son (see p. 35) these drugs were not available, but had the young man received tafenoquine or primaquine in addition to quinine, he would not have suffered a relapse.

Qinghaosu (artemisinin), a Chinese herbal medicine, is both the newest and the oldest in the arsenal of antimalarials. It has been used in China for 2,000 years to reduce fever. It is derived from the leaves of the wormwood *Artemisia annua*. Artemisinin has given way to three more potent derivatives: artemether, artemotil, and artesunate. Because these are eliminated from the body rapidly, they are combined with structurally unrelated and more slowly eliminated antimalarials, called artemisinin-based combination treatments (ACTs), to clear the blood of parasites, both asexual stages and gametocytes. A 3-day course of artemisinin derivative in combination with a slowly eliminated partner drug such as amodiaquine, sulfadoxine-pyrimethamine, mefloquine, lumefantrine, or atovaquone-proguanil is required for optimum cure. ACTs work well against both *P. vivax* and *P. falciparum*.

Regrettably, little is known about the molecular mode of action of many of these drugs or the mechanisms of parasite resistance to them, though it has been contended that multidrug resistance to halofantrine, mefloquine, chloroquine, and quinine is due to mutations. Some antimalarials such as atovaquone, proguanil, Fansidar, Maloprim, and Malarone kill malaria parasites by blocking the synthesis of *Plasmodium* DNA, whereas the target of others, such as mefloquine, quinine, artemisinin, and chloroquine, remains unknown.

The major threat of malaria today is not an increasing range of endemicity but, rather, a rise in the intensity of antimalarial drug resistance. Drug resistance in malaria has been defined as the ability of a parasite strain to survive and/or multiply despite the administration and absorption of a drug in doses equal to or higher than those usually recommended but within the limits of tolerance of the subject. Thus, resistance is a characteristic of the particular parasite strain. First recognized more than 40 years ago with chloroquine in South America and Southeast Asia, drug-resistant malaria poses one of the greatest challenges for controlling morbidity and mortality and successful treatment is dependent on prompt and accurate diagnosis. In an attempt to overcome or delay the emergence and spread of drug-resistant strains, combination therapy (e.g., Fansidar, which combines sulfadoxine and pyrimethamine; Fansimef, which utilizes Fansidar plus mefloquine; Maloprim, which is pyrimethamine plus dapsone; Malarone; and the ACTs) has been recommended by the WHO since 2001. Since 1991 GlaxoSmithKline Biologicals has been developing LAPDAP, a combination of lapudrine or chlorproguanil with dapsone. LAPDAP, released in 2002, costs less than a dollar per tablet and provides a cheaper alternative to the more expensive antimalarials. In addition to resistance, there is another treatment constraint: cost. Chloroquine, costing about 8 cents per treatment, was very cheap, whereas a course of treatment with the newer drugs such as mefloquine may cost 10 times as much as chloroquine; halofantrine may cost 20 to 30 times as much as chloroquine; a course of treatment with Malarone costs even more; and ACTs cost $1.35 to $2.40 per adult treatment course. A third constraint for the drug-based treatment of malaria is the reluctance of pharmaceutical companies to invest their capital to develop antimalarials in the belief that profits will not be sufficient to enable the companies to recoup the very large investments that have to be made in research to discover and test such drugs.

Impact

One of the Four Horsemen of the Apocalypse, Pestilence, in the form of malaria, has long been a companion of War (442). On 11 June 323 BC, Alexander the Great, having conquered most of the known world, fell ill with chills and fever. On 13 June, at age 33 Alexander died, and as a consequence his incursions into the subcontinent of India were aborted. It is widely believed that he died of malaria, although there are other possible causes. Malaria repelled foreign invaders from sacking ancient Rome, and Caesar's campaigns were disrupted by malaria. Frederick Barbarossa's army in the 12th century was also prevented from attacking Rome because it was felled by "bad air." Malaria (then called ague) was prevalent in the fens of England and in colonial America, debilitating the farming population. Indeed, in some instances it was the major obstacle both socially and economically to the growth and development of the American colonies. In the time of the Revolutionary War, malaria played a role in several critical battles, and during the Civil War there was a high incidence of malaria among the troops from the North stationed in the South, with an estimated one-half of the white troops and four-fifths of the black troops contracting malaria annually. Malaria plagued the French and British engaged in battle during World War I and again in World War II. In 1943 Sir William Slim, the British Field Marshal, said, "For every man evacuated with wounds we had one hundred and twenty evacuated sick." Malaria, the most significant disease problem facing the Allies, prompted one writer to claim "that the triumph over malaria was one of the most important victories of allied forces in the Southwest Pacific." Again, during the Korean and Vietnam wars malaria was "The Great Debilitator," with *P. vivax* causing the greater morbidity in the former war and *P. falciparum* causing the greater morbidity in the latter. This prompted the military to provide better antimalarial measures and also stimulated the search for new drugs. The problems facing medical planners in the future, however, will be even greater, because the next time around multidrug-resistant strains of *P. falciparum* and insecticide-resistant mosquitoes will have to be overcome.

Today, malaria still ravages many countries, targeting the indigenous peoples and slowly and inexorably killing them. This is especially true in the rainy and low-lying agricultural areas where transmission is high. In the time it takes you to read this paragraph, 100 children will have died from malaria. Malnourishment increases the susceptibility to malaria and a variety of other diseases, leads to lower productivity, and puts a further strain on the already fragile health care system. Sadly, in Africa the Horsemen of the Apocalypse have as their principal target the children; one-third of infected (and untreated) children will die from malaria. Thus, it is a certainty that malaria will continue to devastate nations unless something is done to control it: in addition to insecticide-impregnated bed nets and insecticide spraying, a protective vaccine(s) and/or effective drugs at an affordable price are sorely needed.

3

The Immunity Alphabet

It is in the warm humid places in sub-Saharan Africa where malaria is endemic that "most people are almost continuously infected by *P. falciparum* and yet the majority of infected adults rarely experience overt disease. They go about their daily routines of school, work, and household chores feeling essentially healthy" (132). Such vigor in the face of a continuing onslaught of *Plasmodium*-infected mosquitoes is a clear indication the adults are immune. The goal of developing a vaccine targeted at the blood stage of the malaria parasite is to mimic that same condition in infants and young children so that they too may enjoy lifelong immunity with natural boosting by exposure to the blood-sucking, parasite-carrying *Anopheles*.

At the turn of the 20th century, not only was knowledge about human immunity to malaria scanty, but most of it was incorrect. The prevailing ideas suggested that immune Africans were able to resist infection because they threw off noxious odors, they sweated profusely, and their skins were as thick as hides. It was far easier then (as it is even today) to describe the immune condition than to understand the remarkably effective and diverse mechanisms that produce it. To "read" the immune system, it would be essential to know its "alphabet"—the building blocks of comprehension (319, 461). Over the past century, much has been learned about the alphabet of immunity; through investigation, reading, and comprehension, it has provided grist for the mills of vaccine producers (vaccinologists) who are searching for protection against a wide variety of potential assassins, including malaria parasites.

A Is for Antibody

Diphtheria is a horrible childhood disease (442). The infected child is pale, feverish, and nauseated, and the nasal discharge is blood tinged. A sticky gray-brown membrane (made of white cells, red cells, dead epithelial cells, and clots) adheres to the palate, throat, windpipe, and esophagus, impairing breathing. If disturbed, the membrane can bleed and break off, blocking the airways of the upper respiratory tract. The throat is sore, the tonsils are inflamed, the individual vomits and suffers with headaches, speech is thickened, and the breath is foul smelling. Within 2 weeks this upper respiratory infection can spread to the heart, leading to heart failure, circulatory collapse, and paralysis. During epidemics 20 to 50% of untreated children may die (46).

In 1883 two German microbe hunters, Edwin Klebs and Friedrich Loeffler, working in Robert Koch's laboratory in Berlin, discovered the cause of diphtheria by using throat swabs taken from an infected child; they were able to grow the bacterium in pure culture and named it *Corynebacterium diphtheriae*. Then in 1888 Emile Roux and Alexandre Yersin of the Pasteur Institute in Paris found that the broth in which the bacteria were growing contained a substance that produced all the symptoms of diphtheria when injected into a rabbit, and at high enough doses it killed the rabbit. This killing was due to a poison, a toxin. In humans it is this toxin, a neurotoxin so powerful that a single molecule can kill a cell, which produces the deadly consequences of heart muscle degeneration, neuritis, paralysis, and hemorrhages in the kidneys, adrenal glands, and liver. The diphtheria bacteria produce the toxin because they themselves are infected with a virus; for the toxin to be synthesized, iron molecules must be present nearby—in body tissues or culture medium.

In 1885, Arthur Nicolaier discovered the tetanus bacillus *Clostridium tetani*. In 1889 Shibasaburo Kitasato in Koch's laboratory at the Institute for Infectious Diseases in Berlin was able to grow the bacillus in pure culture, and in the culture fluid he found a powerful toxin that differed from that of diphtheria. Then in 1890 two critical discoveries related to immunity were made: Emil von Behring and Shibasaburo Kitasato, both working with Robert Koch, found that if small amounts of tetanus toxin were injected into rabbits, the rabbit serum contained a substance that would also protect another rabbit from a subsequent lethal dose of toxin. (Serum [plural sera] is the straw-colored liquid that remains when blood is allowed to clot and then the clotted red material is removed.) The serum capable of neutralizing the toxin was called "immune serum." Behring found that the same situation also occurred with diphtheria toxin. In effect, toxin inoculation could protect the rabbit against a disease. Further, the immune protection seen by Behring and Kitasato could be transferred from one animal to another by injection of this immune serum. In other words, it was possible to have passive immunity; i.e., protection could be provided by another animal that had been immunized with the active foreign substance when its serum was transferred (by injection) to a nonimmunized animal. (Passive immunization, or "immunity on loan," can be life-saving in the case of toxins such as bee, spider, or snake venoms or tetanus since it is possible to counteract the deadly effects of the poison by injection of serum containing the appropriate antitoxin; passive immunity has been demonstrated in bird, monkey, and mouse malarias and has been shown to have clinical benefit in human malaria. See p. 111.)

Behring and Kitasato called the toxin-neutralizing substance in immune serum "antitoxin," meaning "against the poison." Antitoxin, found in immune serum after a toxin is injected into humans, rabbits, horses, chickens, mice, guinea pigs, monkeys, rats, or goats, is simply one kind of antibody. The implications of the research on immune sera for treating human disease were obvious to von Behring and others in the Berlin laboratory. Without waiting for the identification of the active ingredient in immune serum, the German government began to support the construction of factories to produce different kinds of immune sera against a wide range of bacterial toxins.

Treatment of human diphtheria cases with antitoxin serum (or, more simply, antiserum) was so effective that the death rate dropped precipitously. In some cases mortality was halved. For his work on antitoxin therapy, von Behring received the 1901 Nobel Prize. Initially, it was believed that all bacterial infections could be treated with antitoxin therapy, but unfortunately, relatively few kinds of bacteria produce and secrete toxins, so the therapeutic application of immune serum remained limited. Indeed, despite advances in serum therapy and vaccination, neither diphtheria nor whooping cough (which also produces a toxin) has been eradicated. Today, however, diphtheria and whooping cough are less of a problem because of childhood immunization (the diphtheria-pertussis-tetanus [DPT] shot); also, if the infection is acquired, it can be treated with antibiotics.

At about this time, Paul Ehrlich (1854 to 1915) came on the scene in Koch's laboratory (314). Ehrlich was born in the city of Strehlen, Silesia, then Germany, now a part of Poland. He had an undistinguished high school record and preferred mathematics and Latin to German composition. He studied at the University of Strasbourg, where his tutor, Professor Waldeyer, allowed him to experiment with various kinds of dyes. In short order, Ehrlich discovered a new type of cell—the mast cell. (Later, it would be found that mast cells secrete the chemical histamine, which makes nearby capillaries leaky and in some cases provokes an allergic response. Drugs designed to blunt this effect are called antihistamines.) One day, a country physician from Wollstein, Robert Koch, visited the laboratory. It is said that Koch asked Waldeyer who the dye-covered young man was and he replied, "that is 'little Ehrlich' he is very good at staining but he will never pass his examinations" (314). Nevertheless, pass he did, graduating as a doctor of medicine at age 24. His thesis was entitled "The value and significance of staining tissues with aniline dyes."

Because certain dyes stained only certain tissues and not others, it suggested to Ehrlich that there was chemical specificity of binding. He wrote: "curative substances—a priori—must directly destroy the microbes provoking the disease; not by an 'action from distance', but only when the chemical compound is fixed by the parasites. The parasites can be killed only if the chemical has a specific affinity for them and binds to them. This is a very difficult task because it is necessary to find chemical compounds, which have a strong destructive effect upon the parasites, but which do not at all, or only to a minimum extent, attack or damage the organs of the body. This we call chemotherapy" (314).

Ehrlich's first position was as a house physician at the Second Medical Clinic of the Charity Hospital in Berlin. He attended the sick but did not feel comfortable knowing that he could do little to cure them. In 1882, he attended a lecture given by Koch in which he announced his discovery of the "germ" of tuberculosis, and he emphasized how difficult it was to stain the bacteria. That very night, Ehrlich rushed back to his laboratory to experiment with dyes and stained some tubercle bacteria. He placed the slides on a warm stove to dry; however, the cleaning woman whose job it was to get the stove going did not notice Ehrlich's slides, and by the time he got to the laboratory the stove was red hot. To his surprise, the slides were not worthless but instead were beautifully stained. When Koch was shown the stained slides, he was very pleased by Ehrlich's discovery. The method was elegant and simple, and the colors were so intense that anyone could see the tubercle bacteria under the microscope. During his time at Charity Hospital, Ehrlich contracted tuberculosis from a patient. Parenthetically, in 1887, when Ehrlich experienced the clinical symptoms of tuberculosis, he stained his own sputum and demonstrated that he in fact was infected. In despair he left for Egypt to recover, which he did, and upon his return to Berlin in 1890, Koch, who was now the Director of the newly founded Institute for Infectious Diseases, invited him to join the Institute.

Ehrlich's work on immunity at the Institute was begun independently of Emil von Behring and brought to light the fact that immunity against other protein poisons, particularly ricin, could be developed (37). He followed this up quantitatively and found that the potency of the immune serum could be progressively increased by appropriate methods. When Behring's discovery of antitoxins was announced, Ehrlich started to apply the methods used for increasing immunity to ricin, beginning with tetanus and then going on to diphtheria, which was practically more important. This was in sharp contrast to Behring, who had not succeeded in making practical use of his discovery of antitoxin. Drawing on his broad experimental experience, Ehrlich established the principle that there was a quantitative (1:1) relationship between the amounts of toxin neutralized and the specific amount of antitoxin needed to do so. The potency or strength of an antibody, called its titer, is reflected by the degree of dilution that must be made to have a specific amount of antigen that will bind to an equal amount of antibody. (The higher the titer, the greater the potency of the serum; put another way, a high-titer immune serum can be diluted to a greater degree to achieve the same neutralizing effect as that of a weaker immune serum.) Ehrlich's principle of titer was not only of theoretical interest, but it also enabled antitoxins (and later other therapeutic biologics such as insulin and other hormones) to be provided in standardized amounts and strengths (442).

In the autumn of 1901, trial treatments of Behring's serum on children with diphtheria had been carried out without success. In fact, as Ehrlich noted, "it was impossible that they should be, for his serum contained considerably less than 1 unit per cc and could not produce any action. Therefore, I was asked to make experiments with my serum, which immediately gave good results; my sera, obtained from goats had 30 units per cc, and one serum had almost 100 units. It was in these circumstances that von Behring asked me to collaborate with him, and indeed, this may have been necessary for him, for in spite of all the means which the Hoechst Laboratory placed at his disposal he made no progress. When we began working together, he showed me a bottle containing 5 quarts of diphtheria toxin. He believed this would be sufficient for 50 years of immunizations, however, it would hardly have been enough for one horse!" (314).

Ehrlich was convinced of the necessity of using progressively higher doses of a very active toxin to obtain the desired increase in immunity.

Initially, Behring did not accept Ehrlich's point of view or his methods; after more than 6 months of experimenting, however, he finally did succeed in producing active sera. It was Ehrlich who at a critical time straightened out the confusion and put Behring on the right track that enabled him to achieve success and to be covered with honor and glory (including the first Nobel Prize); Ehrlich, however, was given no credit.

From the start, Ehrlich's interactions with Behring were antagonistic. He said: "Behring now puffs himself up like a peacock, calling himself von Behring, but he is a fraud. Behring did not understand the quantitative nature of the binding of antitoxin to toxin, nor had he systematically investigated how to obtain the highest levels of antitoxin. It was my work that allowed the development of an exact measure of the potency of diphtheria serum, which could be used for therapy. And, what did he do to thank me? He cheated me. The factory which was to produce the serum prepared a contract that was to include both of us, but Behring contrived to leave me out . . . he built himself a castle and changed his name to von Behring . . . he was rich, but apparently not influential enough since I never received an appointment as the Director of the State Serum Institute. I think of that as a dark period and the way Behring tried to hide our scientific partnership, but my revenge has come. He has not gotten far without me since our separation. Everything is blocked now: his work on plague, cholera, glanders, streptococcal infections. He makes no progress with diphtheria—only hypotheses. He wanted to be the 'all-highest' who would dictate his laws to the entire world and who, in addition could earn the most money, but thank God he did not have the necessary brain power. His mountain laboratory with its halls, ponds, meadows etc. was conceived by him as a fortress of science; but, it is fortunate for the free investigator that his plan has been destroyed; instead of it being a temple where he could be worshipped . . . it is just a normal laboratory. Away with the mammonization of science!" (314).

Ehrlich clearly recognized the power of immunization: "The marvelous effect of an antibody in a serum, as you know, is due to the fact that in no case has it an affinity for the substances of the body. It flies straight onward, without deviation, upon the parasites. The antibodies, therefore, are 'magic bullets' too, that find the target by themselves. Hence, they're astonishingly specific in effect, and the advantage of serum therapy over chemotherapy. The difference between serum therapy and chemotherapy is: we have, in active and passive immunization, a powerful weapon which has already shown its effectiveness in many infectious diseases and will always do so. What makes serum therapy so extraordinarily active is the fact that the protective substances of the body are the products of the organism itself, and here again we may speak of them as 'magic bullets' which aim exclusively at the dangerous intruding parasites which are strangers to the organism, but the 'bullets' do not touch either the organism itself or its cells. Therefore, obviously serum therapy, whenever it can be carried out, is superior to any other mode of action. However, we know a number of infectious diseases where serum therapy does not work at all or only with much loss of time. I call attention especially to malaria, and to African sleeping sickness, and perhaps a number of diseases caused by spirochetes such as syphilis. In these cases, chemotherapy instead of serum therapy must come to aid the treatment" (314).

Ehrlich also found that although the diphtheria toxin lost its poisoning capacity with storage, it still retained its ability to induce immune serum. He called this altered toxin (nowadays usually produced by treatment with formalin) "toxoid." Even today, toxoids made from diphtheria, pertussis, and tetanus toxins are used as the childhood vaccine DPT, and this is the standard means of inducing immunity to these diseases. The Schick test for susceptibility to diphtheria also involves toxoid: a small amount of diphtheria toxoid is injected just under the skin surface; failure to react (by an absence of a red swelling—inflammation—at the site of injection) indicates a lack of protective immunity (442).

Ehrlich observed that the serum made against one toxin did not protect against another toxin. Thus, the antitoxin made by the animal was specific. Moreover, since this was true for almost any foreign substance, he concluded that antibodies were specific for the foreign material against which they were produced. The foreign substances, such as a toxin or toxoid, that were antibody generators were called antigens. Any material that is foreign to the body, be it a bacterium or its toxin, a virus such as smallpox, a protozoan such as the malaria parasite, a piece of tissue from another individual, or a foreign chemical substance such as a protein, DNA or a sugar polymer, a polysaccharide, can be an antigen.

Ehrlich also formulated a theory of how antigen and antibody interact with one another. It was called the "lock-and-key" theory. He visualized it in the following way: the tumblers of the lock represented the antigen, and the teeth of the key represented the antibody. One specific key would work to open only one particular lock. Another way of looking at the interaction of antigen with antibody is the way a tailor-made glove is fitted to the hand. One such glove will fit only one specifically shaped hand. Similarly, only when the fit is perfect can the antigen combine with the antibody. Yet, despite the fact that antigens are large molecules, only a small surface region is needed to determine the binding of an antibody to it. These antigenic determinants, called epitopes, can be thought of as a small patch on the surface of the much larger molecule of antigen (442).

G Is for Globulin

The study of the protein components found in serum began in the mid-1800s, when it was discovered that addition of water to serum resulted in a white flocculent precipitate. In 1862 the precipitate was named globulin, from the Latin "globus" meaning "sphere," because of the tiny round particles of precipitate (the fraction that did not precipitate was named albumin from the Latin "albus," meaning "white"). At the turn of the 20th century it was possible to separate the globulin into additional fractions. One of the fractions, the euglobulin or "true globulin," could be precipitated from the globulins by addition of a saturated solution of ammonium sulfate, and the remainder, which was water soluble, was called pseudoglobulin or "false globulin." The primary objective of serum fractionation was to determine the ratio of albumin to globulin since this was thought to be of clinical significance, especially in malaria, where the ratio favors globulin (381). (Over time, however, it would be shown that such a ratio had little clinical value in malaria.)

During the 1930s, when high-speed centrifuges became available, the serum components could be separated for the first time into heavy (19S) and light (7S) globulin fractions and albumin (4S); the method, however, was cumbersome and the equipment was costly. In 1937, a much simpler method, electrophoresis, was invented by Arne Tiselius in Sweden, who took advantage of a characteristic of proteins: they carry an electrical charge whose magnitude varies for different proteins. In electrophoresis, when a polarized current is passed through the salty solution of serum, the components separate into several components that may be labeled according to their mobility in the electric field. When this technique was used with serum, it was found that the fastest component was albumin and the slower ones were the globulins; the fastest globulin was named alpha, and the slowest one was named gamma. A year later, when the American Elvin Kabat joined Tiselius and they compared immune and normal serums by electrophoresis, they found antibody to be principally in the gamma globulin fraction. The antibody-containing gamma globulin, produced in response to an antigen, is also called immunoglobulin (381).

There is frequent confusion in the minds of those unfamiliar with immunity concerning the relationship of the presence of an antibody and protection. Unfortunately, more often than not, detectable antibody (or an elevated gamma globulin level) is neither an indicator nor a measure of protective immunity. Although there are instances when antibody levels do correlate with the degree of protection, there are many exceptions to this seeming relationship. Indeed, as yet there is no immunological test for malaria that directly and invariably measures the functional immune status against malaria. Consequently, a best test for putative protective malaria antigens cannot be achieved in the laboratory with a test tube; instead, it requires that protection be assessed in a living susceptible host. This continues to be a significant hurdle for the evaluation of malaria vaccine candidates and makes vaccine testing cumbersome, expensive, and lengthy.

By the 1950s, there was an explosive development of the technique of electrophoresis when a simple commercial apparatus became available and serum samples could be applied to paper; by the late 1950s, paper electrophoresis was replaced by electrophoresis using agar gel, agarose, starch gel, acrylamide gel, and cellulose acetate strips. The electrophoretic patterns and their numerical evaluation have become more informative than the much simpler albumin-to-globulin ratios. After 1965, when the amino acid composition of immunoglobulin was determined, it became possible to divide it into three more abundant classes (immunoglobulin G [IgG], IgA, and IgM) and two lesser abundant classes (IgD and IgE) (381).

Each immunoglobulin molecule has a Y shape made up of four amino acid chains each consisting of a pair of light (220-amino-acid) polypeptide chains (denoted L) and a pair of heavy (440-amino-acid) chains, denoted H, held together by sulfur-sulfur bonds. In 1959, Rodney Porter working at St. Mary's Hospital in London was able to dissect the Ig molecule using the enzyme papain from the papaya plant (377). He found that the enzyme produced two fragments: one, named Fab, contained all of the antigen-binding activity and was at the two tips of the Y, whereas the other, called the constant or Fc fragment, did not bind antigen but did bind complement; it is the stem of the Y. The five classes of immunoglobulin differ in the amino acids of their heavy chains, designated γ for IgG, μ for IgM, α for IgA, etc. Putting all this complexity together in a simplified way, we can say that the outstretched "arms" of the Y-shaped IgG contain the Fab fragment and that the stem of the Y contains the Fc fragment. The first half of the Fab fragment, consisting of a small number of amino acids (15 to

20), called the variable region, is the actual binding site for the antigen. The variable region forms a shallow cavity large enough to make contact with 15 to 20 amino acids from a protein or six sugar residues on a polysaccharide. Thus, although antibodies recognize the three-dimensional shape, it is not the shape of an entire bacterium or malaria parasite or even a virus that is recognized; it is the shape of just a small portion of its surface molecules (known as the antigenic determinant or epitope). Today, Ehrlich's lock-and-key theory of antigen-antibody reaction can be described in molecular terms: it is the geometric fit of an antigen's epitope with its complementary determining region of the antibody. Therefore, even the smallest constituent of a malaria parasite may be recognized by a large number of antibodies—each, as it were, "looking at it" from different directions and "seeing" different antigenic determinants. One consequence of the smallness of the antigenic site is that similar shapes occasionally occur by chance on completely different pathogens, so that an individual antibody stimulated by one stage of a parasite may be found to cross-react with stages from a completely unrelated parasite or foreign cell.

IgM, called macroglobulin because of its large size (having a molecular weight of 10^6 and found in the 19S fraction by ultracentrifugation), consists of five IgG molecules bound together like a cartwheel and has five times as many antigen-binding sites as IgG (374). It is the first antibody to appear in the blood when a naive subject is infected with malaria; it occurs at about 1.5 mg/ml of serum, is broken down in a matter of 5 to 7 days, and cannot cross the placenta. It is able to immobilize and clump microbes very efficiently. IgG, constituting about 75% of the serum immunoglobulins (at about 9 mg/ml of serum), has a molecular weight of ~150,000 (found in the 7S fraction) and is the second antibody to be found in the blood, appearing 3 to 4 days after malaria parasites appear in the blood; it persists for 3 to 4 weeks in the blood and is capable of crossing the placenta. This maternal IgG is essentially gone by the time an infant is 6 months of age, and thereafter the child must make its own IgG (98). IgE is the antibody of allergies, and IgA is the antibody found in the intestinal tract; neither plays a role in malaria infections.

I Is for the Immune System

Being immune means that the body is able to react specifically to a foreign material as a result of previous exposure. To be effective, the immune response has to be swift the second time around. So what are the essential properties of the immune system involved in protection against a foreign invader? First, it must be able to distinguish foreign substances—"non-self" from "self." Second, it must be able to remember a previous encounter; that is, there must be memory. Third, it must be economical; that is, the immune response should not be continuous but should be turned on and off as needed. And fourth, the immune response must be tailored to interact with a specific antigen (334, 374).

A key to understanding the protective mechanisms of the immune system is an appreciation of the role of the body's bloodstream and the lymphatic system (442). Blood consists of a salty protein-containing fluid, the plasma, in which float red blood cells and white blood cells. (When plasma clots, the remaining fluid, now deficient in clotting molecules, is serum.) The red blood cells carry oxygen and carbon dioxide and are not involved in immunity. The white blood cells, on the other hand, are a critical part of the immune system. White cells, or leukocytes, come in two varieties: granule-containing granulocytes (called—depending on whether they stain with <u>eosin</u> [an acidic red dye], a <u>bas</u>ic methylene blue dye, or both—<u>eosin</u>ophils, <u>bas</u>ophils, and neutrophils respectively), and the non-granule-containing cells called monocytes and lymphocytes.

The "plumbing" of our bloodstream does not exist as an entirely closed system of "pipes." It leaks fluid but not cells. This is because blood, headed for the tissues and organs, is pumped by the heart into the arteries. The outgoing arteries branch into smaller and smaller vessels the farther they are from the heart, and finally they ramify into the smallest of vessels, the capillaries. Fluid, consisting essentially of plasma with much less protein and without blood cells, seeps out of the capillaries into the tissue spaces, bathing the cells. Cells are microscopic manufacturing plants, and to accomplish their energy-demanding tasks they contain an internal furnace that burns fuel (glucose) and dumps ashes (carbon dioxide and other waste products). The cell's bathing solution (called interstitial fluid) serves as a waterway for carrying fuel to and removing ashes from the cells. But what happens to all the fluid that has leaked out into the tissues? How does it return to the bloodstream? The fluid cannot return from the direction from which it came because the pumping of the heart creates an opposing hydrostatic pressure. The only means by which the fluid can return to the bloodstream is by an auxiliary drainage system that collects the seepage and returns it. (If the drainage is faulty or limb movement is limited, fluid accumulates in the tissue spaces, especially in the lower limbs by gravity, and the result is swollen ankles and feet!) The drainage system consists of permeable "pipes" (lymph vessels) of the lymphatic system. Just as the capillaries are widely distributed throughout all the tissues of the body, so too are the lymph vessels. Lymph vessels pick up the fluid from the tissue spaces (and such movements are aided by muscular movements of the limbs) and return it to the blood circulation upstream from the heart by a "connector pipe" (the thoracic duct located in the region of the collarbone), and in this way there is a complete circulatory system.

Mammals have co-opted the lymphatic system for another purpose: immunologic defense. To mount an effective defense, the immune system has three sets of components: a recognition system to identify the nonself invader, a disposal system designed to eliminate that invader, and a communication system that coordinates the recognition and disposal elements. Disposal is taken care of by phagocytes (literally "eating cells") that have recognition molecules built into their surface as well as the serum protein complement and antibody. Some phagocytes are able to kill their target cell from the outside (extracellular); when this involves antibody it is called antibody-dependent cellular toxicity. In other cases killing is carried out by two specialized types of cells—the natural killer (NK) cell and the cytotoxic T lymphocyte (CTL). Much of what the immune system does is under the influence of signals from other cells, signals that tell the cells to divide, cease division, migrate, secrete antibody, and change function. These signals are delivered in two ways, either by interaction with surface molecules through contact of one cell with another or by soluble molecules that act at a distance similar to the way hormones act. These soluble signaling molecules, called cytokines, are sometimes also called chemokines because they induce cells to move (chemotaxis) or are referred to as interleukins because they cause an <u>intera</u>ction between <u>leuk</u>ocytes (374).

By placing clusters of immune tissues at "checkpoints" or "surveillance stations" along the lymphatic routes, the body is able to examine the fluids (now called lymph in the lymph vessels) for invaders. The surveillance stations, called lymph nodes, are scattered throughout the body but tend to be clustered in the region of the groin, the armpits, the chest, and the spleen, where they act as filters. The filter is composed principally of two kinds of white cells: macrophages and lymphocytes. Our body contains 10^{10} to 10^{12} lymphocytes, equivalent in mass to the brain or the liver! In addition to the "fixed" lymphocytes in the lymph nodes, some lymphocytes and macrophages are free to roam throughout the body as they circulate in the bloodstream.

N Is for Natural Immunity

Natural or innate immunity serves as a first line of defense against nonself antigens (334, 374). It involves cells (neutrophils, macrophages, dendritic cells, and NK cells), preformed molecules (natural antibodies and complement), and inflammatory cytokines (including gamma interferon [IFN-γ] and tumor necrosis factor [TNF]). Innate immunity acts rapidly (hours to days), preventing invading pathogens from spreading throughout the body, and it is critical to our wellbeing until acquired immunity can develop, which may take several days to weeks. In malaria, innate responses limit the initial phase of malaria parasite replication and effectively control the first wave of infection. In this way the virulence of the malaria infection is reduced, there is less chance that an individual will die early, and the possibility for mosquito transmission to the next person is increased.

Inflammation, an innate response to infection typically characterized by pain and fever, can be most unpleasant. It has a beneficial purpose, however: increased blood flow and infiltration of fluid allow the accumulation of white blood cells (lymphocytes and phagocytes), complement, and cytokines in the affected area where there is a nonself invader (334). The bouts of fever, nausea, headaches, and other malaria symptoms—so typical of inflammatory responses seen when other pathogens such as bacteria and viruses are present—provide convincing evidence that innate immunity is operational when a person becomes a victim of *Plasmodium*.

One can speculate that innate immune mechanisms are triggered when the density of malaria parasites crosses a predefined threshold. This leads to oscillations in the number of blood parasites between a lower level (at which point there is no immune response) and a higher level (at which time the innate response is triggered), resulting in a partial clearance of malaria-infected red cells (481). These density-dependent mechanisms limit the growth of all blood stage parasites, irrespective of the kind of *Plasmodium*, suggesting that similar nonself molecules can trigger innate immunity. The precise nature of these molecular triggers, however, remains incompletely known, although the usual suspects such as malaria pigment (hemozoin) have been implicated.

Macrophages and dendritic cells ingest nonself pathogens and are able to do so in the absence of a specific antibody. The macrophage, the "large eating cell," was discovered in 1882 by the Russian Elie Metchnikoff (1845 to 1916). Metchnikoff was able to get some little particles of

carmine inside starfish larvae, which are transparent. Now with his microscope he could watch the crawling, flowing cells in this starfish ooze toward the carmine particles and eat them. He mused: "These wandering
cells . . . eat food, they gobble up carmine granules . . . but they must
eat up microbes too! Of course . . . the wandering cells are what protect
the starfish from microbes! Our wandering cells, the white cells of our
body . . . they must be what protects us from invading germs . . . they
are the cause of immunity to diseases . . . if my theory is true" (118).
Metchnikoff speculated that when a thorn from a rose was put into the
body of the larva, the wandering cells would react. To his delight, under the
microscope he saw the tip of the thorn surrounded by a cluster of phagocytes that appeared to want to "eat" the intruding foreign object. It was after these observations that Metchnikoff said, "Here is why animals can
withstand the attacks of microbes" (118) and he became a microbe hunter.

Later, in 1887, when Metchnikoff had established himself at the Pasteur
Institute in Paris, he noted that macrophages in vaccinated animals were
more active than those in unvaccinated animals. He formed the theory that
all protective immunity was due to the activity of phagocytes. Much of his
subsequent experimental work was directed toward establishing the principle that phagocytes—including macrophages and their precursors, the
blood-borne monocytes (elevated in numbers in the disease infectious
mononucleosis)—are key elements in this process. For his work on phagocytes and immunity he shared the 1908 Nobel Prize with Paul Ehrlich. Although his "solely the phagocyte" theory of immunity was flawed, he was
not entirely wrong in his opinions about the fundamental role of phagocytes
in immunity.

The bone marrow-derived dendritic cells have a maze of long exten-
sions, giving them an enormous surface area for capturing, processing, and
presenting antigen to other cells (563). Although dendritic cells superficially
resemble nerve cells, they are more closely related to the macrophages.
Dendritic cells depend on external signals for their activation and maturation, and in turn they secrete a variety of different interleukins as well as the
cytokines TNF and IFN-γ. The macrophages and dendritic cells have protein families of pattern recognition receptors that can be expressed on their
outer surface (259). The members of one of these families, the Toll-like receptors, are capable of recognizing dozens of different pathogen-associated
molecular patterns (PAMP). PAMP molecules include the lipopolysaccha-
ride in the cell wall of gram-positive bacteria, viral coat proteins, the my-
colic acid and lipoarabinomannan in the cell wall of tubercle bacteria, fla-

gellin (a protein that makes up the bacterial flagellum), glycosylphos-phatidylinositol (GPI) (the molecule anchoring surface proteins to the cell membrane), and hemozoin-associated parasite DNA. Purified and synthetic extracts from microbes containing some of these materials exert potent adjuvant (from the Latin "adjuvare," meaning "enhance" or "help") effects on the immune system (20, 372). Once a PAMP is fitted into the docking station of the dendritic cell or macrophage (that is, when it is bound to the receptor), cytokines are released; these make the smaller blood vessels leaky, and neutrophils and lymphocytes (of the T-cell kind, which will be considered shortly) become active and attack the invader.

NK cells are bone marrow-derived lymphocytes that are widespread throughout the body, being present in the spleen and lymph nodes as well as in the peripheral blood (374). They can be stimulated by cytokines such as the phagocyte-derived interleukin-12 (IL-12) to secrete IFN-γ, TNF, and several macrophage inflammatory proteins. NK cells are a major source of IFN-γ during the early phase of the malaria infection, and the early production of IFN-γ may determine the outcome of a blood stage infection (481). A robust IFN-γ response measured in the test tube (in vitro) has been associated in humans with a lower susceptibility to infection as indicated by a reduced risk of fever and clinical malaria. Once NK cells have identified their target, they adhere to it and give it a "chemical kiss of death." The chemicals involved are proteins such as perforin and protein-digesting enzymes called granzymes; the perforin punches a hole in the cell, allowing the granzymes to enter and destroy the cell's constituents. In healthy nonimmune volunteers infected with *P. falciparum* by mosquito bite, granzyme, IFN-γ, and interleukins can be found in the serum even before parasites are detected in a stained blood smear. This innate response very early in infection has also been demonstrated by in vitro experiments in which peripheral monocytes produced TNF and IFN-γ and interleukins 10 h after being exposed to infected red cells (481). In a ménage à trois among NK cells, macrophages, and malaria-infected red cells, production of IFN-γ was dependent on the production of IL-18 by the macrophages and NK cell secretion of IL-8, which allowed the recruitment of other types of white blood cells. Hemozoin is also able to aid in the recruitment of macrophages via secretion of IL-8 by NK cells.

The GPIs of blood stage malaria parasites are unique, containing a longer sugar moiety (glycan) and lipid. As a consequence, GPIs react with receptors on the macrophage, inducing it to produce interleukins (particularly IL-1 and IL-6) that activate other lymphocytes to secrete TNF, a cytokine first discovered because it could cause certain tumor cells to shrivel and die. Production of TNF is associated with parasite clearance and resolution of fever. Of some interest is that a chemically synthesized GPI glycan (and a mimic of that found in blood stages of *Plasmodium*) can be a potent activator of the innate immune system; it may hold promise as an antitoxic vaccine (see p. 278).

A considerable amount of the work on innate immune responses to malaria has been carried out using model systems, especially mouse malarias (see chapter 11) (286). Various combinations of rodent *Plasmodium* species and inbred mouse strains or rats have been used to mimic human malaria infections. No single rodent model, however, replicates all of the features of the pathologic changes or the immune response in humans. Despite important differences between human and mouse immune systems including differences in NK and dendritic cells, accumulating evidence suggests that the innate immune system is crucial to protective immunity in both mouse and human malarias (481). For example, in the absence of NK cells, parasite levels in the blood were higher during an acute infection with a mouse malaria (*P. chabaudi*) and there was a recurrence of parasites in the blood during the chronic phase. Early production of IFN-γ by NK cells is correlated with a spontaneous resolution of the infection in malaria-infected mice, whereas a lethal infection occurs in the absence of IFN-γ. Clinical studies of humans support these observations (481).

In malaria-naive individuals there is a coordinated increase in the level of the proinflammatory cytokines (including IFN-γ, IL-12, and IL-8) in the serum at the time when parasites leave the liver and first appear in the blood. This increase in cytokine production is corroborated by test tube studies in which parasitized red blood cells were shown to induce TNF, IL-12, and IFN-γ production within 10 h in monocytes obtained from naive donors. In the case of NK cell activation by *P. falciparum*, IL-12 and IL-8 are required but are not sufficient for optimal production of IFN-γ; direct contact between NK cells and malaria-infected red cells is required. In addition, NK cells from individuals exposed to malaria specifically recognize and then kill *P. falciparum*-infected red cells, suggesting that NK cell activation requires two signals: one cytokine mediated and the other involving direct contact between the NK cell and the malaria-infected red cell. Neither the antigen on the red cell surface nor the NK receptors responsible for this are known, however (481).

The innate immune response can be potentially harmful to the infected individual, and inflammatory mediators have been repeatedly implicated in the severity of the disease, giving rise to the widely held conclusion that severe malaria is in part an immune-mediated disease. Indeed, in the mouse malaria parasite *P. berghei*, overproduction of IFN-γ and TNF as well as IL-12 was associated with pathologic changes. In human malaria, TNF has been associated with an increased risk of cerebral malaria in African children and low levels of IL-2 and IL-8 in serum were linked to severe disease (481).

Eleanor Riley (born in 1956), currently Professor of Immunology and Head of the Immunology Unit within the Department of Infectious and Tropical Diseases at the London School of Hygiene and Tropical Medicine, grew up near Oxford, the daughter of a veterinarian and a medical social worker. After enrolling as a student of veterinary science at the University of Bristol, she became enthralled by immunology and cell biology, undertook an intercalated BSc in Cell Pathology, and, on completion of her veterinary training (1980), joined the graduate training program in pathology at the New York State School of Veterinary Medicine at Cornell University in upstate New York.

Her Ph.D. in immunology and veterinary parasitology at the University of Liverpool (1985) was supposed to prepare her for a research career in tropical diseases of livestock, and she applied, unsuccessfully, for a post to work on trypanosomes in Nairobi, Kenya. By chance, on the same page of *Nature* magazine that week was an advertisement for an immunologist to work on malaria at the Medical Research Council Laboratories in The Gambia, West Africa. She took up her post in The Gambia, under the direction of Brian Greenwood, and began a series of epidemiologic studies attempting to find associations between specific antibody or T-cell responses to malaria antigens and resistance or susceptibility to malaria infection and/or clinical disease which have underpinned much of her subsequent work. After 5 years in The Gambia, and a brief sojourn at the University of Stockholm, she moved to the University of Edinburgh in 1990 as a Wellcome Trust Senior Research Fellow, where in addition to maintaining strong links with The Gambia, she initiated epidemiologic research programs in conjunction with Kojo Koram at the Noguchi Memorial Institute for Medical Research in Legon, Ghana, and with Tom Egwang at MedBiotech Laboratories in Kampala, Uganda. These studies, combined with an emerging line of research on murine malaria infections initiated by Fakhreldin Omer in her laboratory in Edinburgh, led to an abiding interest in the mechanisms by which the proinflammatory immune responses that are required to control and kill blood stage parasites are regulated to prevent immune-mediated pathology. Most recently, this has led to a major interest in the innate immune response to malaria, particularly the role of NK cells.

Appointed as a full professor in Immunology at the London School of Hygiene and Tropical Medicine in 1998, she was asked—alongside her erstwhile mentor Brian Greenwood—to help establish the cross-departmental, interdisciplinary Malaria Centre to bring together basic scientists, clinicians, public health scientists, and policy makers from across the School to enhance the effectiveness of the School's research efforts toward improved malaria control. In 2000, funding from the Bill and Melinda Gates Foundation established the Gates Malaria Partnership and, with a program grant from the British Medical Research Council, allowed the establishment of the Joint Malaria Programme, a multidisciplinary research collaboration, based in Moshi in Northeastern Tanzania and involving the Kilimanjaro Christian Medical College, the Tanzanian National Institute for Medical Research, and the Centre for Medical Parasitology at the University of Copenhagen in Denmark. Today, the Joint Malaria Programme works from purpose-built research facilities in Moshi, Korogwe, and Tanga, hosting basic research programs on entomology, immunology, and parasite and host population genetics; carrying out applied research designed to improve the diagnosis and treatment of malaria; and conducting clinical trials of antimalarial drugs, insecticides, and vaccines, in this case the RTS,S vaccine developed by Walter Reed Army Institute of Research (WRAIR) and GlaxoSmithKline (GSK) (see p. 255).

In addition to limiting an initial infection, a robust and rapid proinflammatory response by the innate immune system might act to control reinfection; too potent a response, however, could promote the development of severe malaria either directly or by amplifying the effects of acquired immunity. Indeed, one of the hallmarks of acquired immunity to malaria is the ability to control the levels of the proinflammatory cytokines such that they facilitate clearance but do not trigger pathologic changes (481). In support of this hypothesis is the observation that overproduction of IFN-γ and TNF as well as IL-12 in infections with the rodent malaria parasite *P. berghei* is associated with pathologic changes. It is thus important to have a balance in cytokines, but how this is achieved in natural infections is still not known. Would a malaria vaccine be able to achieve such a critical balance? We simply do not know.

A link between dendritic cells and virulence during a malaria infection has been proposed. One study showed that the adherence of intact malaria-infected red cells to dendritic cells inhibited their maturation and subsequently inhibited their capacity to stimulate T cells (346). Further, for the induction of antibody responses, the interaction between dendritic cells and B cells may be critical in that dendritic cells provide naive B cells with signals essential for their proliferation and survival. Hemozoin also paralyzes dendritic cells: hemozoin-fed monocytes were prevented from maturing into dendritic cells, with the effect being mediated by malaria pigment—DNA binding to Toll-like receptor 9. *P. falciparum* may thus escape immune clearance by impairing the dendritic cells critical for producing the necessary proinflammatory cytokines (521).

How might studies of innate immunity in malaria contribute to the development of a protective vaccine? The innate immune system is intimately involved both in the initial phases of a malaria infection and in the outcome of acquired immunity. This could have important implications for vaccine design and immunotherapy. One could visualize a scenario where the enhancement of the innate immune response would limit the severity of an infection or modify the course of the subsequent adaptive immune response by "setting the stage" and/or acting synergistically with an antimalarial drug to facilitate parasite clearance from the blood, thereby augmenting the effects of partially effective vaccines (560). A possible strategy for enhancing the immunogenic properties of malaria antigens might be to combine them with cytokines, microbial products, or synthetic compounds that activate the innate immune system. Many current formulations of malaria vaccines use aluminum hydroxide (alum) as an adjuvant because it is one of the few approved for human use; it might not always be appropriate, however, given that it stimulates an IgG response in mice and is unable to induce CTL responses. By contrast, unmethylated CpG (cytidine phosphate guanosine) derived from bacteria and recognized by Toll-like receptor 9 induces a cytokine response characterized by production of IL-12 and IFN-γ. CpG is immunostimulatory with other microbial and peptide-based vaccines and enhances the efficacy of a blood stage vaccine based on a crude antigen preparation delivered in alum in the mouse malaria *P. chabaudi*; Improved vaccine efficiency has also been seen in *P. yoelii* and a merozoite surface protein (481). However, there is no evidence for this with any human malaria vaccine candidate.

Complement, discovered by the Belgian Jules Bordet (1870 to 1961), consists of a naturally occurring complex of proteins in serum (313, 374, 540). In healthy individuals it is quiescent, but during the course of a malaria infection (or infections involving other pathogens) it can become activated. At the time of Bordet's birth, Pasteur was engaged in studies of fermentation and spontaneous generation, Lister had just begun to develop antiseptic surgery, Koch was still an unknown physician, and Laveran had not yet found the malaria parasite. But by the time Bordet had graduated with a Doctor of Medicine degree at age 22, several immunologic milestones had already been reached: Behring had shown that animals immunized against the bacteria causing diphtheria and tetanus produced antibodies and had demonstrated that these antitoxins could protect against these diseases, Pasteur had been able to protect against disease by attenuation, and Metchnikoff had shown the importance of phagocytosis. Clearly, immunotherapy was now possible as a practical solution to disease control. In 1894 a grant from the Belgian government enabled Bordet to work in Metchnikoff's laboratory at the Pasteur Institute in Paris. In that same year, Richard Pfeiffer (1858 to 1945), a student of Robert Koch, made an important finding. In 1883 Koch had examined fecal samples from patients with cholera and had isolated the "germ" of cholera. It was a comma-shaped bacillus that moved by vibrating, and hence he named it *Vibrio cholerae*. When Pfeiffer introduced the vibrios into a guinea pig that previously had been immunized against cholera, he found that the vibrios were unable to swim and that they soon disappeared. The same thing happened in an unvaccinated animal if the vibrios were introduced after being added to serum from an immunized animal; in the absence of the immune serum, however, the vibrios killed the guinea pig. Since the serum from the immunized animal had no effect on the cholera toxin, the immunity that had developed was clearly different from the immunity to the toxins of diphtheria and tetanus.

In 1895 Bordet examined the cholera vibrios under the microscope and was able to show that the immune serum ruptured (lysed) the bacterial cell wall; this killed the vibrios (319, 461). When the serum was heated to 56°C for 30 min and then added to vibrios in a test tube, the capacity for killing was lost; the serum, however, was still able to clump the vibrios. Thus, the antibody's killing activity was heat sensitive whereas the antibody itself was not. Clearly, Bordet reasoned, normal serum contains a heat-sensitive accessory factor that acts together with the heat-stable antibody to exert its effect. This accessory factor, whose concentration did not increase in the blood serum after immunization, was named complement. Three years later Bordet discovered that when red blood cells from one animal species were injected into another species, they were destroyed through a process similar to lysis of bacteria, a process called hemolysis.

(This is also what happens when there is incompatibility in blood types during transfusion and is the reason why blood other than that from humans cannot be used for transfusion.) This discovery allowed Bordet to develop a laboratory test called complement fixation, which would permit a determination of the presence of antibodies to a pathogen in serum. Bordet remained at the Pasteur Institute for 7 years, and when he returned to Brussels he became the Director of a similar Institute in Belgium. For his work on complement, Bordet received the Nobel Prize in 1919.

Once considered to be a single substance, complement is now known to consist of 30 proteins and makes up approximately 15% of the globulin fraction (540). There are two major pathways of complement activation: the classical and alternative pathways. The classical pathway, with its proteins designated C1 through C9, is activated by a combination of antibodies produced during acquired immunity, natural antibodies (IgM), and other molecules such as C-reactive protein and serum amyloid P protein. In the classical pathway the binding of C1 to an antigen-antibody complex initiates the cleavage of the other complement components, producing a cascade of products; it is involved in acquired immunity. The alternative pathway, a part of the innate immune system, does not involve antibody and is self-starting by certain substances such as the bacterial cell wall, cell membrane fragments, and DNA. It begins with the spontaneous breakdown of the fourth component (called C3) and its deposition onto the surface of the pathogen. Once deposited, C3 binds more and more C3 to amplify the cascade. Once initiated, the process is like a chain reaction where one component activates the next until the final (ninth) component becomes activated; this attaches to the surface of the invader, a small hole is formed, and the pathogen dies as the cellular contents stream out.

There is significant activation of the alternative pathway in the preclinical and early clinical stages of human experimental malaria, even when blood smears are unable to detect parasites. This occurs despite the presence of factors on the surface of red blood cells that ordinarily protect against complement-mediated lysis. Children with severe malaria have greater amounts of C3 on the surface of *P. falciparum*-infected red cells, and this, in conjunction with increased IgG deposition and stimulation of macrophages by TNF, leads to removal of infected red blood cells. Further, the factors that protect red cells against complement attack are also reduced (363, 399). Together, these processes contribute to the excessive anemia frequently seen in children with severe malaria.

B Is for B Lymphocytes

After invasion of the body by a pathogen and an initial battle, not all of the pathogens are destroyed. Some of the debris from a pathogen enters the lymph vessels to be carried to the regional lymph node, which swells. These nonself antigens will now trigger the production of antibodies. Antibodies are made by lymphocytes called B lymphocytes or B cells. B cells are essentially little antibody factories, and when they are stimulated by recognition of the correct antigen the synthesis and secretion of antibody molecules are switched on. The discovery of B cells came about in an interesting way (442). It all began with a study of immunity in chickens. In the 1950s, Bruce Glick was a graduate student who sought to discover the function of a lymphoid organ, called the bursa of Fabricius, located at the lower end of the chickens' digestive tract. (The bursa is named after the 16th century anatomist Hieronymus Fabricius, who discovered it.) Glick surgically removed the bursa from some chickens and found no difference between chickens with a bursa and those lacking it. Then another graduate student in Glick's department needed some chickens to demonstrate the production of antibodies. He used Glick's chickens, both those with a bursa and those without one. After the chickens were injected with antigen, he discovered that the chickens lacking a bursa made no antibody. Subsequently it was found that the cells of the bursa (which is essentially a lymph node) that were involved in the production of antibody were lymphocytes. In recognition of their location, they were named B cells after the first letter of bursa. (In mammals, which lack a bursa, the B cells are made in the bone marrow, which fortunately also begins with a B.) The bursa work was submitted to *Science* magazine for publication and was rejected because it was "uninteresting." Later it was published in the journal *Poultry Science*, where it went unnoticed for several years. Glick and his colleagues then made another interesting observation regarding the chickens that lacked a bursa: although they could not make antibody, they could reject skin grafts and were able to overcome a viral infection. This suggested that although the bursa played a role in antibody-based immunity, there must be additional mechanisms that could protect against infection or foreign materials. As will soon be revealed, this additional immune response would involve another kind of lymphocyte, called the T lymphocyte, or T cell.

How do B cells make antibody? On the surface of the B lymphocyte is an antibody molecule that serves as a receptor for the foreign antigen

(374). Once the antigen is bound to the surface, the B cell changes its appearance from the dull-looking lymphocyte, which is almost all nucleus, to the large plasma cell containing a cytoplasm rich with the organelles involved in protein synthesis, ribosomes and endoplasmic reticulum. The plasma cell is triggered to divide asexually, and the population of identical offspring from such division is a clone. Division to produce an adequate number of plasma cells may take several days since the division rate is about 10 h per division. Plasma cells can manufacture large quantities of antigen-specific antibody (up to 100,000 molecules per min), which is secreted into the bloodstream for up to 4 to 5 days, and then they die. Different plasma cells are able to synthesize 10^{15} different antibodies, each having a specificity for the antigen that triggered its division in the first place. The specific antibody produced can protect in several ways: it can neutralize the antigen; it can clump (agglutinate) malaria-infected red cells, bacteria, and free virus in the blood; it may lyse (disintegrate) virus- and malaria-infected cells or bacteria; it may block virus or merozoite entry into cells; or it may make the pathogens more attractive so that they are more readily eaten by phagocytes, a process known as opsonization. Because the antibody is in the liquid fraction (plasma or serum) of the blood, this type of immunity is called humoral immunity after the four humors (black and yellow bile, phlegm, and blood) that the ancient Greeks and Romans believed were able to control our behavior. (We still use these terms when we describe someone who is sluggish as "phlegmatic" or when we say that we are in a bad or good humor.)

Not all the B cells differentiate into plasma cells; some remain quiescent and persist as memory cells. When the memory cell reencounters the same antigen, it undergoes rapid division, plasma cells are produced, and antibody is synthesized and released. This rapid response is due to the memory of an encounter with a nonself antigen that occurred naturally in the past, and it is induced artificially when one is immunized with booster shots.

For many diseases caused by bacteria or viruses, a single infection protects the individual from attack for life. For example, it takes several decades for levels of antibody to diphtheria and tetanus toxin to decline to the nonprotective region, and with the smallpox vaccine (see p. 79) antibody levels remain stable for up to 30 years after vaccination. With DPT and poliovirus vaccine 95% of children vaccinated have protective antibody for 4 years after a booster vaccination. Malaria, however, is a very different kind of microbe: antibody responses are lower, antibodies are short-lived, and more often than not in the absence of parasites, malaria-specific antibodies diminish. Thus, while antibody levels and B cells are crucial for protection against malaria, the specific antibody-producing cells appear to be short-lived and there are fewer and/or short-lived memory B cells able to be fully reactivated. In individuals infected with *P. falciparum*, within 3 to 9 months after exposure only 50% were responders to malarial antigens; in the same period the antibody levels had declined by 66% (1a).

C Is for Cell-Mediated Immunity

The other part of the immune system, the one Glick and his colleagues found not to be bursa related, involves another set of lymphocytes, the T-cell variety, and it is concerned with nonhumoral immunity (334, 374). In the 1960s it was discovered that the bone marrow and thymus are the "master organs" of our (and other mammals') immune system. The bone marrow is the site where all the different cell types of the blood and the immune system are made, and it is here that the blood stem cells also reside. Stem cells are primitive, unspecialized cells that have no function other than to divide and to make other cells. However, when a stem cell divides to produce two daughter cells, the offspring are unlike: one daughter cell is an identical replica of the stem cell, but the other can grow and become a very specialized cell. In this way, stem cells are self-renewing and are able to give rise to a variety of specialized cell types, including B and T lymphocytes. A T lymphocyte arises in the bone marrow but migrates via the bloodstream to a gland in the neck called the thymus. The thymus was so named because in the 2nd century the physician Galen, in an imaginative mood, thought the shape of the gland resembled the leaves of the thyme plant. (The thymus glands of young beef and lambs have another name: butchers and restaurants call them sweetbreads.)

The thymus is made of lymphoid tissue. It is located just above the heart. In children the gland is rather large, but it tends to atrophy and shrink with age. It is the place where the T lymphocytes from the bone marrow establish residence for some time and where they are selected. While in the thymus, some T cells become either helper T cells (also called $CD4^+$ T cells) or killer T cells (also called $CD8^+$ T cells). The killer T cells are also referred to as CTLs. Some of the progeny of these T cells leave the thymus and, by way of the bloodstream, reach and settle in other lymphoid organs such as the spleen, the tonsils, and the variously distributed lymph nodes, where they are activated by dendritic cells that bring and present pathogen-derived antigens. After activation, CTLs patrol the body in search of altered cells (cancer cells), foreign cells (grafts of tissues), or cells compromised by a pathogen (such as liver stage forms). Because they are capable of attacking and destroying any cell in the body that appears to be nonself, they are aptly called killers. But in the attack the CTL does not direct its action against the pathogen itself but, rather, against the cell (no longer recognized as self) in which the pathogen is hiding. Killing may occur in one of two ways. In one mechanism, CTLs have a T-cell receptor molecule that recognizes a foreign antigen on the surface of infected (or transformed) cells and, after attachment, the CTL punches in a "security code" that then triggers a self-destruct mechanism in the nonself cell. In effect, the CTL simply tells the pathogen-infected cell to die. The other mechanism is for the CTL to punch a hole in the cell by using a pore-forming molecule and then inject the cell with a lethal cocktail of destructive enzymes. The CTL response is strong within 5 days after exposure to a foreign antigen, peaks between days 7 and 10, and then declines over time. (Antibody formation usually peaks after the CTL response.)

What roles do T cells play in protective immunity? Macrophages, the first line of defense in cell-mediated immunity, are the "foot soldiers" of the immune system, and they are also the immune system's cellular "vacuum cleaners" or "filter feeders." They ingest, digest, and then regurgitate the foreign pathogens or even the dead and dying pathogen-containing cells they have eaten. In this way, they are able to display the broken bits of antigen on their outer surface in association with another molecule called the major histocompatibility complex (MHC), which occurs in two forms, MHC I and MHC II.

MHC II molecules are found only on a few specialized cell types, including macrophages, dendritic cells, and B cells, all of which are "professional antigen-presenting cells." MHC I molecules are found on almost every nucleated cell of the body and generally interact exclusively with cytotoxic CD8$^+$ T cells (CTLs). The combination of MHC II and antigen provides information that is offered to other lymphocytes expressing either the CD4 or CD8 T-cell receptor on their surface. Once bound, the T cells are triggered to divide and hence to form a clone. Rather than secreting antibody, however (as would a B cell), the T cells release chemical messengers called cytokines. The cytokines then activate T lymphocytes or other cell types including macrophages, or they can trigger the B cells into activity, so that they begin the process of antibody synthesis and secretion.

In addition, as with B cells, memory T cells are left after an encounter with a foreign antigen, and so the response to a second encounter is swifter and more specific. The CD4 helper T cell is the keystone that maintains the connection between the two branches of the immune system: cell mediated and humoral. The humoral branch produces antibodies that prevent invasion of cells, whereas the members of the cell-mediated branch search out and destroy nonself cells. Although CTLs are capable of killing any cell in the body (except red blood cells), they do so only if the cells express MHC I on their surface. This allows infected cells to signal their plight to the CTL. Once this information is received by the CTL, the infected cell is killed.

The goal of vaccination is to provide protection against a foreign invader, and by understanding the immune mechanisms by which protection is achieved it may be possible to design novel vaccines as well as to improve existing ones. The immune system is of necessity complex because it has to protect against a very large army consisting of many different kinds of invaders. The previous discussions have described the two immune systems: innate and acquired. Both are necessary for survival. The former, activated within minutes or hours after invasion by a malaria parasite, consists of specialized cells such as the phagocytic neutrophils, macrophages, and NK cells, as well as different products, such as the cytokines and C-reactive proteins of the complement pathway. The innate system is involved in limiting the early blood stages of a malaria infection (481). The activated antigen-presenting dendritic cells produce the IL-12, which interacts with NK cells that, in turn, secrete large amounts of IFN-γ; the dendritic cells also activate T helper cells, which promote adaptive immune responses that, if successful, control or eliminate the parasite.

The adaptive system, thought by some to have been "superimposed" on the innate system during evolution, differs from the innate system in two critical properties—greater specificity and memory. In general it takes days or sometimes longer for the adaptive immune system to begin to act. There are three components in this system: two classes of lymphocytes (T and B), as well as specialized antigen-presenting cells. The T lymphocytes are often said to be involved in cell-mediated responses, although actually many of their responses are mediated by secreted factors. Of the two main classes of T cells, the CD4 cells help B cells make antibody and the CD8 cells are engaged mainly in recognizing and killing infected cells.

During a blood infection, a cascade of immune responses is set in motion. Early on, dendritic cells are activated to produce IL-12 and/or

TNF-α, as well as interacting with NK cells that produce the first wave of IFN-γ. Monocytes and macrophages are activated by the IFN-γ; this leads to increased phagocytosis of infected red cells, and parasites hiding within the red cell are killed by nitric oxide. CD4 T cells, also activated by dendritic cells, induce B cells to begin the process of producing antibody, resulting in increased opsonization of infected blood cells; this leads to a temporary reduction in the numbers of blood parasites. In parallel with increased numbers of blood stage parasites, hemozoin is ingested; this will act to not only paralyze the monocytes but also to adversely affect dendritic cell maturation. In *P. falciparum*, dendritic cell contact with infected red cells also paralyzes their function. In consequence, there is reduced secretion of IL-12 and increased secretion of IL-10, a growth factor for B cells. As the infected red cells switch to another surface antigen, there will be a second wave of parasite multiplication; this leads, in turn, to dendritic cell and NK cell activation; the IL-10 counteracts the effects of IFN-γ and TNF-α on other cells, and the level of immunoglobulin directed to the new variant begins to increase. This serves to control the infection; however, antigenic variation permits the parasite to escape the newly established immune response and the cascade begins again (520). In effect, a chronic blood infection is established. In the infected liver cell, CD8 T cells (again primed by dendritic cells) that recognize parasite-derived peptides associated with the cell surface MHC I are activated and secrete antimicrobial cytokines such as IFN-γ and TNF-α and cell-destroying factors such as perforin and granzymes, all of which may contribute to the elimination of the infected liver cell and kill the developing parasite (362).

Natural acquired immunity, as well as the practice of vaccination, is dependent on the coordinated activity of the complex components of the immune system. It is the primary goal of vaccination to harness the power of the adaptive immune response such that the enhanced numbers of antigen-specific lymphocytes directed against the parasite will be sufficient to provide protection following subsequent challenge. It is hoped that the more we know about the immune responses to a natural malaria infection, as well as the more we understand about the character of the immune systems themselves, the better able we will be to develop a protective malaria vaccine.

4

From Milkmaids to Vaccines

Thanks to vaccination smallpox has been eradicated in nature, but at one time it was one of the most devastating of all human diseases. The cause of smallpox is a virus, one of the largest of all the viruses, so large that it can actually be seen under a light microscope. Much of its detailed structure, however, can be visualized only by using an electron microscope that can magnify objects 100,000 times. The outer surface of the smallpox virus resembles the facets of a diamond, and its dumbbell-shaped core contains about 200 genes, 35 of which are thought to be involved in virulence (442).

Most commonly the smallpox virus enters the body through droplet infection by inhalation. It can also be acquired by direct contact or through contaminated fomites (inanimate objects) such as clothing, bedding, blankets, and dust. The infectious material from the pustules can remain infectious for months. The virus multiplies in the mucous membranes of the mouth and nose. In the first week of infection there is no sign of illness; however, during this time the virus can be spread by coughing or by nasal mucus. The virus moves on to the lymph nodes and then to the internal organs via the bloodstream. Here it multiplies again. It then reenters the bloodstream. Around day 9 the first symptoms appear: headache, fever, chills, nausea, muscle aches, and sometimes convulsions. The person feels quite ill. A few days later a characteristic rash appears. The individual is infectious a day before the rash appears and remains infectious until all the scabs have fallen off. Many patients die a few days or a week after the rash appears. Not infrequently there are complications from secondary infections. The infection results in destruction of sebaceous (oil) glands of the skin, leaving permanent craterlike scars known as pockmarks (442).

There are two varieties of the smallpox virus: *Variola major* and *Variola minor*. *Variola major*, the deadlier form, frequently killed up to 25% of its victims, although in naive populations the fatality rate could exceed 50%. *Variola minor*, a milder pathogen, had a fatality rate of about 2% and was more common in Europe until the 17th century, when it mutated to the more lethal form. In the 17th century, smallpox was Europe's most common and devastating disease, killing an estimated 400,000 persons each year. England suffered most severely in the second half of the 17th century; between 1650 and 1699 there were 20 deaths a week. In 1660 in London alone, more than 57,000 died of smallpox in a population of 500,000: a mortality rate of 10% (442). It is estimated that smallpox killed an estimated 400,000 Europeans each year during the 18th century and was responsible for a third of all cases of blindness; over 80% of infected children died of the disease. It is estimated that during the 20th century smallpox was responsible for 300 million to 500 million deaths, more than the deaths due to all the wars during that same period. In the early 1950s there were ca. 50 million cases of smallpox in the world each year, and in 1967 the WHO estimated that 15 million people contracted the disease and 2 million died in that year.

Smallpox and Cowpox

The farmers of Gloucestershire in England knew from experience that if a person contracted cowpox, he or she was assured of immunity to smallpox (375, 442). In milk cows, cowpox appears as blisters on the udder that clear up quickly and do not develop into a serious illness. Farmers who come into contact with cowpox-infected cows develop a mild reaction with the eruption of a few blisters on the hands or lower arms. Once exposed to cowpox, neither cows nor humans develop any further symptoms. Early on it was noted that hardly any milkmaids or farmers who had contracted cowpox showed any of the disfiguring scars of smallpox, and most milkmaids were reputed to have fair and almost perfect skin, as suggested in the nursery rhyme:

> Where are you going, my pretty maid
> I'm going a-milking sir, she said
> May I go with you my pretty maid
> You're kindly welcome, sir, she said
> What is you father, my pretty maid
> My father's a farmer, sir, she said
> What is your fortune, my pretty maid
> My face is my fortune, sir, she said.

In 1774 a cattle breeder named Benjamin Jesty accidentally contracted cowpox from his herd, thereby immunizing himself. He then deliberately inoculated his wife and two children with cowpox. They remained immune even 15 years later when they were deliberately exposed to smallpox. Edward Jenner (1749 to 1823), an English physician, did not know what caused smallpox or how the immune system worked, but nevertheless he was able to devise a practical and effective method for producing immune protection against attack by the lethal smallpox by taking advantage of the local folk tale. On 14 May 1796, Jenner took a small drop of fluid from a pustule on the wrist of Sarah Nelms, a milkmaid with an active case of cowpox. He smeared the material from the cowpox pustule onto the unbroken skin of a small boy, James Phipps. Six weeks later, Jenner tested his "vaccine" (from the Latin "vacca," meaning "cow") and its ability to protect against smallpox by deliberately inoculating the boy with material from a smallpox pustule. The boy showed no reaction—he was immune to smallpox. In the years that followed, "poor Phipps" (as Jenner called him) was tested for immunity to smallpox about a dozen times, but he never contracted the disease. Jenner wrote up his findings and submitted them to the Royal Society, but to his great disappointment his manuscript was rejected. It has been assumed that the reason for this was that he was simply a country doctor and not a part of the scientific establishment. In 1798, after a few more years of testing, he published a 70-page pamphlet, *An Inquiry into the Causes and Effects of* Variola Vaccinae, in which he reported that the inoculation with cowpox produced a mild form of smallpox that would protect against severe smallpox. He correctly observed that the disease produced by vaccination would be so mild that the infected individual would not be a source of infection to others, a discovery of immense significance.

It is Jenner, however, not Jesty, who is given credit for vaccination, because he carried out experiments in a systematic manner over a period of 25 years to test the farmer's tale. Jenner wrote: "It is necessary to observe, that the utmost care was taken to ascertain, with most scrupulous precision, that no one whose case is here adduced had gone through smallpox previous to these attempts to produce the disease. Had these experiments been conducted in a large city, or in a populous neighborhood, some doubts might have been entertained; but here where the population is thin, and where such an event as a person's having had the smallpox is always faithfully recorded, no risk of inaccuracy in this particular case can arise."

The reactions to Jenner's pamphlet were slow in coming, and many physicians rejected his ideas. Cartoons appeared in the popular press showing children being vaccinated and growing horns. But within several years, some highly respected physicians began to use what they called the "Jennerian technique" with great success. By the turn of the century the advantages of vaccination were clear, and Jenner became famous. In 1802 Parliament awarded him a prize of £10,000, and in 1807 he was awarded £20,000. Napoleon had a medal struck in his honor, and he received honors from governments around the world.

By the end of 1801 Jenner's methods were coming into worldwide use, and it became more and more difficult to supply sufficient cowpox lymph or to ensure its potency when it had to be shipped over long distances. To meet this need, Great Britain instituted an Animal Vaccination Establishment in which calves were deliberately infected with cowpox and lymph was collected. Initially this lymph was of variable quality, but when it was found that addition of glycerin prolonged preservation, this became the standard method of production. The first "glycerinated calf's lymph" was sent out in 1895. With the general acceptance of Jenner's findings, the hide of the cow called Blossom, which Jenner used as the source of cowpox in his experiments, was enclosed in a glass case and placed on the wall of the library in St. Georges's Hospital in London, where it remains to this day.

Why does vaccination work? Variolation, or inoculation with variola major, can protect against death from the disease; however, because it is the virulent form of the virus it may result in the death of 1 to 3% of those who have been variolated; further, the variolated person is infectious. Vaccination, on the other hand, involves inoculation with the agent of cowpox, named by Jenner *Variola vaccinae*. The exact relationship between the virus used in vaccines today and Jenner's cowpox virus is obscure. Indeed, in 1939 it was shown that the viruses present in the vaccines used were distinct from contemporary cowpox as well as smallpox. Samples of Jenner's original vaccine are not available, so it is not possible to determine what he actually used; current thinking is that the currently available strains of vaccine are derived from neither cowpox nor smallpox, but from a virus that became extinct in its natural host, the horse.

The two viruses, *Variola vaccinae* and *Variola major*, are 95% identical genetically and differ by no more than a dozen genes. For human protection to occur, there must be cross-reactivity between the two viruses such that antibodies are produced in the vaccinated individual that can neutralize the variola major virus should the need arise. Fortunately, this does happen. Recovery from a smallpox attack gives lifelong immunity, but vaccination ordinarily does not. The advantages of vaccination over variolation are twofold. First, the recipient of a vaccination is not infectious to others; second, death is very rare.

Smallpox is a disease of historical interest because it was certified as eradicated by the World Health Assembly on 8 May 1980. This feat was accomplished 184 years after Jenner had introduced vaccination. Smallpox is the first and only naturally occurring disease to be eradicated by human intervention. The eradication program began in 1967. By 1970 smallpox had been eliminated from 20 countries in western and central Africa; by 1971 it had been eliminated from Brazil, by 1975 it had been eliminated from all of Asia, and in 1976 it had been eliminated from Ethiopia. The last natural case was reported from Somalia in 1977. Eradication of smallpox was possible for three reasons: (i) there were no animal reservoirs; (ii) the methods of preserving the vaccine proved to be effective; and (iii) the vaccine was easily administered. With this kind of success there has been a move to eradicate other infectious diseases, particularly measles, polio, and malaria. So far, however, eradication has not been achieved for any of these diseases and smallpox remains the only disease ever eradicated by immunization.

Using Crippled Microbes

Louis Pasteur was small in stature and sturdily built, and his face appeared to be carved from granite; with his high forehead and grayish green eyes with a deep look, he seemed both serious and sad. After his stroke in 1868 he moved only with difficulty and dragged his left foot, and a crippled left hand made him dependent on others for carrying out most of his experimental procedures. The atmosphere in the Pasteur laboratory was severe; smoking was not permitted, his assistants were expected to keep strict hours, they were treated more like servants than colleagues, and they all worked in silence. Pasteur spent hours in silent observation, his eye glued to the microscope (147, 367). Indeed, as Roux remarked, "Nothing escaped his myopic eye and jokingly we said he saw the microbes grow in the broth." He was authoritarian and demanding, and he would focus on only one thing at a time, becoming completely absorbed by the subject he had chosen to study. He trusted only himself, and he alone kept the records of the experiments by writing down the information provided by his assistants. In 1875 he attempted to induce immunity to cholera (147, 367). He was able to grow the cholera bacteria that cause death in chickens and to reproduce the disease by injecting healthy chickens with the bacteria. The story is told that he placed his cultures on a shelf in his laboratory exposed to the air, went on summer vacation, and on his return injected chickens with an old culture; the chickens became ill, but to his surprise they recovered. Pasteur then grew a fresh culture of the cholera bacteria with the intention of injecting another batch of chickens, but the story goes that he was short of chickens and used those which he had previously injected. Again he was surprised to find that the chickens did not die from cholera but instead they were protected from the disease. To the prepared mind of Pasteur, this turn of events suggested that aging and exposure to air (and possibly the heat of summer) had weakened the virulence of the bacterium, that such an attenuated or weakened strain might have protected against disease, and therefore that attenuated strains might be used for protection in the future.

In 1878 Pasteur spoke before the Academy of Sciences and described his experiments with the anthrax bacillus, never once mentioning the name of Robert Koch, who had already demonstrated the germ 2 years earlier. In 1880 he once again presented to the members of the Academy his results with a chicken cholera vaccine. In that talk he casually mentioned that Jean-Joseph Henri Toussaint, a professor at the Veterinary School in Toulouse, was the first to isolate the microbe that caused chicken cholera. For 9 months Pasteur kept secret the method by which he attenuated the virulence of the cholera microbe, and never once did he mention the summer accident of 1875. Finally, he disclosed that attenuation of the cholera microbe involved simply culturing it for 2 to 3 months exposed to atmospheric oxygen. He then went on to describe how a "live attenuated" cholera vaccine was able to protect fowl against a lethal challenge. Pasteur theorized that the live attenuated vaccine protected the chickens by "depleting the host of a certain element that healing does not reconstitute and that the absence of which hinders the development of the microbe when re-inoculated a second, third or fourth time" (187). Ever bold, he claimed that it would be possible to attenuate the virulence of all microbes simply by growing them under special conditions.

After Pasteur's success in creating a vaccine against chicken cholera, the next major step came in 1886 from the United States. Theobald Smith and Daniel Edward Salmon published a report on their development of a heat-killed cholera vaccine that immunized and protected pigeons. Two years later, these two investigators claimed priority for having prepared the first killed vaccine. Although their work appeared in print 16 months before a publication by Chamberland and Roux (who worked silently in the laboratory of Pasteur) with identical results, the fame and prestige of Pasteur were so widespread, and so bright was the aura surrounding those who worked in Pasteur's Institute in Paris, that the claim of "first killed vaccine" by Smith and Salmon was lost.

Flush with success at having produced a cholera vaccine, Pasteur began to develop a vaccine for anthrax. In this his competitor was not his archrival the German Koch but a fellow Frenchman. On 12 July 1880, Henri Bouley (a veterinarian) read before the Academy a report from his friend Toussaint which described his anthrax vaccine, produced by killing the bacteria by heating for 10 min at 55°C, that had been used on 8 dogs and 11 sheep. Of the eight dogs, four injected with the killed vaccine survived a series of four successive injections of virulent live anthrax. All unvaccinated dogs died after the first injection. Similar results occurred with the sheep. When Pasteur heard of Toussaint's results from Bouley, he was astonished. He wrote, "it overturns all the ideas I had on viruses, vaccines etc. I no longer understand anything . . . it makes me want to confirm it to my own satisfaction" (123, 124). Pasteur could not, would not accept with grace that he was wrong in his concept of immunity, namely, that the biological activity of a living, if attenuated, microbe was critical to deplete the host of essential nutrients.

By August of 1880 Toussaint had changed the method of killing the anthrax bacteria, and instead of heat he had resorted to carbolic acid (phenol), which Lord Lister had used as an antiseptic. In February 1881 Pasteur announced he had an attenuated anthrax vaccine produced by the same technique he had used for attenuation of the cholera microbe, i.e., exposure to atmospheric oxygen. A month later, when Pasteur reported successful results of using his anthrax vaccine in sheep, a veterinarian, Hippolyte Rossignol from Pouilly le Fort, a farm town 40 miles outside Paris, challenged him. Rossignol prodded Pasteur to test the vaccine at his farm. Pasteur had to accept the challenge or else his claim of priority over Toussaint would have been damaged. Further, there were rumors that Pasteur was seeking to profit financially from his "secret remedies" against livestock diseases. The flamboyant Pasteur "welcomed" the challenge, and a public trial was conducted at Pouilly le Fort in June of 1881. There were more than 200 observers including government officials,

politicians, veterinarians, farmers, cavalry officers, and newspaper reporters. Of 50 sheep in the trial, half were vaccinated on 5 May and 17
May and the remainder were left unvaccinated. On 31 May all the sheep
were challenged with a virulent culture of anthrax bacilli. By 3 June all the
unvaccinated sheep were dead. The London correspondent of the Paris
Journal des Debats wrote: "When Pasteur spoke, when his name was mentioned, a thunder of applause rose from all benches from all nations. An
indefatigable worker, a sagacious seeker, a precise and brilliant experi-
mentalist, an implacable logician and an enthusiastic apostle, he has produced an invincible effect on every mind." Pasteur's fame spread
throughout France, Europe, and beyond (123, 124, 147).

There is, however, another side to the dramatic success of Pasteur's
attenuated anthrax vaccine. Chamberland in Pasteur's laboratory had
been experimenting with a dead vaccine prepared by chemical treatment
with potassium dichromate. In small-scale tests it worked. Only Pasteur
and his collaborators knew the nature of the vaccine used for his famous
trial at Pouilly le Fort. Indeed, it is now known from an examination of
Pasteur's laboratory notebooks that Pasteur used Chamberland's "dead"
vaccine, not the live attenuated one that he emphasized was critical to
success against chicken cholera (187). At the time, however, Pasteur, not
Toussaint, received credit for developing the first successful vaccine
against anthrax. Toussaint subsequently published only two more scientific papers before his death in 1890 at age 43 after suffering a mental
breakdown. In France Pasteur was recorded as the inventor of the vaccine
against anthrax; Koch, however, persisted in denigrating Pasteur's contributions and hailed Toussaint as the real inventor. Koch also criticized Pasteur's incomplete disclosure of the process used for attenuation. We do
not know how many in Pasteur's laboratory knew the true nature of the
anthrax vaccine used at Pouilly le Fort, but we do know that Pasteur's
concept of attenuation of microbes by exposure to atmospheric oxygen
was wrong. Pasteur acknowledged that there were differences between
the published method and the technical knowledge required to produce
the vaccine, and he and his collaborators Chamberland and Roux never
filed a patent on the production process. (The 1844 Patent Law in France
prohibited patenting pharmaceutical products, including those for veterinary use.) Nevertheless, Pasteur was able to create a monopoly for the production and distribution of the vaccine by keeping secret the laboratory
methods and technical devices used for attenuation as well as the test standards used for establishing the virulence of the vaccines (76). In addition,

there were formal contracts for the exclusive rights to exploit the vaccine and control was exercised over the commercial use of Pasteur's name and later the trade name. The anthrax vaccine manufactured commercially by Pasteur was produced at first in a small production laboratory in Rue d'Ulm; later, a separate commercial unit at Rue Vauquelin was established with Chamberland in charge. In 1882, 245,000 sheep were vaccinated, and between 1882 and 1887 some 2 million were vaccinated. By 1887 the three inventors renounced their share of the profits for the benefit of the newly established Pasteur Institute, thus yielding a substantial income for the Institute. The Pasteur anthrax vaccine, however, was not perfect: not only did it require two immunizations administered within a 2-day interval, but also it required a particular degree of virulence; there were accidental deaths involving the vaccine, and no protection was achieved if the manufactured vaccine was less virulent. Declines in virulence required selection of new isolates, and complex tests were required to measure the degree of attenuation. Associated with losses in virulence were problems of stability stemming from storage and transport time. These factors served to justify Pasteur's position that vaccine production should remain under the inventors' control, and by limiting the development of special laboratory facilities elsewhere the inventors prevented exploitation by others, in effect creating a monopoly. By 1886 a commercial enterprise, La Compagnie de Vulgarisation du Vaccin Chambonneux Pasteur, was established to market the vaccine outside of France. This gave legal recognition to the Institut Pasteur monopoly since the contracts specified the scope of the concession to a licensee, determined the distribution of the product (spores and culture fluids) and the transfer of knowledge between the Pasteur laboratory and the licensee, fixed the price, set the amount of royalties to be paid to the inventors, and confirmed the Pasteur laboratory's technical primacy.

In humans the dose necessary to kill half the persons infected with anthrax is estimated to be 2,500 to 55,000 inhaled spores; as few as 1 to 3 spores, however, may be enough to result in an infection. Natural cases of human anthrax through inhalation of spores is rare in recent history, but in 1979 anthrax spores were accidentally released from a military laboratory in the former Soviet city of Sverdlovsk and 64 people were reported to have died, although U.S. intelligence sources claimed the numbers might have reached 1,000. In the century prior to 2001, a year when there was a deliberate delivery of anthrax spores in mailed letters that resulted in 10 confirmed cases, the incidence of anthrax in the United States amounted to a total of 18 cases. Currently, because anthrax is considered to be an appealing biological weapon, research into the production of a protective anthrax vaccine is under way. The currently licensed anthrax vaccine in the United States is not based on a live attenuated vaccine, as proposed by Pasteur, nor is it the inactivated preparation first developed by Toussaint and recognized only in 1998 by the French government; it consists of a protein, called protective antigen (a central component of the anthrax toxin), adsorbed to aluminum hydroxide with small amounts of formaldehyde and benzethonium chloride added as preservatives (472). Prepared from attenuated cultures of a capsule-free strain of *Bacillus anthracis* and named Biothrax, it is licensed and recommended for a small population of mill workers, veterinarians, laboratory scientists, and others with a risk of occupational exposure to anthrax (541, 564).

Pasteur's final and certainly most famous success was his rabies vaccine (123, 147, 187). Rabies, referred to as mad dog disease or hydrophobia, has always had a hold on the public's imagination. Its victims were described as raging, howling, and having to be controlled by being placed between two mattresses. The treatment used was as horrifying as the symptoms: feeding of cock brains, crayfish eyes, livers from mad dogs, and snake skins mixed with wine; cauterizing the bite wounds with a red-hot poker; or sprinkling gunpowder on the wound and then igniting it (356). Pasteur's goal was to produce a vaccine by attenuation that would protect the individual before the appearance of clinical symptoms. In the case of anthrax and chicken cholera, Pasteur knew the causative agents; he had seen them with his own eyes under a microscope. Rabies was different. The infectious agent was a self-reproducing entity and, although invisible to the naked eye, could still be identified. In 1884, Pasteur's associate Chamberland invented a porcelain filter that made "it possible to have one's own pure spring water at home, simply by passing the water through the filter and removing the microbes." Since the causative agent of rabies was able to pass through the small pores of Chamberland's filter, Pasteur deduced that it had to be smaller than a bacterium; today we know that rabies is caused by a virus (meaning "poison"). The rabies virus would be seen only in the 1950s, after the development of electron microscopes with magnifications of 100,000 diameters.

In the work on a vaccine for rabies, Pasteur's colleague was the young physician Emile Roux. By air drying in a glass jar devised by Roux, the virus in the spinal cords from rabid rabbits were attenuated. A strip of spinal cord was suspended by a thread at the top of the jar containing a hole at the top and another at the lower side, thus allowing air to enter from the bottom opening and to exit from the hole at the top. Cotton plugs placed in the holes allowed air, but not microbes, to enter the jar. Initially Roux used heat to attenuate the virus; at Pasteur's suggestion caustic potash was added to the jar to prevent the strip from putrefying. Over a 12-day drying period, the potency of the rabies virus in the cord diminished. In 1885 a 9-year-old boy named Joseph Meister was severely bitten by what was presumed to be a rabid dog. His grief-stricken mother took young Meister to Paris, where she appealed to Pasteur to help keep her son from dying in agony. Pasteur agreed to treat the boy with his attenuated rabies vaccine. As the world knows, Meister survived and Pasteur became the stuff of legends.

There is, however, another side to the story, and this concerns the ethical issues of the rabies vaccine and the difficult personal relationship between Roux and Pasteur (185, 186). Pasteur and Roux were an odd couple: Pasteur, short and stout, was at age 60 a visibly exhausted and stroke-debilitated man who could no longer work at the bench performing his own experiments, whereas the tall and thin Roux, age 30, worked alone in the evenings. Roux the physician took a clinical approach to experiments and tended to draw limited conclusions from the experimental data, whereas Pasteur the chemist paid greater attention to general patterns, to "signals" amidst the "noise." Where Roux was self-effacing, exceptionally critical, and witty, Pasteur was confident, highly focused, ambitious, driven, serious, and utterly devoid of a sense of humor. Pasteur was concerned with material possessions and financial security, was family conscious and a bellicose patriotic Bonapartist, whereas Roux was a bohemian bachelor pacifist who shunned material comforts and was fiscally generous to a fault. Roux began his career in 1878 when Pasteur, recognizing that work on infectious diseases of animals and humans needed a medical man, hired him as an "animal inoculator." Roux was superb technically and was soon attenuating anthrax cultures and injecting them into experimental animals. The first signs of discord between the two men surfaced in the vaccine trial at Pouilly le Fort in 1881. When some of the vaccinated animals became feverish Pasteur feared the worst, and he attributed this to a careless technique on the part of Roux. At first he thought of sending Roux to face the humiliation of a failed trial; when news came that the trial was to be a triumph, however, Pasteur traveled alone to bask in the applause of the crowd. Initially Pasteur refused to allow Chamberland and Roux to publish the results of their chemically altered anthrax vaccine that had resulted in the trial's success, but when Pasteur allowed publication he stipulated that they were to avoid mention of the method used. More was at play than Roux's resentment of his role as an assigned subordinate to the master to whom all glory redounded. Roux did not subscribe to Pasteur's reckless attitude in the face of ambiguous experimental evidence. This was particularly evident during the rabies vaccination, where Roux's absence was conspicuous. A recent review of Pasteur's 102 laboratory notebooks (187) has revealed that when he undertook to treat Meister he had not achieved "multiple proofs" of the efficacy of his attenuation method on "diverse animal species," as he stated in his paper read to the Academie des Sciences on 26 October 1885 and in which he wrote: "I have succeeded in immunizing 50 dogs of all ages and breeds . . . without a single failure." In fact, at the time of Meister's vaccinations, only 20 of the 40 to 50 experimental dogs had received a full treatment, and none of the dogs had survived as long as 30 days since their last and highly lethal injection. Further, as Pasteur himself admitted, not a single one of these dogs had first been bitten or otherwise inoculated with rabies before being treated by the method used on Meister. These dogs were rendered "immune" by a method precisely the opposite of that used on Meister: the dogs were treated with a rabies vaccine that was step-by-step made with less lethal strains, whereas Meister received injections that began with a less lethal strain and moved to a more lethal strain. It must be emphasized that although later animal experiments vindicated Pasteur's original intuition, at the time of Meister's treatment, he did not have the evidence. Roux did not participate in the rabies vaccine trials (nor in any published account), probably because he did not agree with Pasteur's cavalier attitude in administering a highly virulent rabies strain to a human and because he considered the treatment of Meister to be a form of unjustified human experimentation. Pasteur failed in his ethical conduct (186) on several counts: he did not seek informed consent, he did not show great concern for his subject's right to privacy, and he had not established the safety and efficacy of his method in animal or human testing when he undertook treatment of Meister. Roux kept secret these ethical lapses, the deliberate deceptions, and his resentment of the subordinate role accorded him by Pasteur's high-handed treatment. Shortly before Pasteur's death, Roux did achieve success on his own for the introduction of serum therapy against diphtheria, and he later became Director of the Institut Pasteur, a position he held for 29 years (1904 to 1933); these honors, however, never matched the glory and praise that had been showered on his master, Pasteur.

Attenuation was also used in the development of BCG (bacillus Calmette-Guérin), the vaccine for tuberculosis named after the two investigators at the Pasteur Institute who produced it: Albert Calmette (1863 to 1933) and Camille Guérin (1872 to 1961). It took 231 subcultures of a bovine tubercle bacterium, *Mycobacterium bovis*, in an ox bile-containing medium every 3 weeks for 13 years before the bacterium was attenuated. Different preparations of liquid BCG were used in controlled community trials conducted before 1955; the results of these trials indicated that estimated rates of protective efficacy ranged from 56 to 80%. Since 1975, case-control studies with different BCG strains indicated that vaccine efficacies ranged from 0 to 80%. In young children, the estimated protective efficacy rates have ranged from 2 to 80% for prevention of pulmonary tuberculosis. In the United States BCG is not used, because tuberculosis is not prevalent and the chances are low that infants and young children will become exposed. Another reason BCG is not used in the United States is that it may cause a tuberculin skin test to convert from negative to positive. This can be confusing because the tuberculin skin test is the best available test for a tuberculosis infection, and widespread use of BCG would make the skin test less useful.

It is difficult to minimize the contribution that vaccination has made to our well-being (442). It has been estimated that were it not for childhood vaccinations against diphtheria, pertussis (whooping cough), measles, mumps, smallpox, and rubella, as well as protection afforded by vaccines against tetanus, cholera, yellow fever, polio, influenza, hepatitis B, bacterial pneumonia, and rabies, childhood death rates would probably hover in the range of 20 to 50%. Indeed, in countries where vaccination is not practiced, the death rates of infants and young children remain at that level. In the period from 1870 to 1890, killed vaccines for typhoid fever, plague, and cholera were developed, and by the turn of the 20th century there were two human attenuated virus vaccines (vaccinia and rabies) and three killed ones. Although there were other successes, such as a formalin-inactivated diphtheria toxin (toxoid) in 1923 and BCG in 1927, the golden age of vaccine development really began in 1949 when viruses were first grown in chicken embryos and in human cell cultures. This led to the first live polio vaccine; the Salk formalin-inactivated polio vaccine; attenuated Japanese encephalitis, measles, and rubella vaccines; and to an inactivated rabies vaccine. In the 1970s and 1980s bacterial vaccines were made using the capsular polysaccharides from *Streptococcus pneumoniae*, *Neisseria meningitidis*, and *Haemophilus influenzae*.

The majority of vaccines being developed today use new technologies: DNA vaccines, subunit vaccines (purified proteins or polysaccharides), and genetically engineered and virus-transmissible antigens. The older methods continue to be useful, however, as is the case with the live influenza virus, poliovirus, and yellow fever virus vaccines. A prominent vaccinologist said, "If these new technologies succeed, the golden age will turn to platinum." Indeed, the list of successful protective vaccines is an impressive one (Table 1). There is, however, still no malaria vaccine. The question is: why?

Table 1 Vaccines against viruses and bacteria[a]

Live attenuated microbe is used as vaccine
  Yellow fever (V)[b]
  Rabies (V)
  Polio (oral) (V)
  Measles (V)
  Mumps (V)
  Rubella (V)
  Chicken pox (V)
  Shingles (V)
  BCG (tuberculosis) (B)[b]
  Typhoid (oral) (B)

Killed microbe is used as vaccine
  Polio (injected) (V)
  Hepatitis A (V)
  Typhoid (B)
  Cholera (B)
  Plague (B)
  Pertussis (B)
  Rabies (V)
  Japanese encephalitis (V)

Purified protein or polysaccharide from microbe is used as vaccine
  Hepatitis B (V)
  Influenza (V)
  Human papillomavirus (V)
  DPT[c] (B)
  Streptococcus pneumoniae (B)
  Typhoid (B)
  Haemophilus influenzae (B)
  Neisseria meningitidis (B)
  Anthrax (B)
  Lyme disease

[a]Modified from B. Bloom and P.-H. Lambert (ed.), *The Vaccine Book* (Academic Press, Inc., San Diego, CA), 2002, p. 185.
[b](V), viral disease; (B), bacterial disease.
[c]DPT, diphtheria pertussis tetanus vaccine.

5

Fundamental Findings

The various life cycle stages of *Plasmodium* (Fig. 1, p. 40) provide potential targets for immune (and drug) interventions. Ideally, research on vaccines (and drugs) should be carried out with human plasmodia; however, the availability of human subjects is limited, there are ethical and safety considerations in the use of human subjects, and most *Plasmodium* species that infect humans cannot be grown in vitro or in other animals. Therefore, apart from the use of malaria to treat the symptoms of late-stage syphilis (see p. 285), the early research on protective vaccines and immune mechanisms was carried out using birds, monkeys, and rodents as surrogates for the disease as it occurs in humans.

Only 4 years separated Laveran's discovery of the human malaria parasite *P. falciparum* and the discovery of avian malaria parasites by Basil Danilewsky in 1884. Danilewsky (1852 to 1939), a physician from Kharkov, Russia, examined 300 birds from the Ukraine and found malaria-like parasites in their blood. Neither Danilewsky nor Laveran, however, was immediately aware of the other's findings. The parasites Danilewsky saw were mistakenly believed by him to be different forms of one kind of parasite whereas he actually had discovered several different kinds of blood parasites. Although he observed exflagellation, like Laveran, he also failed to recognize its significance. Nevertheless, despite these gaffes, his finding prompted others to seek out infections in other birds. Of the 43 different kinds of avian malaria parasites, only a few have been found to be useful for laboratory studies. The first of these to be identified with certainty was *P. relictum*, which had been isolated from heavily infected sparrows by Grassi and Feletti in 1892 and which was transmissible to canaries by blood inoculation. (Later, this same parasite would be used by Ross to identify the mosquito stages in *Culex* mosquitoes [see p. 25].) In 1926, Wilhelm Roehl (1881 to 1929), a student of Paul Ehrlich working at the Bayer laboratories in Düsseldorf, used *P. relictum*-infected canaries to screen potential antimalarials and was able to identify the first synthetic antimalarial, pamaquine (plasmochin) (468).

The ties that bound bird malaria to human malaria were the overall similarity in the life cycle of the parasite and the similarity in appearance and behavior of parasites in the blood to those in cases of human malaria as observed in the clinic. This was especially true in the Department of Medical Zoology at the Johns Hopkins School of Hygiene and Public Health, founded in 1916 and headed by Robert W. Hegner (468). Hegner, educated at the University of Chicago and the University of Wisconsin, where he received his Ph.D. in zoology in 1908, suggested that the inability to infect animals with human malaria parasites or to grow them in test tubes (as well as the inconvenience and difficulties in using humans as study subjects) made it imperative to use animal models, particularly avian malarias. In 1913 Eugene Whitmore, one of Hegner's students, isolated a strain of *P. cathemerium* from sparrows in New York, and in 1924, Ernest Hartman, another Hegner student, isolated the same species from sparrows in Baltimore. Both strains were found to be transmissible to canaries. The malaria researchers at Johns Hopkins used female canaries purchased from Baltimore pet dealers since male canaries were colorful and more likely to sing and hence were preferred in the pet trade. Among those who used avian malarias for their doctoral studies at Johns Hopkins and later as faculty in other institutions, in addition to Hartman and Whitmore, were Shulamite Ben-Harel, George Boyd, Reginald Manwell, Reginald Hewitt, Clay Huff, and William Hay Taliaferro and Lucy G. Taliaferro (see p. 92). Remarkably, and despite Hegner's death in 1942, work with avian malarias continues at Johns Hopkins to this day. It does not involve the use of a malaria parasite that infects canaries, however, but uses parasites able to infect domestic fowl. The reasons for abandoning the canary for studying malaria was the bird's small size, the variability in response among the randomly bred canaries, and the small amount of blood that could be taken from a single canary.

Further research on vaccines and immunity to malaria (as well as the biochemistry of the parasites) required hosts larger than the canary. In 1933, a veterinary surgeon in Colombo, Sri Lanka, noted that a number of imported fowl were dying from infection by a *P. relictum*-like parasite. In 1935, Emile Brumpt, working in the Faculty of Medicine in Paris and recognizing that such a parasite would have great value in drug testing and other studies, traveled to the Far East in search of live parasites in fowl but could find none. En route home he stopped in Sri Lanka, visited with the veterinarians there, selected 16 survivors of the original outbreak, and transported them to France. Upon arrival in Paris he inoculated blood from these birds into young chicks. The chicks readily became infected, the infection could be passed from chick to chick by blood inoculation, and it could also be transmitted by several kinds of mosquitoes; he named the parasite *P. gallinaceum*. With great generosity, Brumpt shared this parasite with researchers around the world (468).

P. gallinaceum causes subclinical disease in its natural host, the Indian junglefowl (*Gallus sonnerati*); it became a valuable model, however, because the domestic chicken (*Gallus gallus*) is highly susceptible to infection (mortality rates can be as high as 80 to 90%), it is easily transmitted in the laboratory by *Aedes aegypti* mosquitoes, it has retained its ability to produce gametocytes, and chickens have a much larger blood volume than canaries.

In the United States, Lowell Coggeshall of the Rockefeller Foundation in New York City began to search for a new kind of avian malaria with which he could screen antimalarial drug candidates. He reasoned that it would be cheaper and more efficient to examine birds imported into the United States rather than mounting expeditions to exotic locales. In cooperation with an ornithologist and a pathologist at the New York Zoological Gardens in the Bronx, he examined the blood of specimens from Borneo and Ceylon. In June 1937, a malaria parasite was observed in a drop of blood taken from a Borneo fireback pheasant, *Lophura igniti igniti*. The parasite was infectious to chickens and white Pekin ducklings and was named *Plasmodium lophurae* after its host (94). In spite of subsequent collecting expeditions to Borneo, neither this parasite nor its mosquito vector has ever been identified in the wild. *P. lophurae* soon lost its gametocytes, thus making mosquito transmission impossible, but it nevertheless served as a model for studies of malaria immunology and was a rich source of biochemical investigations until the 1970s.

From 1924 through 1950, William H. Taliaferro (1895 to 1973) and Lucy Graves Taliaferro (1895 to 1984), working initially at Johns Hopkins and later at the University of Chicago, were two of the most influential researchers on malarial immunity in the United States (497). W.H. (nicknamed Tolly) was born in Virginia and received his undergraduate education at the University of Virginia, working not on parasites but on the reactions of the free-living *Amoeba proteus* to food. He continued these studies as a graduate student at Johns Hopkins, carrying out research on the reaction to light by a free-living flatworm. During the summer of 1917 he was an assistant in the Invertebrate Zoology course at the Marine Biological Laboratory in Woods Hole, MA. There he met Lucy Graves, who had just completed her undergraduate work at Goucher College, and he was smitten with her. Courtship and graduate studies, however, were interrupted when the United States entered World War I. Tolly volunteered for service in the U.S. Army and was assigned to the Chemical Warfare Division at Yale University, where he was a private, sergeant, and second lieutenant working on the effect of respiratory gases on dogs. During that same period he received his Ph.D. (1918) from Johns Hopkins. (His thesis was entitled *Reaction to Light by the Eyes of Planaria*.) In 1919, after discharge from the U.S. Army, Tolly married Lucy, and he received a fellowship to work with Hegner in the Department of Protozoology and Medical Entomology at the Johns Hopkins School of Hygiene and Public Health. It was this serendipitous appointment that would lead to a quarter of a century of studies of blood-dwelling parasites—trypanosomes and malaria.

Early on, Tolly studied the trypanosome *Trypanosoma lewisi* growing in the rat. These infections had a characteristic pattern: at first the numbers of trypanosomes in the blood increased due to rapid division of the parasites, reaching a peak after a week or two. Although there followed a "crisis" indicated by a rapid decline in the numbers of trypanosomes, not all the parasites disappeared from the blood, and the trypanosomes that persisted remained constant in number because they were inhibited in their reproduction. These nondividing trypanosomes remained in the blood for some weeks, and then they were cleared from the blood to produce sterile immunity. In 1924 Tolly presented his first immunology paper, showing that crisis and termination of the infection were due to two separate antibodies, one that killed the parasites outright and one, named ablastin, that inhibited reproduction. During this time Lucy was studying *Plasmodium relictum* and was awarded an Sc.D. from Hopkins in 1925 for a thesis entitled *Periodicity of Reproduction, Infection and Resistance in Bird Malaria*. Tolly and Lucy formed a perfect research team—he with ideas and questions and she with experimental data involving inoculations, blood smears, autopsies, and tests galore. Together they graphed, tabulated, and wrote up their results for publication. Their commitment to research was exceptional: when an experiment demanded it, they took the experimental birds home with them, placed them in a bathtub, and were able to take hourly blood samples through the entire night.

The Taliaferros looked in vain for an ablastin-like antibody in malaria—an antibody that would depress the number of merozoites produced and lead to a reduction in the numbers of asexual parasites in the blood (492, 493). Working with *P. cathemerium* in the canary, the Taliaferros found that the majority of merozoites produced at each asexual cycle perished and at crisis large numbers of parasites were destroyed; during the latent period parasite destruction kept the numbers in the blood below levels detectable by microscopy. (This was not an entirely novel finding since Whitmore had shown in 1918 that blood collected during a latent period was infective to other birds for as long as 29 months after infection and Mazza in 1924 had found birds to be infective 4 years and 2 months after infection.) In 1929 Lucy and Tolly found that when infected blood was injected into birds with a latent infection, the parasites were removed from the peripheral blood between 24 and 72 h later, depending on the number of parasite-infected red blood cells that had been introduced. (This confirmed earlier work done by Wasielewski in 1901 and rediscovered by me in 1957!)

At the time of Pasteur's death in 1895, several overseas Pasteur Institutes had been established. The first colonial Pasteur Institute, established in 1891, was in Saigon, and in succeeding years Institutes were established in North Africa in Tunis, Algiers, Tangiers, and Casablanca. Disease was rampant in parts of French colonial North Africa, and especially serious were the deaths due to undiagnosed fevers. In 1893 two professors at the Algiers School of Medicine and Pharmacy wrote to Louis Pasteur and gained permission to establish a Pasteur Institute in Algiers. There was inertia in the early years, but in 1909 Emile Roux (then Director of the Pasteur Institute in Paris) appointed Albert Calmette (then Director of the Pasteur Institute in Lille and codiscoverer of BCG) as Director in Algiers. Later, he was succeeded by two intrepid microbe hunters: the brothers Edmond (1876 to 1969) and Etienne (1878 to 1948) Sergent. Both studied medicine, Etienne in Algiers and Montpelier and Edmond in Algiers and Paris, where he furthered his education in bacteriology, entomology, and protozoology. Edmond was a laboratory assistant to Roux, studying a variety of blood parasites. At the request of Roux, the brothers worked from 1900 to 1910 at the Pasteur Institute in Algeria, carrying out extensive field studies to control the transmission of malaria. But they did more.

The Sergent brothers were able to show that birds were immune to a second acute attack of malaria as long as they had a latent infection, whereas upon complete recovery they could become reinfected. Resistance to reinfection, called premunition, was found to persist for 2.5 years or longer (427, 428). No protective antibodies, however, could be found in birds with a latent infection. Following the path blazed by Pasteur in developing vaccines for chicken cholera, anthrax, and rabies by attenuation (see p. 81–87), in 1910 the Sergent brothers conducted the first malaria vaccine trial using attenuated sporozoites of *P. relictum* (429). Regrettably, the report of this protective vaccine is brief and lacks critical details, but it does reveal a "proof of concept." In 21 cases after sporozoites were held in glass vessels (in vitro) for 12 to 24 h they no longer resulted in a blood infection; when 24 canaries immunized with attenuated sporozoites were challenged by the bite of infected mosquitoes, 7 had a light or no infection, 1 infection was severe, and in 16 there was evidence of immunity. In 1921, the Sergents tried a variety of methods to produce attenuation and achieved positive results in 30% of 24 cases when "old" sporozoites kept in vitro for 12 to 48 h were inoculated into canaries before the introduction of viable parasites and in approximately 20% of the cases in which blood was taken before parasites were apparent in blood smears and inoculated into birds. The Sergents used different doses of sporozoites and found that smaller doses in most cases produced lighter infections with resulting immunity to reinfection. (The optimum dose was two-thirds or three-quarters of a body unit, where a body unit is the total number of sporozoites in a single naturally infected mosquito.) In 1934, the brothers tried to extend their sporozoite work but failed to induce protective immunity when using attenuated blood parasites. Based on their extensive studies with *P. relictum* in canaries, the Sergents theorized (428) that immunity to malaria could result from both premunition (where the body is infected but not sick and the individual cannot be reinfected) and "true" acquired immunity (where there is complete resistance without the presence of parasites after clinical cure). In 1929, Tolly Taliaferro (after a decade of research on immunity to a variety of parasites) critically reviewed the Sergents' immunization studies in his seminal work *The Immunology of Parasitic Infections*. He suggested that the success of their vaccine "was probably due to the injection of a very small number of parasites (with possibly a lowered virulence) that a latent infection was set up without an apparent acute attack" (491).

The availability of *P. gallinaceum* and *P. lophurae* allowed the Taliaferros to expand their studies of immunity by allowing them to challenge animals with the same (homologous) or different (heterologous) kinds of avian malaria parasites (494). In chickens *P. gallinaceum* conferred a marked homologous and heterologous immunity whereas *P. lophurae* conferred a strong homologous but weak heterologous immunity. Repeated infection fortified the immunity against *P. lophurae* but not against *P. gallinaceum*. The Taliaferros thought that this resulted from the amount and the spacing of the immunizing dose, as well as the fact that immunity to *P. gallinaceum* increased over a much longer period. The same patterns were found if the infections were initiated by sporozoites or erythrocytic stages. In 1950 the Taliaferros found evidence of acquired immunity to both *P. gallinaceum* and *P. lophurae*; it appeared progressively during infection with the former and suddenly at the peak of infection in the latter (495). Homologous immunity to reinfection resulted mainly from parasite-killing (parasiticidal) antibodies in both species, which acted on both free merozoites and the parasites within the red cell. Indeed, 97 to 99% of the reinfecting parasites were destroyed within 36 h whereas inhibition of reproduction was only temporarily decreased. The reproduction-inhibiting effects were not apparent when the reinfecting dose was large enough to break down the acquired immunity they ascribed to parasite-killing antibodies. Their conclusion "after the acquisition of immunity after crisis, the few surviving parasites of each brood are apparently resistant to both parasite-killing and reproduction-inhibiting mechanisms, whereas the nonresistant parasites of each brood are removed before observations can be made on degeneration (as evidenced by crisis forms, lowered merozoite mean and lengthening of the asexual cycle)" (495).

My own studies of malaria immunity and vaccines began when I was a graduate student at Northwestern University in Evanston, IL. *P. lophurae* in white Leghorn chickens had been established by my predecessor in the laboratory, the Vietnamese priest Father Hoang Quoc Truong, using blood inoculation. Truong (whose research aside from his 1956 Ph.D. thesis [517] was never published) found that as the chickens aged, they became increasingly resistant to infection, and that after 1 month of age they were able to survive a level of infection as high as 70 to 80%, a level that invariably killed younger fowl. He observed resistance to reinfection by the older birds and extended the work of the Taliaferros by being able to confer passive immunity by using transfusion of plasma from either recovered or older chickens. This reduction in the severity of the infection (that is, immunity), which was dependent on the amount of "immune" plasma transfused into the recipient birds, had already been demonstrated by Manwell and Goldstein with *P. circumflexum*, and they felt "that whatever immune substances are present in serum must be there in relatively small concentration" (312). Father Truong went a step further, showing that the immune substance could also be demonstrated by incubation of parasitized red cells for 1 h with immune plasma, and he showed that immune activity was associated with the globulin fraction. In my work (441) I used paper electrophoresis to separate the serum components and found that the amount of gamma globulin increased by day 4 and decreased to its original value by day 7, and that during this period there was a change in the mobility of the gamma globulin. Further, a rise in the amount of β-globulin on day 4 persisted into the latent period, suggesting the presence of antiparasite antibodies in this fraction. Fifty years later, Williams used electrophoresis to demonstrate (as did I) that when chickens were infected with *P. gallinaceum* there was a change in the proportions of the albumin and globulin fractions and that a sharp drop in these proportions occurred coincident with the increase in the levels of 19S and 7S gamma globulins 8 days after infection (555). These studies suggested, but did not prove, that the protective antibodies were contained in the gamma globulin fraction.

Longenecker et al. (298) provided additional support for the findings of antibody-mediated immunity as had been described by the Taliaferros. When chicks had the bursa of Fabricius removed (hormonally or surgically) on the first day after hatching, their infections with *P. lophurae* were more severe than in control animals that had undergone sham operations; when the birds had the bursa removed on day 18, however, the severity was no greater than in control animals. Severe infections resulted in chickens whose spleens were removed on day 5 (when levels of parasites in the blood had already decreased), whereas removal of the thymus had no significant effect on the severity of infection. Removal of the spleen after recovery from an initial infection resulted in a severe and often fatal secondary infection. In chickens missing a bursa there was a reduction in the level of 7S gamma globulin whereas the 19S component was retained. Longenecker et al. (297) found that the rise in the level of the 7S gamma globulin during a *P. lophurae* infection in normal chickens was decreased during the time of parasite clearance in chickens missing a bursa. The specific infected red-blood-cell-clumping antibodies (agglutinins) detected in recovered birds were not found in the birds missing a bursa that had been infected and reinfected; however, passive transfer of hyperimmune sera to such bursaless and intact birds resulted in a reduced parasite burden in the blood (299).

In an attempt to understand the immunologic basis for premunition, Rank and Weidanz (384) made chickens B-cell deficient by removing the bursa. When eight bursaless and eight bursa-intact birds were each infected with 10^5 *P. gallinaceum*-parasitized red cells, the infections were the same in both groups until day 7, when the severity of infection in the B-cell-deficient birds increased until death occurred on day 11. When the B-cell-deficient birds were rescued from a primary blood infection with the antimalarial drug chloroquine, they were found to be resistant to challenge even though they lacked gamma globulin. This was in contrast to the B-cell-deficient chickens that had been treated with chloroquine but had not been previously exposed to malaria. The conclusion drawn was that "premunition in avian malaria is B-cell and antibody independent, whereas acute infections are controlled by antibody-dependent mechanism of immunity" (384). These investigators went a step further, suggesting that premunition was possibly dependent on T-cell mechanisms. This hypothesis was never pursued in avian malaria, however; what their study did do was to stimulate investigations into the role of T cells in rodent and human malarias (200).

William Weidanz (born in 1935) received his early education in the public schools of Clifton, NJ. In 1956, he was awarded a B.S. degree in pharmacy from the Rutgers College of Pharmacy in Newark, NJ, and in 1958 he earned an M.S. degree in bacteriology from the University of Rhode Island in Kingston, working with Philip Carpenter on antigens of enteric bacteria. In 1961 he received a Ph.D. degree in medical microbiology from Tulane University under the mentorship of Morris Shaffer. His dissertation was on the pathophysiology of *Salmonella enterica* serovar Typhi in chicken embryos. From 1961 to 1963 he was a postdoctoral fellow at the National Institutes of Health (NIH) studying natural antibodies to gram-negative bacteria with Maurice Landy, and then he began his academic career at Louisiana State University in Baton Rouge. As a young Assistant Professor at Louisiana State University he met John Finerty, who was earning his Ph.D. in another department. Finerty was working with *P. lophurae,* and the mere sight of Giemsa-stained malaria parasites piqued Weidanz's interest. At the time he was continuing studies on natural antibodies to gram-negative bacteria in chickens, but he became interested in the work on the ontogeny of the immune system done in Robert Good's laboratory at the University of Minnesota as well as the studies by Bruce Glick on the suppression of B-cell immunity in bursaless chickens. He began to wonder whether these models of immunodeficiency could be used to study the immunology of malaria. In 1966 Weidanz moved to Hahnemann Medical College (Philadelphia, PA), where over the next 24 years he studied immunity to malaria in different immunodeficient animal models. He initiated the Malaria Research Unit at Hahnemann, recruiting both Carole Long and Akhil Vaidya and aiding in the mentoring of James Burns, Jr., toward his doctoral degree. This group subsequently evolved into the Center of Molecular Parasitology. In 1990 he took the Chair of Medical Microbiology and Immunology at the University of Wisconsin—Madison, where he is at present. His current studies are focused on the identification of immune cells and molecules and the mechanisms by which they participate in the immune response to malaria.

Weidanz's contributions to the understanding of malaria immunology have involved the immunosuppressive effect of experimental infection, the function of antibody-independent T-cell-mediated immunity in resistance to malaria, and the use of genetic dissection to study the roles of cells and molecules in the immune response to experimental malaria. Perhaps the best testimony to the importance of his contributions to our knowledge of immunity malaria is the neglect by prominent authors making similar discoveries some years later and declaring them to be novel.

Clearly, the next step in developing a malaria vaccine (even for birds!) would be to identify the parasite antigens that provoked an immune response. As a postdoctoral fellow in William Trager's Rockefeller University laboratory, I used the then fashionable immunologic techniques of Ouchterlony double diffusion and immunoelectrophoresis to demonstrate that one of the major antigenic components of the erythrocytic stages of *P. lophurae* was hemoglobin, that hemozoin was nonantigenic, and that formalin-killed free parasites contained at least four to six parasite-specific antigens (434). However, when it became obvious that we had underestimated the number of antigens and that none of these would serve as a protective vaccine, I left the field of malaria immunity in birds for a decade. During the 10-year hiatus, however, other malariologists, recognizing that mice and monkeys were closer to humans, began to study the immune mechanisms in rodent and primate malarias (see p. 105).

Blood Battles in Birds

In the early 1940s, Wendell Gingrich (a student of W. H. Taliaferro) showed that canaries immunized by a dozen intravenous injections of large numbers of formalin- or heat-killed *P. cathemerium*-infected red blood cells were protected. If the dose of infected red cells was reduced, however, or if the number of immunizing doses was smaller, the vaccine failed to influence the severity of the infection; despite this, most birds did survive (8 of 11). Jacobs (257) showed that when ducklings were vaccinated with insoluble extracts of *P. lophurae* mixed with a *Staphylococcus* toxoid (that had been used to make nonantigenic materials more antigenic), four of six ducklings showed increased protection against challenge. He concluded that the antigens were either too insoluble or partially antigenic, or both, and that the parasite material did not efficiently provoke antibody production. This suggested to Jules Freund, whose name today is remembered most for the practical advantage of his adjuvant (from the Latin word "adjuvare" meaning "to aid") techniques for immunization, that a vaccine might be possible if the potency of the malarial antigens could be enhanced.

Freund was born in Budapest, Hungary, on 24 June 1890. He received his M.D. at the Royal Hungarian University, served in the Austrian Army (1913 to 1914), and then held several public health positions. He emigrated to the United States in 1923 to work at the Antitoxin and Vaccine Laboratory of Harvard University (12). Thereafter, he joined the group of Eugene Opie (the former Johns Hopkins medical student, who, along with William MacCallum, recognized the significance of Laveran's flagellum) in the Department of Pathology at Cornell Medical School in New York. In 1943 Freund became Chief of the Division of Applied Immunology at the newly established Public Health Research Institute of the City of New York. Robert Koch was long dead when Freund began to work on developing an adjuvant, but Koch's work on a tuberculosis vaccine was an essential ingredient that led to a successful demonstration that a malaria vaccine was possible.

For Koch's discoveries of the germs of tuberculosis (1882) and cholera (1883), the Imperial Government of Germany rewarded him in 1885 by appointing him Professor of Hygiene at the newly established Hygiene Institute of the University of Berlin. For several years after his appointment, contrary to his past practice, Koch ceased to do research with his own hands and instead relied on his laboratory assistants to do the bench work. Instead, at age 47 and at the peak of his career as a microbe hunter, Koch spent his time lecturing and serving on commissions. Then, sometime in 1889, there was an abrupt change in behavior. He was back in the laboratory, doing work with his own hands behind closed doors and speaking to no one for days. In 1890, at the Tenth International Congress of Medicine held in Berlin, Koch revealed his secret research in a lecture given in an auditorium that seated 8,000 and was decorated in an opulent manner that simulated the Temple of Zeus. The German Government provoked his world-shattering announcement of a "remedy" for tuberculosis by promising him a new institute with superior facilities, far better than France had provided to his nemesis, Pasteur. The stolid and cool Koch approached the podium, peered over his gold-rimmed glasses, shuffled his papers, and addressed the crowd: "My discovery of the tubercle bacillus led me to seek substances that would be therapeutically useful against tuberculosis. In spite of many failures I continued my quest and ultimately found a substance, which I call tuberculin, which halted the growth of the tubercle bacteria not only in test tubes but also in animal bodies. Guinea pigs, highly susceptible to tuberculosis, even if they already have the disease in an advanced state, when injected with tuberculin have the disease halted." He concluded the lecture by saying, "It is possible to render harmless the pathogenic bacteria in a living body and to do this without disadvantage to the body. Tuberculin is a cure for tuberculosis!" (I. W. Sherman, *Great Scientists Speak Again*, 1985, unpublished). Koch had found that when a glycerin extract of a pure culture of tubercle bacteria was injected under the skin of a guinea pig, a remarkable immunological response occurred. After a delay of 1 or 2 days the guinea pig showed a local or systemic reaction whose severity was dependent on the amount of tuberculin injected and the host's immunity. The site near the injection became red, swollen, tender, and ulcerated; it oozed pus, and the lymph nodes enlarged. Despite the fact that at the time of his announcement of tuberculin therapy Koch had tested it only in guinea pigs, public and political pressure encouraged testing in human subjects. These tests soon showed that the therapy was not only useless but also on occasion fatal. The failure of tuberculin therapy to cure tuberculosis led Koch to become grumpy, quiet, and restless. He offered his microbe-hunting services to countries around the world in the hope of conquering plagues. During sojourns in Italy and Africa, he hunted the malaria parasite and the means by which it was transmitted, and in these endeavors he failed. Ironically, however, his failed "remedy"—tuberculin—in the hands of Freund would find success in trials of a protective malaria vaccine.

Unlike Koch, Freund was not a hunter of microbes. He had trained in medicine and pathology and was interested in tissue responses produced by immunological reactions. As early as 1913 Freund had studied anaphylactic shock, and later he developed a special interest in tuberculin hypersensitivity in guinea pigs injected with Koch's tuberculin. Freund found that guinea pigs sensitized with tubercle bacilli in paraffin oil produced high-titer antibodies to the bacteria. This discovery led to the idea

that the oil might act as an adjuvant that would result in a general enhancing effect. Indeed, when typhoid bacilli, rabies virus, and tetanus toxoid were incorporated as water and oil emulsions and injected into the le-
sion, the trio resulted in enhanced antibody production. Extending these
findings, Freund and coworkers (175) incorporated formaldehyde-killed
P. lophurae combined with a lanolin-like substance, paraffin oil, and killed
tubercle bacteria to vaccinate 2-month-old white Pekin ducklings. (Older
ducks, weighing 2 to 3 kg, were used because the experiment required at
least 2 months from the time of first vaccination to challenge, and since
these were probably age-immune the challenge dose had to be high, i.e., 1
billion to 5 billion parasitized red blood cells, to produce a lethal infection
in control birds.) When the ducks were given three injections 4 weeks
apart and challenged with blood stage parasites by intravenous inoculation 1 month after the last immunizing injection, seven of eight vaccinated
ducks survived, in contrast to a 50% mortality rate in the control birds.
The survivors, however, had a few parasites in the blood.

In a more extensive study, Freund and coworkers (506) found that
when vaccinated ducks received two injections of vaccine none died of
malaria; they also found, however, that four of five ducks developed a low-
grade blood infection and that one of the five had no detectable parasites
in a blood smear. There was also evidence of protection after a single vaccination, and this seemed to persist for several months. Although protection could be achieved without adjuvant, it was of relatively short duration. The work of Freund and coworkers had shown that it was possible
to produce immunity to malaria in the absence of a prior infection. However, immunization did not prevent the animals from becoming infected.
In addition, inoculations had to be done intramuscularly, and protection
appeared to be related to adjuvant-induced local tissue reactions, which at
times were severe. The intramuscular lesions produced by Freund's adjuvant preclude its use in humans; it is clear from Freund's experiments,
however, that vaccination against malaria could be achieved (but only
when a suitable and potent adjuvant was used).

The partially successful production of nonliving and attenuated vaccines titillates the imagination, and although successful vaccination of birds
with whole parasites had already been achieved, there had been no identification of a protective antigen. Then, in 1974, as Araxie Kilejian was working
with *P. lophurae* in Trager's Rockefeller Laboratory (and before Trager was
able to continuously culture *P. falciparum* in vitro), she serendipitously discovered a possible protective antigen (272). Kilejian was trying to isolate nu-

clei from *P. lophurae*. The parasites were broken open by sound waves, and the crude suspension was separated into several fractions by sedimentation in a centrifuge. When each of the fractions was examined under a transmission electron microscope, the nuclear fraction showed a great enrichment of electron-dense granules. When the isolated granules were dissolved in acetic acid and analyzed, they were found to contain an unusual protein with an unusually large amount ($\sim$70%) of the amino acid histidine. It was called the histidine-rich protein (HRP) (271). Although the dense granules (as well as malaria pigment) are discarded when the infected red cells burst and merozoites are released, Kilejian could not countenance that the parasite would expend energy synthesizing a waste product. The localization of radioactive histidine in *P. lophurae* trophozoites and merozoites led her to hypothesize that the HRP was a component of "the polar organelles of merozoites and function in their penetration of erythrocytes" (273). The next "logical" experiment she carried out was to vaccinate ducklings with HRP in the absence and presence of Freund's adjuvant (274). As Freund had found in his earlier studies, Kilejian discovered that three of four ducks in each group showed low-grade infections, indicating "protective immune response without the use of adjuvant." Two weeks after challenge one of the ducks was used to obtain a globulin preparation, and this was used to transfer protection passively. All the ducklings survived, and when they were challenged 2 and 4 weeks later, they showed "total resistance." Here is where I came on the scene. For several years I had tried to identify protective antigens in *P. lophurae*, and the immunologist Vincent McDonald joined me in this pursuit. Vincent was making some progress with a ribosomal protein preparation when I suggested that he compare his "protective" antigens to the HRP. We carried out an extensive study (322, 446), trying as best we could to repeat Kilejian's work precisely, but without success. We claimed that Kilejian's success could be explained by the facts that she used age-immune ducks, that there were too few experimental animals for statistical analysis, and that HRP played no role in protection. We suggested that she repeat the experiments (440); neither she, however, nor any other laboratory did.

Battles outside the Blood

In 1955, Augusto Corradetti (1907 to 1986) of the Istituto Superiore di Sanità in Rome may have been the first to provide evidence for immunological responsiveness to EE stages. When he injected chickens with *P. gallinaceum*-infected red cells, EE stages were found in the brain 11 days later;

by day 21, however, they were no longer seen, and when birds were inoculated directly with tissue containing EE stages these forms were seen between 8 and 25 days postinfection but not later. Removal of the spleen in
recovered birds did not result in renewed multiplication of EE stages (but
did affect blood stages). Eighteen years later (227), evidence suggesting an
immune response to EE stages was also found. Turkey poults infected
with *P. fallax* EE stages obtained from cell cultures initially exhibited an increased number of EE forms in brain smears, but over time these diminished. In 2002 Permin and Juhl (371) infected chickens with different
doses of *P. gallinaceum*-infected blood; although there was evidence of EE
stages in tissue smears in all organs examined from 2 to 4 weeks postinfection, these completely disappeared after 5 weeks, suggesting that immunity to the tissue stages had developed. (In contrast, parasites did persist in the blood, and the severity of the clinical symptoms was related to
the dose given.)

It has also been possible to transfer protection passively by intravenous administration of antisera specific for EE stages of *P. fallax*; the
protective activity was found in the IgG fraction (227). The results suggested that the resistance of turkeys to EE stages was mediated, at least in
part, by antibody. The hyperimmunized birds that had served as serum
donors, however, were fully susceptible to a challenge with blood stages
of *P. fallax*. Thus, the antibodies seemed to be specific for a particular
stage. Indeed, earlier it was noted that serum from turkeys and chickens
that had recovered from a *P. lophurae* blood infection did not affect the appearance of EE forms in serum recipients. Of 20 turkeys vaccinated with
formaldehyde-killed EE stages of *P. fallax*, 18 survived a challenge infection with EE merozoites from culture, whereas 21 of 30 control birds died.
In two further experiments, when birds immunized with formalin-killed
EE stages were challenged with infected red cells, 5 of 6 immunized birds
survived whereas 5 of 7 control birds died, and in the second experiment 4
of 14 immunized birds survived whereas all 28 controls died. These findings suggest that immunization with formalin-killed EE merozoites (with
or without Freund's adjuvant) was able to protect against both red cell and
EE infections after challenge with either EE merozoites or infected red
cells. The effect of immunization on the development of EE stages after
sporozoite inoculation was not determined. In another study, however,
with *P. gallinaceum* in chickens, immunization with EE stages (from
formaldehyde-treated infected brain tissue or from culture with saponin as
adjuvant) had no apparent effect on challenge with sporozoites or blood

stages (395), and the immunized chickens survived longer than did control chickens when subsequently challenged with EE forms.

Monkeying with Vaccines

In 1931, when the Calcutta School of Tropical Medicine ran short of rhesus monkeys customarily purchased from Indian suppliers for its various experimental purposes, they imported long-tailed Malayan kra monkeys (*Macaca irus*= *M. fascicularis*) from Singapore. A blood smear revealed that one of the monkeys harbored a malaria parasite despite an absence of clinical symptoms. When Robert Knowles, the Director of the School, inoculated blood from this monkey into an Indian rhesus monkey (*Macaca mulatta*), the parasites multiplied rapidly and within a few days the monkey died from a fulminating infection. In 1932, John Sinton and H. W. Mulligan reisolated the parasite from another kra monkey originating in Singapore; this parasite was maintained at the Malaria Institute of India (Delhi) in rhesus monkeys and named *P. knowlesi*, after Knowles (464, 552). *P. knowlesi* was exported to laboratories around the world, and for 20 years it was used extensively for studies of chemotherapy, biochemistry, nutrition, immunology, vaccination, and analysis of the requirements for erythrocyte invasion.

In 1932 Knowles and Das Gupta succeeded in experimentally transmitting *P. knowlesi* from monkey to human and then from human to human by inoculation of infected blood. In the early 1930s (and before the advent of penicillin), this monkey malaria was widely used for the treatment of general paralysis of the insane, with tertiary syphilis being one of the main reasons for admission to neuropsychiatric institutions. But it soon became apparent that this infection could become uncontrollable, and after several fatalities its use was discontinued in favor of the less virulent *P. vivax* (see p. 285–287). By 1952, either most laboratories had lost the strain or it had become attenuated, so the following year Singh, Ray, and Nair (464) isolated *P. knowlesi* virulent for rhesus monkeys from a kra monkey living in the Nuri Valley (Malaysia). This isolate, known as the Nuri strain, has since been favored for biochemical, immunological, and vaccine studies.

P. knowlesi is unusual among malarial parasites of primates in that under natural conditions humans and rhesus monkeys as well as baboons can be infected. The first natural infection in a human was reported to have occurred in 1965 in a man who returned to the United States after visiting Malaysia, and there was another report in 1971. In nature, transmission is restricted to *Anopheles lateens*, a mosquito that is equally attracted to monkeys and humans and feeds primarily at dusk in the forest and forest fringe. Recently it was shown that *P. knowlesi* is not as rare as previously thought, and in a particular region of Malaysian Borneo almost all of the 108 malaria cases reported in the study were due to this parasite, which is sometimes fatal (463). Some investigators now suggest that despite its preference for monkeys, *P. knowlesi* should be considered the fifth human malaria parasite (552).

During the 1930s, Eaton and Coggeshall (155), working in the Rockefeller Foundation Laboratories, found that it took between one and five *P. knowlesi*-infected red blood cells to produce an infection and that once the infection was established it was almost invariably fatal unless treated. Furthermore, immunization with parasites killed by heat, freezing and thawing, formalin, or drying produced no resistance to challenge despite there being complement-fixing antibodies. By contrast, repeated reinfection of chronically infected monkeys enhanced the potency of immune serum (97), and when injected into animals with an acute infection the immune serum had a variable but generally depressing effect on the course of infection if administered in daily doses; relatively large amounts of immune serum given shortly before or at the time of injection of parasites had only a minor effect (96). Coggeshall and Eaton (95) showed that the protective effect of immune serum was more marked when the challenge involved 10 *P. knowlesi* parasites than when it involved 1,000 parasites, and protection was more effective when immune serum was incubated with the parasites before injection and given daily during the incubation period and the first stages of infection; however, it was more difficult to protect the monkey with immune serum if the serum was administered after parasites appeared in the blood. Thus, in protection experiments, survival or death of the rhesus monkey was dependent on both the amount of immune serum and the number of *P. knowlesi* parasites used for challenge. In 1999, Biswas (45) confirmed the 60-year-old findings of Coggeshall and Eaton by repeatedly infecting rhesus monkeys with *P. knowlesi*-infected blood and measuring the immunoglobulin levels by a sensitive and specific assay. The first and second infections were cured by chloroquine treatment, and with the third there was self-recovery; in every case IgM was present, there was a substantial increase in the amount of IgG, and there were elevated levels of complement. Again, there was no direct correlation of immunoglobulin levels with the acquired immune response. Clearly, not much new had been discovered.

Michael Heidelberger (1888 to 1991) (158), one of the founders of immunochemistry, was born in New York City and received his B.S., M.S., and Ph.D. in organic chemistry from Columbia University. His initial research was in chemotherapy at the Rockefeller Institute, where, in collaboration with Walter Jacobs, he synthesized aromatic arsenicals for the treatment of syphilis and African sleeping sickness. In 1919 they developed a variant of Ehrlich's "magic bullet," salvarsan, which proved to be effective for the treatment of African sleeping sickness; named tryparsamide, it was successfully tested in the Belgian Congo and became a standard chemotherapeutic treatment for the disease. Subsequently, Heidelberger and Jacobs studied the chemotherapy of pneumococcal pneumonia. Among the compounds they studied was a group of azo dyes containing a sulfonamide group. Finding that the drug did not kill the bacteria directly in the test tube (in vitro), they abandoned the work. Later, Gerhard Domagk, in trying to understand the lack of correlation between in vitro and in vivo antibacterial tests, resorted to in vivo testing of the Heidelberger-Jacobs azo dye. This led to his discovery that the dye protected mice against a streptococcal infection but had no effect on the bacteria in vitro. Later, it would be shown that within the body of the mouse the sulfonamide-containing azo dye was converted to the active antimicrobial, p-aminobenzenesulfonamide (sulfanilamide). Subsequent studies demonstrated that sulfonamides on their own were as effective as the parent dyestuff in protecting mice infected with streptococci. Indeed, 20 years later the work done by Gerhard Domagk in 1933 to 1935 launched the era of sulfa drugs. In 1939, Domagk received the Nobel Prize. Heidelberger often mentioned to his students that he had followed too closely Pasteur's dictum and that his preconceived notion of a drug acting as a "magic bullet" with a direct action on the pneumococcus prevented him from making a seminal discovery and perhaps earning a Nobel Prize.

In 1923 Heidelberger first ventured into immunochemistry, performing studies of the antigenicity of hemoglobin with the Nobel laureate Karl Landsteiner, who had come to the Rockefeller Institute in the previous year. Over the next 50 years, Heidelberger concerned himself with the fact that antibodies were not directed against an antigen as a whole but against particular chemical groupings (later called antigenic determinants or epitopes) on the surface of the antigen and to which the antibody binds. Heidelberger and Landsteiner showed that antigenically distinct hemoglobins (from horses, rats, rabbits, or humans) could be precipitated by the same antiserum. These experiments provided convincing evidence that immunological cross-reactions could occur because an antibody prepared against one antigen reacted with another antigen and that this was due to their having identical or closely related epitopes. During this period at Rockefeller, and before the introduction of sulfa drugs and penicillin, pneumococcal pneumonia was the major cause of illness and death. At the time, the only effective treatment was injection of an antipneumococcal antiserum produced in rabbits or horses by inoculating them with killed pneumococci. The question was: What was the chemical composition of the pneumococcus that made it such a powerful antigen? In 1917 Oswald Avery and Alphonse R. Dochez at the Rockefeller Institute had shown that the virulent form of the pneumococcal bacteria, covered by a slimy capsule, released a "soluble specific substance" that could be precipitated with a type-specific antiserum. It was the capsule that imparted to each pneumococcus its specificity. In 1923 the chemical nature of this substance was characterized when Heidelberger teamed up with Avery. Heidelberger precipitated Avery's specific soluble substance with antiserum, and after purifying the precipitate he concluded that the antigen consisted of nitrogen-free carbohydrate, namely, polysaccharide, a sugar compound containing a string of five or more sugar residues. The reason each type of pneumococcus was different was that they each had a distinct polysaccharide. Heidelberger used these findings to develop a simple protective vaccine against pneumonia, and then he set his sights on malaria.

Armed with Pasteur's finding that attenuated (weakened but living) microbes or repeated doses of their toxic products induced an immune reaction and long-term protection against disease, with Theobald Smith and Daniel Edward Salmon's discovery that even bacteria killed by heat could impart protective immunity, with Behring's and his own demonstration that immunity stemmed from antibodies in the serum, and with the passive-transfer experiments of Coggeshall and Kumm (97), Heidelberger was optimistic that a malaria vaccine could be developed. In 1942 to 1945, as a Professor of Immunochemistry at Columbia University and under a contract recommended by the Committee on Medical Research funded by the Office of Scientific Research, he and a graduate student, Manfred Mayer, prepared a vaccine from *P. vivax*-infected red cells obtained from malaria-infected troops who had returned to the United States from the South Pacific and volunteered to donate blood (215). They used *P. vivax* because it was "safe"—it usually did not cause serious or fatal infections and had been used for decades in the treatment of those with late-stage syphilis (see p. 285–287). Blood containing large, hemozoin-laden parasites was used as the starting material; this was fixed with dilute formaldehyde, lysed by freezing and thawing, and "purified" by centrifugation. Processing 500 ml of blood took 15 to 20 h! Using this material, Heidelberger conducted the first controlled active-immunization studies for vivax malaria. Some 200 volunteer patients suffering from chronic relapsing vivax malaria were divided into three groups. One group received routine therapy, a second group received normal red cell membranes, and the third group received the vaccine, consisting of 2 billion to 4 billion formaldehyde-killed parasites administered by intracutaneous, subcutaneous, and intravenous routes in divided doses over four to five consecutive days. The relapse rate in the three groups was the same (214). A subsequent experiment with late-stage syphilitics found no protection from challenge (216). They also administered formaldehyde-killed sporozoites to healthy volunteer inmates, and again this vaccine had no effect on susceptibility to infection due to the bite of *P. vivax*-infected mosquitoes. Heidelberger et al. concluded that the experiments were "a complete failure" (216a). He returned to working with the pneumococcal bacteria and studied the protective role of antisera against anthrax and the highly toxic castor bean protein ricin, both of which were thought to be in preparation by the Germans for use against Allied troops, and never again worked on a vaccine for malaria.

Heidelberger's disappointing findings about a malaria vaccine were a direct contrast to those of Freund and coworkers, who in 1947 (506) reported that ducks could be immunized against malaria with formaldehyde-killed parasites combined with paraffin oil and killed tubercle bacilli (Freund's complete adjuvant [FCA]). They extended the work with avian malaria to *P. knowlesi* in the rhesus monkey. In a preliminary trial (173) each dose was divided into three or five equal portions and injected into the subcutaneous tissues of the neck, armpits, groin, and neck, and then the monkeys were challenged with infected blood. In one of seven vaccinated monkeys, no parasites were seen in blood smears, and in the others there were fewer than 10 parasites per 100 red blood cells; in the latter case the numbers declined until none could be seen. There were palpable masses at the injection site, although they did not ulcerate. In a more extensive study (174), the vaccinated animals developed low-grade infections and none died from malaria when challenged; the low-grade blood infection was of short duration, and there were no relapses for as long as 6 months. It was not possible to substitute for the killed tubercle bacteria, however, and no protection occurred when peanut oil was used instead.

When killed parasites in saline were injected or when tubercle bacteria were replaced by alcohol-ether extracts of the bacilli or with cholesterol or lecithin, there was no protection. The titer of complement-fixing antibodies was higher in monkeys vaccinated with adjuvant than in those without adjuvant; there was no exact correlation, however, between immunity and the titers of sera. The investigators stated, "Because of the possible application of vaccination to man it seemed desirable to prepare antigen by a method which Heidelberger and associates found successful in obtaining *P. vivax* parasites from human blood" (174). Of the three monkeys vaccinated, two died of malaria (the course of the disease was similar to that in controls) and the third developed a low-grade infection lasting 28 days and death occurred on day 34.

Freund's ability to protect ducks and monkeys against challenge by blood stage malaria parasites and the demonstrated need for an adjuvant to immunize against such infections did not have an immediate impact on the development of malaria vaccines. Indeed, it was 20 years later that Targett and Fulton (498) confirmed Freund's findings by using formaldehyde-fixed erythrocyte-free *P. knowlesi* and infected red cells emulsified in FCA, the latter being more immunogenic. Vaccination was carried out with an interval of 6 to 8 weeks between injections with or without adjuvant, and the monkeys were challenged 6 weeks after the second vaccination. Of the seven animals vaccinated with either parasites or infected red cells with FCA, no blood stage parasites were seen in three whereas three others showed a low-grade infection that persisted for 23 days after challenge; the remaining animal had a more severe infection but recovered. Four weeks later, none of the vaccinated monkeys showed parasites in the blood. Without FCA there was no protection; intramuscular inoculation was found to be more effective in protection than was subcutaneous inoculation. Although there was a marked increase in the level of gamma globulin in serum, this increase was not correlated with protection. Furthermore, an increase in the antibody titer did not correlate with functional immunity in every case, suggesting that much of the gamma globulin increase was nonspecific (499), something I (and others) had found earlier with the avian malarias (see p. 96–97).

The conclusions to be drawn from using bird and monkey malarias are inescapable: protection could be achieved after vaccination with the various stages of the parasite, although in most instances this required the use of an adjuvant. Protection against reinfection was stage specific and more often than not was nonsterilizing. Finally, in spite of the promise of a malaria vaccine, the protective antigens had still not been identified.

6

Dreams about Vaccines

The early vaccine studies with bird and monkey malarias were, for the most part, of limited success. Indeed, during that half century (1910 to 1960), serious reservations about a practical malaria vaccine for humans were being voiced based on clinical, logistical, and economic considerations. Because immunity to malaria in humans develops slowly and incompletely, the assumption was that vaccination would not improve on the immunity that developed in response to repeated severe infections. It was also felt that problems would arise with the acceptability of a vaccine for children within a target population. Finally, some contended that not only would a malaria vaccine be costly to develop, but it could serve only as an adjunct to the inexpensive and effective insecticides and antimalarial drugs that were the basis for the boastful proclamation of "man's mastery of malaria." Full of hubris in 1955, the Eighth WHO Assembly meeting in Geneva, Switzerland, endorsed a policy of global eradication of malaria with reliance on chloroquine treatment and DDT spraying. No mention was made about conducting research that might lead to a protective vaccine. Over time, however, the malaria vaccine that seemed both extravagant and unnecessary would become an intriguing prospect.

By the 1960s, many parts of the world where eradication had once seemed possible were experiencing a resurgence of malaria; in other places economic constraints forced premature relaxation of surveillance; and in some countries the control programs were curtailed because of political turmoil. There was also widespread mosquito resistance to insecticides and parasite resistance to the cheap and once effective antimalarial drugs. It became apparent, even to the WHO, that the Global Malaria Eradication Program was a failure and that eradication could not be achieved. A malaria vaccine, discounted for decades during the "eradication era," became the possible dream when in 1961 Ian McGregor and Sydney Cohen found that IgG from immune Gambian adults had an antiparasitic effect when administered to children infected with *P. falci-parum.*

Ian McGregor (1922 to 2007) completed his medical education at St. Mungo's College in Glasgow, Scotland, and shortly thereafter began his malaria training in the Middle East School of Hygiene at Dimra near Gaza (57). He spent most of his time traveling through Palestine and Transjordan, inspecting malaria control units and organizing training programs. Toward the end of 1948, at his own expense, he enrolled at the London School of Hygiene and Tropical Medicine and received a Diploma in Tropical Medicine. That is where he met B. S. Platt, then Director of the Medical Research Council (MRC) Gambian Research Unit, who invited him to move to Africa. In October of 1949 he left London to take up a post in the Nutrition Center at Fajara—an outpost of the MRC—to study how parasitic diseases contributed to protein malnutrition. In The Gambia, McGregor's energies were devoted to diagnosis and treatment and to efforts to control malaria through application of insecticides. The relatively infrequent episodes of clinical illness in adults led him to posit that this was due to acquired immunity, a view at odds with most clinicians, who looked upon malaria immunity as ineffective and tenuous.

In 1956, McGregor and coworkers (324) reported that newborn infants resident in The Gambia who had received weekly chloroquine showed significantly lower concentrations of gamma globulin than did a group of unprotected children. (The protected group never showed parasites in the blood, whereas the unprotected group was found to be infected by the second year of life.) A follow-up investigation using electrophoresis showed that although the gamma globulin concentration fell for the first 3 to 6 months after birth, there was a subsequent progressive rise with age similar to that in Europeans, but the level in the African children was always higher. These findings convinced McGregor that there was an association between malaria and enhanced levels of gamma globulin in serum. At the time, however, there was no proof that the raised levels reflected a specific antibody response or that the gamma globulin response was protective and responsible for effective immunity. McGregor recognized that definitive experiments would require collaboration with an immunologist. He recruited Sydney Cohen to the project.

In 1954, Sydney Cohen (born in 1921), who had trained as a physician in Witwatersrand University in South Africa, received a Nuffield Fellowship to work at the National Institute for Medical Research (NIMR) in London to study the distribution and metabolism of plasma proteins, especially gamma globulin. After 6 years at NIMR he moved across town to join Rodney Porter's Department of Immunology at St. Mary's Hospital, where purification and the chemical structure of gamma globulin were being pursued. When Cohen studied the synthesis of gamma globulin in Europeans and adult Gambians exposed to malaria, he found that the former group synthesized up to 80% less than the latter, suggesting that the Gambians were making more gamma globulin as a protective response to malaria (98). McGregor arranged to collect a pool of serum from healthy Gambian volunteers and sent this to London, where Cohen carried out the fractionation. The 7S gamma globulin from serum from adult Gambians, judged to be pure by electrophoresis, as well as the purified 7S gamma globulin fraction and serum minus the 7S component from adult Gambians, was provided to McGregor. In addition, the 7S gamma globulin fraction from serum of British blood donors was prepared as a control. The therapeutic effect of these fractions in young Gambian children suffering from acute clinical *P. falciparum* and *P. malariae*, as well as in untreated children, was assessed. The fractions were administered intramuscularly at 8- to 24-h intervals for 3 days with a total dose equivalent to 10 to 20% of a child's own gamma globulin. By day 4 the parasite density did not increase, and by day 9 no parasites were seen in the blood in 8 of 12 children. Protection by passive transfer was limited, however, lasting only 3 months. Unlike the other two 7S fractions, only the 7S gamma globulin fraction from immune adults reduced both the levels of parasites in the blood (asexual but not sexual stages) and severity of illness. This provided the first reliable experimental data to support the view that humans repeatedly exposed to malaria-infected mosquitoes could develop an immunity capable of restricting clinical illness and parasite blood density and that this immunity could be passively transferred to nonimmune individuals (children) via gamma globulin (101).

Clearly, high levels of gamma globulin were being maintained in people living where malaria was endemic due to repeated exposure to infection. (It was also found that West Africans treated with antimalarials had reduced levels of gamma globulin, and that when West Africans were resident in the United Kingdom for some years there was a fall in gamma globulin concentration and synthesis.) These observations led to the proposal that, at least in theory, vaccination against malaria could be feasible. When McGregor and Cohen showed that the 7S fraction from serum from adult Gambians had the same therapeutic effect in Tanzanian children infected with *P. falciparum*, their results suggested that West and East

African strains of malaria had antigenic similarities and that a vaccine prepared against parasites from one region of Africa might be effective against parasites from other regions. This emboldened them to begin a hunt to identify the blood stage antigen responsible for protection against falciparum malaria.

In the early 1900s, Britain's colonial involvement in Africa resulted in considerable interest in the endemic diseases of Africa that had led to its being called "the white man's grave" and to the admonition: "Take care and beware of the Bight of Benin, of forty goes out only one comes in." African sleeping sickness, caused by *Trypanosoma brucei rhodesiense* and *T. b. gambiense*, was of particular concern since after infection the trypanosomes in the blood invaded the central nervous system, leading to the onset of coma referred to as "sleeping." The trypanosomes elicited a variety of antibodies, some of which clumped (agglutinated) them but did not result in complete clearance of the parasites from the blood, and often the infection became chronic. In 1909, Paul Ehrlich and coworkers, while studying the chemotherapy of such infections, had demonstrated a most remarkable feature concerning African sleeping sickness. They found that when mice were treated with a subcurative dose of trypan red and parasites appeared in the blood, the relapsing trypanosomes were no longer agglutinated by antiserum to the original infecting strain. The trypanosomes were altered in their antigenicity!

Following up on these observations was James (Jim) Williamson (1918 to 1993), who began his long career in chemotherapy in 1945 with E. M. Lourie at the Liverpool School of Tropical Medicine, where he received his Ph.D. in 1949. After a sojourn at the West African Institute for Trypanosomiasis Research in Nigeria (1954 to 1960), where he studied the chemotherapy of the African trypanosomes, he joined NIMR. Convinced of the central role of chemotherapy in the control of parasitic diseases, he initiated a research program designed to use studies of the genetic basis of antigenic variation to understand the means by which trypanosomes were able to overcome the host's immune response, as well as to characterize the antigens chemically. At NIMR he was joined by Neil Brown.

Neil Brown (born in 1929) was brought up in North London, was educated at the Southgate Grammar School, and graduated from Imperial College, London, in 1953. After 2 years at the London School of Hygiene and Tropical Medicine, he joined May & Baker Ltd., becoming one of the team that subsequently discovered and developed isometamidium, which has been used for over 40 years for the treatment of cattle trypanosomiasis in Africa. Resigning from May & Baker, he was recruited by Frank Hawking, Head of Parasitology at NIMR, to screen for compounds active against trypanosomes and malaria parasites. Finding drug screening extremely tedious, he asked Sir Charles Harrington, Director of NIMR, for permission to investigate variable antigens of trypanosomes. This permission was granted for 3 years with no strings attached. At NIMR, Jim Williamson and Neil Brown set about identifying and characterizing the variable trypanosome antigens (62). At the time Brown completed his Ph.D. (University of London, 1963), a post became available at NIMR for screening for a malaria vaccine. He was convinced that the only way to tackle the malaria vaccine problem was to pursue basic research, which he immediately set about doing.

Ivor Brown (born in 1938) collaborated in these experiments. Ivor (no relation to Neil) was born and brought up in South London, educated at the Haberdasher's Aske's School at New Cross, and graduated with a degree in microbiology and general science from Bristol University (1961). His first appointment was as a Scientific Officer in the research facility of Glaxo Laboratories (now GSK) at Greenford in West London, where he worked with the head of the research group, Morris Berenbaum, who was interested in how anticancer drugs worked. The primary immune response in mice was used as a model of the massive cell division potential associated with most tumors. Drugs or treatments (such as irradiation) were given in single doses before, at the same time as, or after the antigenic stimulus, to find when in the cycle they were most effective against dividing cells. Several good papers came out of this work. With this grounding in immunology, Ivor Brown applied successfully in 1964 for a research position at NIMR (Mill Hill) in the malaria immunology group led by Neil Brown. Malaria immunology was a new venture for Ivor, but he knew from his attendance at Immunology Society meetings that there was an abundance of immunological talent at Mill Hill, and so the environment was particularly attractive. Neil Brown and other members of the Department of Parasitology provided a good grounding in malaria over the first 6 months or so, but things really became fascinating when the Monroe Eaton papers were found (153, 154). The group at NIMR was particularly fortunate to have a potential source of rhesus monkeys that had been used for testing polio vaccines, and the results with *P. knowlesi* translated into the now-classic studies of antigenic variation.

In 1966, Ian McGregor arranged for Peter Trigg and Ivor Brown to spend a few months in The Gambia. Trigg was working on in vitro culture of the blood forms of malaria, and there was thus an opportunity to look for antibody-induced clumping of in vitro-grown schizonts of *P. falciparum*. At the very outset, however, there were huge problems with the competing blood group system, and given the short time available they could obtain no firm indications of schizont clumping. If only they had tried other approaches, but they had neither the time nor the resources and so work with *P. falciparum* was abandoned. At the time of their discovery of antigenic variation, there were the expected doubts about what had been found with *P. knowlesi*; others, however, later confirmed the results. The initial reservations were related to the use of blood-induced infections only and to the relevance to human malaria infections in the wild and natural human immunity. Despite these concerns, the work provided a huge impetus to malaria research, and in 1968 Ivor Brown was awarded a Ph.D. for his work on antigenic variation. In 1970, when his contract ended, he transferred to the Leprosy and Mycobacterial Research group at NIMR, and in 1973 he moved to St. Mary's Hospital Medical School, joining the Bacteriology Department under Alan Glynn. In 1998 the unit became the Medical School component of Imperial College, and Ivor moved with it to the South Kensington campus and the new Medical School in 2001. Although some malaria work was done, he concentrated more on the mycobacteria, responsible for the diseases tuberculosis and leprosy, particularly (with Charles Easmon and Tobin Hellyer) the agent of bird tuberculosis found in AIDS patients and later (with Douglas Young) the human pathogen *M. tuberculosis*. Over the years, the application of molecular genetics in all fields has produced significant advances in our understanding of antigenic variation, but it is especially troubling that the development of an effective vaccine has remained elusive.

Following the observations of Monroe Eaton (153), who had shown that serum from monkeys infected with *P. knowlesi* could agglutinate schizont-infected red cells, Neil Brown and Ivor Brown used this schizont-infected cell agglutination reaction (SICA) to show that SICA was species and stage specific, that it resulted from a parasite antigen on the surface of the infected red blood cell, and that relapse parasites from a single isolate differed in their SICA antigens (153, 154). In a series of seminal papers over the following years, Brown and Brown demonstrated that a chronic malaria infection was maintained by the serial expression of different antigenic types and that immunity to infection was associated with the development of variant-specific antibodies (60). Neil Brown hypothesized that antigenic variation in malaria resulted from the selection of a minor parasite population and that antibody acted as a signal for the induction of antigenic switching (58, 59). His hypothesis was based on immunizing rhesus monkeys with Freund's incomplete adjuvant (FIA) plus *P. knowlesi* bearing a particular SICA-induced high-titer, non-parasite-killing antibody. When immunized and naive monkeys were challenged with 10, 100, or 1,000 *P. knowlesi* cells expressing the same SICA antigen to which the immunized monkeys had been sensitized, the infections in sensitized and control monkeys were nearly identical in terms of the time at which parasites appeared in the blood and subsequent blood profiles, yet the sensitized monkeys contained parasites that were of a different SICA.

When Brown et al. (61) discovered that spontaneous relapses were usually well controlled in the immune host but that the relapse variant was fully virulent on inoculation into nonimmune monkeys, they suggested that there was an additional immunity that was not variant specific (59). Not only were they able to confirm the observations of Freund (173) that monkeys immunized with FCA developed an immunity that transcended the parasite's capacity for antigenic variation, but also they discovered that after monkeys were vaccinated by infection and cure or by administration of dead parasitized cells in FIA and then challenged with the same variant, there was an intense but short-lived infection that was eliminated in a few weeks (61). This immunity did not depend on the presence of a chronic infection, and unlike the immunity usually occurring in persons with chronic infections it was effective against other strains of *P. knowlesi*, although not against other simian malaria parasites (*P. inui* and *P. cynomolgi*). In 1972, Butcher and Cohen (65) wrote: "Despite the wide antigenic variability . . . within individual strains of *P. knowlesi* . . . the occurrence of cross-immunization between variants is encouraging from the point of view of vaccine production." The challenge ahead was to identify that strain-transcending protective antigen.

In addition to the malaria work by Brown and Brown at the NIMR, there was a team headed by Sydney Cohen at Guy's Hospital Medical School. When he learned of the NIMR experiments and read the conclusion of the Brown and Brown (60) report: "This variation can explain the chronicity of the disease and the inconsistent results with artificial vaccines . . . Some degree of immunity transcending antigenic variation also occurs, but its potential value as a possible basis for the development of a vaccine is uncertain," Cohen was more than irritated. Indeed, the passive-transfer experiments he had carried out with McGregor in The Gambia had raised the hope that a malaria vaccine was possible, and for Cohen immunology was all about specificity—destroy that, and the possibility for a protective vaccine would be severely diminished. Cohen could not accept that parasites could change their antigens, and he went on the offensive, countering Brown's contentions and arguing that even if antigenic variation did occur in *P. knowlesi*, it could be downplayed as a special case and was not relevant to the development of a protective vaccine for humans. Neil Brown, however, was the kind of person who would not "bend the knee" to a dismissive peer such as Cohen, despite the latter's eminence as Professor of Chemical Pathology and Chairman of the MRC Tropical Medicine Research Board. This disagreement led to a serious falling out, and bitter arguments between them persisted for years.

Rejected at first by some in the malaria immunology community, Brown and Brown's assertion for antigenic variation gradually accumulated evidence over the years. Much of this would come from the painstaking work of colleagues Neil Brown inspired, especially Stephen Phillips working with *P. chabaudi* and Chris Newbold working with *P. knowlesi* and *P. falciparum*.

Stephen Phillips (born in 1942) was raised near Manchester on the edge of the Peak District. After completing his schooling at Manchester Grammar School (1964), he enrolled in the Department of Zoology at University College London (UCL). The Head of Zoology at UCL, Sir Peter Medawar (who, along with Macfarlane Burnet, had received the Nobel Prize for Physiology or Medicine in 1960 for their work on transplantation biology and immunological tolerance), stimulated Phillips' interest in immunology, and on the staff in the Department of Zoology at that time was Keith Vickerman, who introduced him to the excitement of parasitology. After graduating in 1964, Phillips moved to the Department of Animal Pathology at Cambridge University. During his Ph.D. training in Cambridge, he had the privilege of interacting with the immunologists in the Department of Pathology, led by Robin Coombs (of the Coombs test). A contemporary in the Department of Pathology, Dick LePage, was working on antigenic variation in the African trypanosomes (antigenic variation in the African trypanosomes as a phenomenon was well established at that time), and his interest in the phenomenon prompted Phillips to search for immune evasion mechanisms in babesias (*Babesia rodhaini* in rats) as well as investigating the nature of the protective immune response to the parasite. Most of LePage's work was never published but formed the basis of others' work. LePage and Phillips often met in the Eagle pub for a pint of ale and reflection on the day's outcomes. Sadly, the Eagle never gave them the kind of insights and inspiration as it did to Crick and Watson a decade or so before.

Phillips completed his Ph.D. in 1968, working on the immune response of rats to *B. rodhaini*, and in the process he showed that the parasite could undergo antigenic variation under immune pressure and that this was indeed a mechanism of immune evasion. He then spent the best part of 6 years at NIMR. The Director of NIMR in 1967 was Sir Peter Medawar (who had previously been at UCL), and that may or may not have been a factor in Phillips' passing the interview. The head of Parasitology at NIMR was Frank Hawking, father of the physicist Stephen Hawking. He was a somewhat eccentric individual who gave all who worked in his laboratory enough stories to liven up any dinner party should the need arise. The transition from investigating *Babesia* immunity to investigating malaria parasites was easy. The question was, how did antigenic variation fit into knowledge of immune mechanisms to the asexual red cell stages of malaria? Was there a variant transcending immunity that might limit the duration of the chronic infection, and was it possible to vaccinate animals to induce a variant transcending immune response? Was antigenic variation a phenomenon in all or most of the malaria parasites, not least in their natural host, since the rhesus monkey was not the natural host of *P. knowlesi*?

In 1972 Phillips moved to take up a lectureship in the Department of Zoology at the University of Glasgow, where he has spent the rest of his academic career, giving almost 15 years in total as Head of the Department of Zoology and later Head of Infection and Immunity, with periods as Vice Dean and Dean of Science in between. At Glasgow his interest in antigenic variation continued. In 1982 he demonstrated that the process occurred during chronic infection in mice infected with a cloned *P. chabaudi* AS strain, looked at the effect of mosquito transmission on variant expression, and showed that different cloned variants of *P. chabaudi* AS strain were able to change their surface antigens in vivo at different rates. (This is still the only study where the rate of change between variants of a malaria parasite has been measured in vivo, where it matters.) More recently, the Phillips laboratory has identified a multigene family, the *cir* gene family, that is most likely to be responsible for the antigenic variability of the parasite.

Neil Brown was against rushing into the field to work on human malarias when, in his view, much basic research using animal malarias still needed to be done in the laboratory; to a degree this was in opposition to the wishes of his senior colleagues. With hindsight he had a point when one considers what has come out of all the studies of immunity to *P. falciparum* in terms of the nature of the protective mechanisms and the antigens involved. Gallons and gallons of blood have been collected from many thousands if not hundreds of thousands of Africans of all ages, and we still do not know how the asexual blood stages are controlled in vivo and eventually removed by the immune response. Notwithstanding Brown's view, Phillips went off for short working visits to The Gambia, first when still at NIMR and later when at Glasgow, to work on immunity to *P. falciparum.* In 1972 Peter Trigg and Phillips published the first observations of the effects of serum from immune adult Gambians on the growth and invasion of *P. falciparum* in vitro, noting that it was after schizogony when the merozoites were released that immune serum had its effects. When this work was submitted for publication, one reviewer thought they had not placed sufficient emphasis on similar work by Sydney Cohen's group at Guy's Hospital working with *P. knowlesi!* Trigg and Phillips noted in their paper that with the field isolate chosen for the cultures with a collection of immune sera, some of the sera from adult Gambians had antiparasitic activity in vitro but others showed no activity. Some 4 years later, Iain Wilson and Phillips showed that in children the protective activity of serum collected from the child after he or she had recovered from an acute infection had its greatest activity against the same parasite population (frozen while the child recovered from the acute attack) and had little or no activity in vitro against parasites from another child. Clearly, there was antigenic variability between wild isolates of the parasite in children and an element of the immune response was population specific. This explains why infection with a different parasite population at a later date could induce another acute attack.

As a graduate student at Cambridge, Phillips had looked after and bred his own experimental animals, washed all glassware, and done so with no technical assistance. Indeed, at the time there was a University edict that research students receive no technical assistance. NIMR provided technical assistance, its own animal breeding unit, and a stimulating and very challenging research environment—what a change in circumstances, and what a great place to be! In particular, inbred rats were available, and this enabled an investigation, using adoptive transfer experiments, into the possible role of cell-mediated immune mechanisms in controlling the asexual stages and the role of individual cell types. In the past 15 to 20 years, this has been more easily investigated with the availability of techniques to specifically eliminate the function of certain genes, called "gene knockout" techniques. One observation made at NIMR, using an adoptive transfer protocol, was that an acute primary *P. berghei* infection in rats could suppress the immunological responsiveness of *P. berghei*-primed lymphocytes (i.e., an acute infection could suppress the immune response to the parasite itself) and that this was another mechanism of survival of the malaria parasite in the semi-immune host. Some 30 years later, it has been most enjoyable for Phillips to pick up this observation again and to be part of some very exciting work with colleagues at the University of Strathclyde showing the effect of *P. chabaudi* asexual blood stages and hemozoin on the immune function of dendritic cells and the consequences of impaired dendritic cell function downstream on lymphocyte migration into B-cell areas of the lymph node.

Thirty-five years ago, Phillips noted that in rats the immune response to *P. berghei* declined rapidly, within a few days, as the blood infection was controlled. Later, this was confirmed to occur in mice immune to *P. chabaudi* AS—protective immunity is short lasting. Jean Langhorne and her colleagues, in a recent excellent review (286), have concluded that important gaps in our understanding of basic immunology of malaria remain. For example, they say that we do need to know why immunity in humans to malaria is so ephemeral. As a self-described "grumpy old man," Phillips is occasionally irritated when work done nearly 40 years ago (by himself and others) has clearly never been read by young scientists who are busily reinventing the wheel. Neil Brown was an example of avoiding the lemming tendency of jumping on the latest bandwagon for a quick publication but ultimately not taking the field further forward. Indeed, Phillips laments that "there is still surprisingly little known about the key aspects of malarial immunity. I still believe, however, that an investment in research on the basic aspects of the immune response to the malaria parasite and yes using suitable animal models will provide the leads we need."

By the time Neil Brown retired from the laboratory in 1994, there was a general recognition of his insight and acknowledgment by his peers. Brown's final triumph would come with the genomic era, when Barnwell and coworkers (31) convincingly proved that the parasite had the genetic capacity to express successive SICA variants, that the SICA antigens were high-molecular-weight polymorphic parasite-encoded proteins uniformly distributed on the surface of the infected red cell; and that the spleen of the host plays a critical role in expression of SICA.

Antigenic variation by malaria parasites is now accepted as an important immunological phenomenon. Furthermore, antigenic switching in vivo is an attractive parasite evasion strategy since it does not occur until after the appearance of a specific antivariant antibody and consequently the malaria parasite makes more efficient use of its variant repertoire. Even more striking, antigenic variation suggests an active signaling process between host and parasite. Antigenic variation is important for an understanding of the nature of *P. falciparum* infections, which are chronic, in which immunity to infection is acquired slowly and is rarely if ever complete, and where severe disease on a per-infection basis is rare (see p. 292). Its impact on vaccine development remains a problem still to be reckoned with. Can antigenic variation, the parasite strategy of immune invasion, be surmounted or not? The wrangling continues.

In 1965 Cohen placed an advertisement in the *Observer* for a research assistant. Geoffrey Butcher (born in 1940), who had obtained his B.Sc. at Kings College London in 1962, had been awarded a Ph.D. in 1970, and then had spent a year in a course for teachers at Bristol and 2 years (1963 to 1965) teaching at Lagos Anglican Grammar School in Lagos, Nigeria, responded and was hired because of his experience in Africa. At the time, the team (Cohen and Butcher!) at Guy's Hospital did not have access to human malaria parasites, and so they began work in Cyril Garnham's laboratory at the London School of Hygiene and Tropical Medicine with *P. knowlesi* as a surrogate for *P. falciparum*. At the outset, the idea was to repeat the experiments that Cohen and McGregor had carried out in The Gambia with passively transferred antibody, but this time using rhesus monkeys infected with minimal doses of *P. knowlesi* parasites. When it was discovered, however, that it took far too much antibody to get any effect (66), an in vitro method had to be developed to assess protection. At first, even culturing *P. knowlesi* for one cycle in the Harvard medium (based on a 1946 analysis of monkey plasma) proved to be a problem: the parasites would neither multiply nor invade. When it was discovered that the rhesus monkeys at Guy's were fed pellets plus cabbage whereas those at NIMR were fed pellets and fresh fruit, the monkeys at Guy's were placed on a diet that included fruit instead of cabbage. The result was that the monkeys looked healthier, their red cells were the normal discoid shape, and the parasites grew. (Presumably the animals were deficient in vitamins E and C when eating their previous diet and were under oxidant stress.) Now sera could be taken from drug-treated animals challenged with a large parasite dose to obtain immune serum and see what effects it had in culture. The findings were interesting—immune serum had little influence on intracellular growth of parasites but did inhibit merozoite invasion and eliminated the succeeding cycle of parasite development; this suggested that the antibody combined with free merozoites to prevent reinvasion of red cells (100). Later studies showed the protective antibody was associated with IgG and IgM (99). (Apparently Neil Brown had done the same experiment, but because the antibody also caused massive schizont clumping, the merozoite invasion assay was not pursued.)

The focus of malaria vaccine research at Guy's, begun under the direction of Sydney Cohen and continued since 1986 by Graham Mitchell (born in 1948), was then and still is on the antigenic composition of the merozoite. Mitchell trained as a zoologist (he obtained his B.Sc. in 1969 at Exeter University), joined the group at Guy's in 1970, and worked for his Ph.D. on the immune response to merozoites and on their use as a vaccine in *P. knowlesi* infections, in close collaboration with Geoffrey Butcher and under Cohen's supervision. His thesis was entitled *Experimental Studies on Acquired Malarial Immunity*. Dave Dennis, the technician who invented a method to isolate merozoites by using polycarbonate sieves (125), and Lawrence Bannister were also early members of the team, and together they produced viable merozoites for electron microscope studies during invasion (and to show antibody on merozoites) as well as to provide material for use as a vaccine.

In 1975 Mitchell, Butcher, and Cohen (336) reported a spectacular finding: when six animals were vaccinated twice intramuscularly with freshly isolated merozoites emulsified in either FCA or FIA and challenged with the same variant, no parasites were found in the blood in three and the remainder had a low-grade infection (maximum, 1.5%) that persisted for 6 to 11 days. Six other monkeys challenged with a variant different from the one used in vaccination with FCA showed a low-grade infection that terminated in less than 2 weeks, and the remaining two animals died. After the initial challenge, all the surviving immunized animals were resistant to challenge for up to 16 weeks. Merozoite vaccination with FCA was required for resistance to challenge with a different variant but not for resistance to challenge with the same variant. Inoculation of blood from vaccinated animals after clearance of blood parasites into naive monkeys did not result in an infection, indicating that the vaccination had induced sterilizing immunity. Merozoites frozen in liquid nitrogen provided comparable protection, as did freeze-dried merozoites stored for up to 20 weeks at 4°C; the latter, however, gave somewhat less consistent protection.

The work, hailed in a 1975 *British Medical Journal* (9) article entitled "Malaria vaccines on the horizon," was responded to by Paul Silverman of the University of New Mexico (see below), who was clearly miffed (460). Silverman questioned the claim of a broader spectrum and higher degree of immune response and was incensed at the reference that his group's work was "only marginally successful." He went on to state that the vaccine trials carried out under his supervision over the previous 5 years and involving over 200 rhesus monkeys had proven to be 90% effective and that adjuvant 65 (emulsified peanut oil) when combined with BCG was as effective as FCA, which was not acceptable for use in humans.

In 1965 the WHO had invited a group of scientists and health educators to its headquarters in Geneva to suggest innovative ways to rescue the failing Global Malaria Eradication Program. Paul Silverman was an attendee, and although at the time he had no "hands-on" experience with malaria, he pressed his case for the feasibility of a malaria vaccine. One of the other attendees, Lee Howard, Head of the Health Division of the U.S. Agency for International Development (USAID), listened to Silverman's arguments and was intrigued. At the meeting's end, Silverman rearranged his flight home so that he could sit next to Howard, continue the discussion, and persuade him of the benefits of a malaria vaccine. Indeed, Silverman went so far as to say that by lifting the burden of disease through a protective vaccine USAID would be able to reduce and perhaps even eliminate its continuing economic support programs in areas where malaria was endemic (127, 451).

Paul Silverman (1924 to 2004) served in the U.S. Army during World War II, received a bachelor's degree from Roosevelt University, in Chicago, and completed an M.S. thesis on trypanosomes at Northwestern University in 1950. Under the impression that he was being rejected from admission to several medical schools because of his membership in the Progressive Party (and had given the controversial Communist Paul Robeson a ride to an event), he moved to Israel, where he began research on leishmaniasis under the direction of the eminent parasitologist Saul Adler. In 1953 Silverman moved to the United Kingdom; in 1955 he earned a Ph.D. in parasitology and epidemiology from the Liverpool School of Tropical Medicine for his work on identifying the mode of transmission of the beef tapeworm, showing that gulls carried the infectious stages from sewage treatment outlets to meadows full of cows. At the time of Silverman's doctoral work, the maintenance and cultivation of parasites outside the host (in vitro) were held in high regard by many parasitologists, who were convinced that in vitro cultures would reveal the biochemical nature of parasitism as well as the basis for host-parasite specificity. Silverman's initial work on vaccination began at the Immunoparasitology Center of Allen and Hanburys Ltd. (acquired in 1958 by GlaxoSmithKline) in London, where he developed nonliving vaccines to protect against tapeworm infections as well as roundworms of cattle, sheep, and horses. One of these, patented in 1959, was a nonliving vaccine for the parasitic roundworms *Haemonchus* and *Trichinella*, produced by in vitro incubation of third-stage infective larvae, removal of the larvae, and freeze-drying of the product excreted by the larvae (so-called exoantigens).

In the 1950s, Senator Joseph McCarthy launched a campaign to rid the United States of subversives who he claimed were strengthening communism and infiltrating the government at all levels. As a consequence, the American consulate suspended Silverman's U.S. passport until he named his "radical" friends from Chicago. Unable to travel, he remained in the United Kingdom until 1968, when there was a change in the U.S. political climate. With his travel privileges restored, Silverman, age 39, returned home to join the Biological Sciences faculty of the University of Illinois in Champaign-Urbana. His experience with nonliving worm vaccines led him to propose that it would be possible to develop a protective vaccine against other parasitic infections, including malaria. Since sterile immunity, well documented in virus and bacterial infections, had not been observed with parasites, Silverman's notion was contrary to the prevailing view of the majority of parasitologists and immunologists.

Silverman's proposal was for a vaccine that would target malaria transmission and pathology by a combination vaccine to two stages: the sporozoite and the asexual blood forms. At this time, there were formidable obstacles to achieving these goals. There was no method for continuously growing the blood stages of *P. falciparum* (or, for that matter, any other species infecting humans) in the laboratory for use in tests for protection, none of the human malarias had been adapted to infect primates, and there was no practical way to mass produce sporozoites of sufficient purity to be used as a vaccine. Silverman dismissed these hurdles as being "simply technical problems" that could be overcome by a major, well-funded research program. Several weeks later he was invited to Washington, DC, and presented his plans to the Health Division of USAID. USAID, having no in-house scientific expertise, sent Silverman's proposal to an ad hoc panel of malaria researchers. They concluded that it was not feasible to develop a practical malaria vaccine for humans and that consequently USAID should continue its business of distributing food and insecticides, providing advice, and assisting with economic programs and should not invest in vaccine research. USAID ignored the expert advice and in 1966 elected to support a $1 million contract with the University of Illinois under the direction of Silverman. The project was designed to determine the feasibility of developing a vaccine against human malaria and involved testing of sporozoite and erythrocyte antigens and in vitro cultivation of these stages.

In 1972, when Silverman became the Vice President for Research at the University of New Mexico, the USAID contract moved with him and a subcontract was established with Rush Memorial Institute to develop a monkey-human malaria model system. Following a 1974 Malaria Vaccine Workshop where the current status of research on vaccines was reviewed, a "road map" for more specific research approaches and priorities was established as well as for the means to increase interest and cooperation among scientists in the area. The priorities included a continued emphasis on erythrocyte antigens including testing in monkeys, use of adjuvants, the continuation of basic studies of sporozoite immunization but with a deemphasis of sporozoite antigens vis-à-vis red blood cell antigens, and consideration of attenuated strains. The agency placed an advertisement in *Commerce Business Daily*, and four projects on in vitro cultivation at Rockefeller University, University of Hawaii, and New York University were selected for funding. When Silverman accepted an administrative position as Vice President for Research and Graduate Studies at the University of New Mexico, the USAID contract remained there, but the project was then under the direction of Karl Rieckmann, a German physician who had moved to Australian New Guinea after World War II, where he had been involved in malaria control programs and had emigrated subsequently to the United States. Three years later, Silverman left New Mexico and malaria vaccine research to become Provost for Research at the State University of New York. He continued in other administrative posts as President of the University of Maine (1980 to 1984); Senior Scientist at the Lawrence Berkeley Laboratory (1984 to 1990), where he headed the Biotechnology Research and Education Program and strongly advocated for funding of the Human Genome Project; Director of Scientific Affairs at Beckman Instruments (1990 to 1994); and finally Associate Vice Chancellor for Health Sciences at the University of California at Irvine. When he died in 2004 from complications from a transplant to correct a blood cell disorder, there was still no protective vaccine either for monkeys or humans.

At the University of Illinois, Silverman's objective was to immunize humans, and he had set his sights on using prisoners at the Statesville Prison at Joliet, IL, where antimalarial studies had been carried out during the 1940s; it was neither practical nor feasible (and certainly not ethical), however, to carry out human trials with an unproven vaccine. So enormous and overly optimistic were Silverman's ambitions that he was convinced that a vaccine which protected rhesus monkeys against *P. knowlesi* malaria could be easily adapted to protect humans against their own malarias. Shortly after receiving funding, Silverman's group realized that the sporozoite antigen was too technically challenging to be practical as a component of the vaccine. Instead, they concentrated on the blood stage vaccine, with the goal of producing a cleaner antigen by removing the parasite, intact, from the erythrocyte and then purifying it of red cell contaminants. Silverman hired an osteopathic physician, Lawrence D'Antonio, who previously had worked at WRAIR and successfully used a French pressure cell (FPC) to remove *P. berghei* and *P. knowlesi* from their respective host red cells and had then cleaned up the preparation by washing, centrifugation, and column chromatography to produce a partially purified parasite fraction for use in a serologic test. Electron microscopy revealed the preparation to have contaminants, and although it might be useful for serology, it was certainly not suitable for vaccinating humans. Crude as it was, this was the best antigen preparation then available.

In 1970, when Silverman was about to present some preliminary results on the FPC technique at a meeting, the USAID office in Washington informed him that he could not present his paper because D'Antonio had filed for a patent to cover the method, which for the moment was therefore not government property, even though D'Antonio's research and salary were paid for by the U.S. government through WRAIR and Silverman's USAID contract. The FPC method was eventually returned to government ownership; however, a pattern of patenting malaria vaccines had been initiated and persists to this day (127). Subsequently, D'Antonio left the University of Illinois and joined the faculty of the Philadelphia College of Osteopathic Medicine; he never again worked on malaria vaccines.

In an attempt to simulate the immune status of children, the Silverman group vaccinated juvenile rhesus monkeys with partially purified parasite fraction and Freund's adjuvant (the first dose with FCA and the second dose with FIA given 4 weeks apart and followed by challenge 4 weeks later); six of nine monkeys survived (462). Even when the antigen was freeze-dried it gave 75% protection when used with an adjuvant such as BCG or FCA (421). The studies with freeze-dried antigen were extended to show that the amount of antigen could be reduced from 1 mg to 250 µg. With the lower dose, 11 of 18 vaccinated monkeys were protected and 16 of 18 vaccinated animals developed a delayed-type hypersensitivity reaction; only 10 of those positive reactors survived challenge, however, once again demonstrating a lack of correlation between DTH and protection.

By the late 1970s, the claims and counterclaims of successful vaccination on both sides of the Atlantic, principally between the Cohen and Silverman groups, led to questions whether the schizont or the merozoite was better as an antigenic source. A comparative trial was contemplated despite the fact that in all of the previous vaccine trials with blood stage vaccines in which some degree of protection was shown, the toxic FCA was required, the "pure" preparations still were not free of host cell contamination, the degree of protection was less than desirable for use in human trials, and in most cases the vaccinated animals became infected. At the time Cohen, the Chairman of the WHO Scientific Group on the Immunology of Malaria, was asked by USAID to visit New Mexico to assess progress. It was suggested by a group of consultants that to resolve the differences between the American and British vaccines, there should be a direct comparison of the antigens in a single laboratory. By the time of this trial (1977), the USAID-University of New Mexico malaria vaccine program was under the direction of Karl Rieckmann, who had been taken on earlier by Silverman in the hope that he would conduct vaccine trials with human subjects. Without a suitable vaccine for humans, however, Rieckmann was to supervise a trial using rhesus monkeys, not human subjects. A total of 32 monkeys were used in the study; 8 controls and 24 rhesus monkeys were divided into groups of eight to test three antigenic materials: a frozen-thawed schizont preparation prepared by the methods of Neil Brown, a freeze-dried preparation as described by the D'Antonio-Silverman FPC method, and fresh merozoites prepared by Mitchell, who had been dispatched by Cohen to New Mexico. Each preparation was emulsified in FCA, monkeys were inoculated intramuscularly twice within a 6-week interval, and then they were challenged with a variant different from the one used for vaccination. All controls developed severe blood infections and died within 12 days. Of those that received the frozen schizonts, six died while the two surviving monkeys had a blood infection that lasted 10 to 14 days with 0.01 and 2.7% parasites in the blood. Four of the eight monkeys that received the D'Antonio-Silverman preparation died, and the remaining animals had low-grade blood infections (0.07, 0.4, 0.9, and 1.2%) that persisted over 9 to 15 days; only two of the eight monkeys that had received the merozoite preparation survived.

Clearly, none of the vaccine preparations was as successful as reported previously. Silverman blamed the poor showing of his vaccine on storage problems and countered that a fresh preparation might have produced better results, and the British claimed that, like good wine, their vaccine did not "travel well" and the prolonged period over which the merozoites were collected to obtain sufficient antigen for the second immunization may have contributed to the lower degree of protection. In a subsequent study, a fresh merozoite preparation was prepared at Guy's, sent to New Mexico in a frozen state, and then stored in liquid nitrogen until use; of eight monkeys vaccinated, four survived a challenge with blood parasites and the remainder died. The results of this head-to-head competition were presented at the NMRI/USAID/WHO Workshop on the Immunology of Malaria held in Bethesda on 2 to 5 October 1979.

Upon peer review of these experiments, the USAID-supported University of New Mexico group lost support from the scientific community and its USAID contract was terminated. The USAID, however, did not lose faith in the development of a malaria vaccine. Indeed, in 1977 the agency advertised for contractors. Of 25 responses to the advertisement, two projects were selected for funding in 1978, increasing the total number of contracts to six. And 2 years later additional contracts were funded by USAID, bringing the collaborating network to 10 USAID projects. The group at Guy's, joined by the immunologists Judith Deans and Alan Thomas and the molecular biologist Andy Waters, persisted in the goal of antigen isolation from merozoites, and ultimately the *P. knowlesi* work would mature into the discovery of the apical membrane antigen, AMA-1, currently being tested as a blood stage vaccine candidate (see p. 181).

Blood stage vaccine research conducted for 35 years (1945 to 1980) primarily used the rhesus monkey-*P. knowlesi* model and could hardly be considered a success; it did, however, provide for a better appreciation of the nature of the immune response to a naturally induced infection as well as to artificial immunization; it established the role of antigenic variation in immune evasion. Still, there was much left to be understood. Identification of the appropriate antigens of the experimental malarias that could provide protection through vaccination was totally lacking. A serious problem that had emerged was how antigen could be presented to the immune system so there would be maximal stimulation without the use of draconian adjuvants such as FCA and saponin. And for some there still remained a nagging question: could a malaria vaccine successful in protecting monkeys be adapted to protect humans?

Despite all these experimental efforts with birds and monkeys, the fact remained that, after half a century, a protective malaria vaccine was not on the horizon. At a minimum, a vaccine for human use would require an immunogen that could be prepared in large amounts in a reproducible manner and would be free of microbial and host cell contaminants, pyrogens, toxins, and other potentially dangerous antigens. The desirable vaccine would have to remain stable over a reasonably long period and not undergo serious deterioration; its potency would have to be measured in terms of efficacy in vivo and/or in vitro; and it would have to be efficacious in an appropriate animal model before being used in humans. And, as Neil Brown has said, manipulation of what is essentially an autoimmune reaction would require special care.

7

Promises, Promises, Promises

It was 1964, and the WHO-sponsored global eradication campaign was in tatters. Chloroquine-resistant malaria was spreading across Southeast Asia and South America, and anopheline mosquitoes were showing resistance to DDT. There was an urgent need to prevent the undoing of all that had been achieved during the previous 20 years. No quick fix would be possible, however. Investigations into the biology, biochemistry, and immunology of human malarias had been almost totally neglected for decades because the world's health workers were convinced that eradication was on the horizon. There was no convenient way to study human malarias since in the United States, Great Britain, and Europe infected patients were lacking, the parasites infecting humans could not be grown in the laboratory, and suitable hosts to which human malarias could be adapted were unavailable. The strategic needs for malaria control were self-evident: find suitable primate hosts for the human malarias, discover the means for growing the parasites continuously in vitro, and recruit a cadre of well-trained scientists to direct their studies toward the biochemistry and immunology of malaria parasites in the hope of discovering a protective vaccine.

Monkey Business

By the 1960s, the widespread resistance of *P. falciparum* to chloroquine led to a new and intensive search for novel antimalarial drugs. One of the critical needs of the drug discovery program was a convenient and relatively inexpensive laboratory animal in which to test the putative "magic bullets." At the time, none was available and the only hosts were the unsuitable and endangered species of larger apes. Nonhuman primates would first be successfully employed for drug testing; however, the rather strict host specificity of human malarias posed a major constraint to the use of nonhuman primates for studies of immunity, so that only later would they be used for investigating vaccines. The solution to the problem of finding suitable nonhuman primate hosts came from several research institutions; one of these was the Gorgas Memorial Laboratory (GML) in Panama, and the other was Stanford University, Palo Alto, CA. The GML, founded in 1921, was named after William C. Gorgas, the U.S. Surgeon General who had "conquered the mosquito" and by so doing enabled the United States to effectively control malaria and yellow fever, thus allowing completion of the Panama Canal. Martin D. Young, with a long and distinguished career in the U.S. Public Health Service working mostly on the chemotherapy of malaria, assumed the Directorship of the GML in 1964. He initiated programs to determine whether Panamanian monkeys would serve as suitable models for the study of human malaria and of mass antimalaria drug administration. Forthwith, the GML established a colony of night owl monkeys (*Aotus trivirgatus*).

Coincident with the work on *Aotus*, squirrel monkeys (*Saimiri sciureus*) were found to be susceptible to *P. vivax* (118), and a few years later Young and Rossan (571) induced *P. falciparum* and *P. vivax* infections into the now endangered Panamanian squirrel monkey, *S. orstedi*. Breeding of *S. sciureus* was found to be relatively easy, and as a consequence squirrel monkeys could serve as excellent hosts for asexual blood stages; however, unlike *Aotus*, the poor development of infective gametocytes in squirrel monkeys made sporozoite passage impossible. Nevertheless, in general, both *Saimiri* and *Aotus* have been considered convenient surrogates for human malaria, and the WHO has promoted preferential use of these two primate hosts. Indeed, much of what is known about immune responses to different antigens and the efficacy of adjuvants comes from vaccine trials using these New World monkeys. All too frequently, however, the admonition by Gysin (1998) that "it should be kept in mind that experimental approaches with any nonhuman host can furnish a wealth of data, but these data may not necessarily reflect reality in man; therefore, care should be exercised before making extrapolations" (206) is ignored by those anxious to test the latest vaccine candidate and grab the prize of being the first developer.

Attempts to induce human malaria infections in nonhuman primates began in 1934 with work in Panama by the Taliaferros (see p. 92), who used the black howler monkey (*Alouatta villosa*); however, the infections could not be maintained by serial transfer of infected blood (496). Thirty-two years later, Martin Young, James Porter, and Carl Johnson, working at the GML and supported by funds from the U.S. Army Medical and Research Command, reported that they were able to transmit *P. vivax* from a patient to owl monkeys (570). The infection was transmitted by the bite of *Anopheles albimanus* mosquitoes, using two self-experimentation volunteers (Johnson and Porter), and subsequently their blood infections were used to infect other night owl monkeys. In succeeding years, further studies using nonhuman primates were conducted at the GML (568, 569).

Richard Rossan (born in 1929) had his first encounter with malaria parasites during the summer of 1949 at the University of Illinois, Champaign-Urbana, while taking an Introductory Protozoology course taught by Richard Kudo, a world-famous protozoologist. Coincident with being awarded an M.S. degree in 1951, Rossan received a draft notice due to the Korea conflict. He opted for a 4-year commitment in the U.S. Navy, a decision that was to affect his entire research career. Told that Clay Huff at the NMRI was looking for assistance, Rossan was interviewed, hired, and spent the next four years in Huff's laboratory working on the drug treatment of *P. fallax* in turkeys and the effect of drugs on the EE stages. After his discharge from the Navy, Rossan returned to graduate school and received a Ph.D. degree under the direction of Huff's old friend Leslie Stauber at Rutgers University, in New Brunswick, NJ. His thesis described the changes in the serum proteins of animals infected with *Leishmania donovani*, and his Ph.D. research was his introduction to immunology. In 1960, Rossan began an intense learning process in the use of rhesus monkeys as hosts for infections of *P. cynomolgi* at Christ Hospital Institute of Medical Research, Cincinnati, OH, under the direction of Leon H. Schmidt. Rossan was able to demonstrate that the "bodies" seen in liver biopsy specimens from rhesus monkeys infected by sporozoite inoculation were, indeed, the preblood stages that had been reported by Garnham and Shortt for *P. vivax* in humans in 1945. In 1963, the Schmidt group moved to Davis, CA, to set up the National Center for Primate Biology. At Davis a wide variety of nonhuman primates were evaluated as hosts of monkey and human malarias, with the goal of screening antimalarial drugs. About this time, Quentin Geiman at Stanford University reported that the night owl monkey, *Aotus trivirgatus*, from Colombia, would support the development of *P. falciparum*. This was a significant breakthrough in malaria research, as now a small primate host would be available for chemotherapy and immunity studies. An extensive series of immunological studies with monkey malarias was done by Alistair Voller, a visiting scientist from the Nuffield Institute of Comparative Medicine in London. Antigenic variation was found (535) in *P. cynomolgi bastianelli*; cross immunity was demonstrated between two species, *P. c. bastianelli* and *P. c. cynomolgi*, but not between *P. c. cynomolgi* and *P.c. ceylonensis*

(536). The final part of the study involved an investigation of immunity and antigenic variation of *P. knowlesi* in *Macaca mulatta*, which showed that immunity to one antigenic variant was effective against subsequent challenge with different variants (534), work that was pursued by Neil Brown at NIMR (see p. 116–117), not by Rossan.

In 1969, Rossan accepted what he believed was to be a 2-year contract as Staff Scientist at GML; however, what he did not realize was that he would spend 28 years there. During his early years at GML, *Saimiri sciureus* was used as a host for blood-induced *P. falciparum* infections and sporozoite-induced *P. vivax* infections, including the demonstration of EE stages. Martin Young retired in 1974, and at about the same time Leon Schmidt retired from his position at Sloan Kettering Laboratory, Birmingham, AL. Schmidt had worked for many years on the U.S. Army's contract, using the Colombian *A. trivirgatus* monkey experimentally infected with many different strains of *P. falciparum* and *P. vivax*. To continue the program of seeking new and effective drugs against *P. falciparum* strains resistant to chloroquine (and other drugs), the U.S. Army funded a program using Panamanian *A. trivirgatus* monkeys at GML with Rossan as Principal Investigator. The first 18 months or so of the contract was devoted to demonstrating that the indigenous monkey could replace the Colombian monkey as a suitable host. In May 1989, Rossan was appointed Acting Director of the laboratory, during an economic embargo placed against Panama by the United States. On 20 December 1989, the Just Cause invasion to capture Manuel Noriega began. For 4 days Rossan was unable to go to the laboratory, and he became extremely concerned about the more than 400 monkeys in the colony. Fortunately, one of the animal caretakers stayed at the laboratory to provide food and water to the monkeys. Following the invasion, the almost $2 million congressional appropriation for the GML was cancelled, with the Laboratory being given to the Panamanian government under the terms of the original charter. To continue the drug evaluation program, additional funds were obtained from the U.S. Navy vaccine program; the grant was administered by a Panamanian medical supply company, with rent for the rooms at the laboratory being paid by the Panamanian government. The drug evaluation continued, and then a vaccine component was initiated.

In 1964, Quentin M. Geiman (1904 to 1986) at Stanford University, encouraged by the U.S. Army Medical and Research Command, attempted to cultivate *P. falciparum*. Geiman had experience in the in vitro culture of malaria parasites; however, this had been gained during World War II. In

1944 the U.S. Office of Scientific Research and Development was created, with a Board for the Coordination of Malaria Studies "to better understand the mode of action of antimalarials in order to protect U.S. troops in Southeast Asia, North America and the Pacific." A project to cultivate the malaria parasite *P. knowlesi* in vitro and to study its metabolism was sponsored at Harvard under the leadership of Eric Ball (1904 to 1979), who was Acting Head of the Department of Biological Chemistry (1943). Ball was joined by Geiman, then an Assistant Professor at the Department of Comparative Pathology, who had received his Ph.D. from the University of Pennsylvania in 1934 for studies with amebas. After Geiman left Harvard in 1955, he joined Stanford University, where he maintained a research program on pathogenic amebas. Despite his lack of recent experience with malaria, in 1964 he was the only member of the World War II Harvard project who was available and interested in the cultivation of malaria parasites. However, Geiman found it difficult to grow *P. falciparum* in vitro, unlike *P. knowlesi*. Further, apart from nine cases of human malaria and two samples of infected blood, there was no steady supply of falciparum parasites available in the San Francisco Bay Area, and storage and freezing of blood caused changes in the parasite before in vitro culture began. It was the 1966 report by Young et al. (570) that prompted Geiman to try to circumvent the shortage of human cases and to investigate the possibility of adapting *P. falciparum* to night owl monkeys. Night owl monkeys obtained from a local dealer were infected with blood from a soldier who had returned from Vietnam; later, blood from a female patient who had contracted an infection in Uganda was used to infect a monkey. Infections were then transferred from these owl monkeys to other owl monkeys by inoculation of infected blood (184). At Stanford the former line was named FVO (Falciparum Vietnam Oak-Knoll) and the latter was named FUP (for Falciparum Uganda Palo Alto). In 1980, the FUP strain, previously adapted to the *Aotus* monkey, was adapted to *S. sciureus* from Guyana (207).

In 1965, to assist him in the cultivation of *P. falciparum* at Stanford, Geiman recruited Wasim A. Siddiqui. Siddiqui had received a Ph.D. in 1961 from the University of California at Berkeley for studies of amebas, returned to India briefly, and then spent 1963 to 1966 working at the Rockefeller University with William Trager studying the nutritional requirements of *P. lophurae*. At Stanford, and with support from a U.S. Army Department of Defense contract, Siddiqui used commercially available media for the short-term cultivation of *P. falciparum* in *Aotus* red blood cells, ostensibly to produce high yields of malaria-infected red cells for isolation of antigens to be used in vaccination studies (364). In 1970 Geiman retired from Stanford, and shortly thereafter Siddiqui moved to the University of Hawaii. From 1970 to 1975 the NIH and a U.S. Army contract supported Siddiqui's work on in vitro cultivation of *P. falciparum* and the use of *Aotus* monkeys to grow malaria parasites for vaccine purposes.

The *Aotus* genus is composed of nine species widely distributed in Central and South America; although these owl monkeys were not found to be naturally infected with human parasites, some species are highly susceptible to experimental infection (219). Each species can be distinguished by appearance, chromosome pattern (karyotype), and geographic distribution. For malaria research, it is the differences between the species in susceptibility to infection with a human *Plasmodium* parasite that is significant. For example, *A. lemurinus griseimembra* was susceptible to nine strains of *P. falciparum*, and infection with the St. Lucia strain was highly reliable and reproducible; although this parasite does not develop well in culture, it is able to produce high levels of infective gametocytes (219). In contrast, *A. nancymai* resisted or recovered quickly from infection with FUP and Malayan Camp strains; Collins et al. (110) found *A. griseimembra* from Colombia to be equally susceptible to Panamanian strains of *P. vivax* and *P. falciparum*, and 12 *A. griseimembra* from Peru were equally susceptible to *P. vivax* and *P. falciparum* (110). Taylor and Siddiqui (501) found that Colombian *A. griseimembra* died following challenge with the FUP strain whereas Peruvian *A. nancymai* generally survived. Espinal et al. (164) found that *A. griseimembra* from Colombia was susceptible to the FCB-1 strain of *P. falciparum*, and in naive *A. griseimembra* the FVO strain causes a very reproducible and fulminating blood infection whereas the FUP strain produces less severe but longer-lasting blood infections. Because susceptibility to death after infection was so variable among the various *Aotus* spp. (102, 103, 219), it has certainly influenced the interpretation of the findings when vaccine candidates were tested for efficacy.

Although *A. griseimembra*, *A. vociferans*, and *A. lemurinus* are susceptible to sporozoite infection, some of these monkeys require removal of the spleen for adaptation to and maintenance of long-term infections as well as for the production of gametocytes (219). Because spleen-intact *A. griseimembra* monkeys exposed to intravenous inoculation of *P. falciparum* sporozoites reliably experience blood infections, it has been suggested that this host-parasite combination is the most suitable for testing pre-blood-stage vaccines. Clearly, the various host-parasite combinations with differing immunological responses can affect the outcomes of vaccination trials, and some combinations would be useless for immunology studies. The problem with the use of New World monkeys is their expense and scarcity and the difficulties in handling them in the laboratory. Also, there is no agreement as to which *Aotus* (or *Saimiri*) species-parasite strain combination is best for testing vaccine efficacy, nor is there consensus on how vaccine efficacy and/or protection should be measured.

The Culture Club

In 1975, USAID advertised for contractors to develop a method for the in vitro cultivation of *P. falciparum*. This resulted in funding projects at the Rockefeller University, the University of Hawaii, Parke-Davis Co., and New York University. Within a year, the Rockefeller University project made a significant breakthrough. The scientific consultants who had advised USAID that a malaria vaccine would require the cultivation of *P. falciparum* had been correct, but they could not have foreseen how serendipity would play a role. William Trager, who had spent a lifetime studying the nutrition of malaria parasites, was specifically invited by USAID to apply for funding. He was not optimistic, having abandoned research in this area 5 or 6 years earlier. Short-term in vitro growth of *P. falciparum* and *P. vivax* was reported as early as 1912 by Bass and Johns, using glass vials and a medium supplemented with glucose; however, subsequent attempts to reproduce this system met with failure (261).

From 1947, Trager worked with *P. lophurae* and maintained infected duckling red blood cells in vitro by using the Geiman-Harvard rocker dilution method. The culture system consisted of red cells suspended in the nutrient "Harvard" medium, gently rocked to simulate blood flow and gassed with humidified 5% CO_2– 95% air. Under these conditions, parasite growth was less than optimal and reinvasion rates were low so that continuous culture could not be achieved. When Trager reviewed the state of malaria parasite cultivation, he lamented that he had failed to reproduce the growth of *P. falciparum* seen by Bass and Johns and wrote, "we are not very close to a method for the continuous cultivation of the erythrocytic stage of any species of malaria parasite" (510). However, in that same review he commented briefly on preliminary experiments he had conducted using specially designed flow vials.

In 1969 Trager decided to abandon the Geiman-Harvard rocker dilution method and to substitute a flow vial system in which the culture medium would move gently over a settled layer of infected red blood cells. His reasoning was that since *P. falciparum*-infected red cells spend most of their 48-h developmental cycle attached to the walls of the post-capillary venules, agitation of the infected red cells might be detrimental to parasite growth and invasion. In his proposal to USAID, Trager specifically asked for support of a postdoctoral fellow experienced in the cultivation of intracellular parasites.

James B. Jensen (born in 1943), who received his Ph.D. in 1975 from Auburn University, was the person selected. Shortly after Jensen's arrival at Rockefeller University in January 1976, the two planned their approach to cultivating *P. falciparum*. First, they selected commercially available culture media high in glucose; second, they abandoned the bicarbonate buffer system (since it was clearly inadequate to control the lactic acidosis); third, they decided to compare parasite growth in the rocker flasks and flow vials; and fourth, they elected to modify the gas mixture. In February 1976 they tested the suitability of commercial media by using *P. falciparum*-infected red cells removed from an *Aotus* monkey infected with the FVO strain of *P. falciparum* (obtained from Trager's former postdoctoral researcher Wasim Siddiqui, who at the time was working at Stanford with Geiman). The cells were washed, diluted with human AB red cells, suspended in 15% human serum, and placed in flow vials, and a variety of tissue culture media were screened. The newly developed RPMI 1640 medium was found to be superior to all others tested. They also changed the gas mixture from 5% CO_2–95% air to one high in carbon dioxide and low in oxygen: 7% CO_2–5%O_2–88% N_2. Under these conditions, and with a settled layer of red blood cells, it was possible to maintain the parasites for 24 days by adding fresh uninfected red cells every 3 to 4 days (261, 512–514). The numbers of parasites increased dramatically in the flow vials when fresh cells were added to a diluted sample of infected red cells, but attempts to maintain parasites in the rocker flasks failed time and time again. In the meantime, Jensen decided to take some of the infected red cells and place them in small flat glass dishes, invented by Koch's young coworker Julius Petri, with a variety of media (such as RPMI 1640 and Dulbecco's modified Eagle's medium, Ham's H-12, MEM, and Medium 199). When he placed the petri dishes containing the infected red cells into a 5% CO_2–air incubator, the parasites died out after 2 to 3 days. It was then that Jensen decided to employ a candle jar instead of the CO_2 incubator—a method he had used to grow the cells for the cultivation of various coccidian species (and when the CO_2 cell culture incubators in the virology laboratory at Utah State were unavailable to him). Jensen located a large glass desiccator, placed his petri dish cultures inside, and, after lighting a candle, closed the stopcock when the flame went out; the cultures were incubated at body temperature for several days. At first Trager was dismayed to observe Jensen's use of a 19th century technology, not far removed from the days of Robert Koch; however, when he was shown the Giemsa-stained slides, Trager was convinced that Jensen was onto something. In the summer of 1976, Milton Friedman, a graduate student in the Trager laboratory who was working in Ian McGregor's MRC laboratories in The Gambia, arranged for a sample of human blood infected with *P. falciparum* to be sent to New York. This sample was diluted with RPMI 1640 (which turned out to be the best of the commercial media), placed in petri dishes, put into a candle jar, and incubated. The line grew very well (much better than FVO) and became FCR-3, one of the most widely used strains. Later, other lines would be established using similar methods, and the impact of continuous cultivation of *P. falciparum* would be phenomenal, allowing for the sequencing of the *P. falciparum* genome, allowing the identification and isolation of blood and insect stage antigens, and providing a means for identifying vaccine candidates.

Although *P. falciparum* grew well in Trager's flow vials, the method would not have been practical for growing these parasites outside of laboratories such as those at Rockefeller University: the flow vials were handmade by an expert glass blower, were cumbersome to set up, used a lot of medium, produced very small samples of infected red cells, and required expensive peristaltic pumps. The greatest value of the candle jar method is that it can be used in laboratories almost anywhere in the world where there is an incubator, a candle, petri dishes, and a desiccator (261). Indeed, this "19th century technology" would, in short order, energize research on malaria vaccines.

The Possible Dream

The stated objective of the 1977 WHO-sponsored Workshop on the Biology and In Vitro Cultivation of Malaria Parasites was "to review the status of research and to identify areas in which efforts and progress were most needed to further expansion and acceleration of chemotherapeutic and immunologic research." The venue for the Workshop was the campus of Rockefeller University, and it was a celebration of the achievement of William Trager and James Jensen. Their work would allow laboratories around the world to have access to *P. falciparum* and to require only human blood donors, not monkeys. Trager, who had been at Rockefeller for his entire career (since 1933) studying various aspects of *Plasmodium* nutrition, now basked in the glory of his accomplishment. A diligent culturist, he had never been interested in a malaria vaccine, but now he believed that such a vaccine could be developed using the in vitro culture methods he had pioneered. As he walked across the Rockefeller campus, Trager announced to several of those who were within earshot, "You know, using in vitro cultures, there will be a protective vaccine in five years." I knew Trager quite well, having been a postdoctoral researcher with him, and after leaving his laboratory I had studied the biochemistry of *Plasmodium* and done some work on antigenic analysis, so I was astounded at his declaration. This was uncharacteristic of the meticulous and self-effacing man I knew. I countered, "You cannot be serious; it will be decades, if ever, before there is a practical vaccine." However, he was serious, as was another who walked with us, an immunologist, who put a finger across her lips, saying, "Shush. We must promise a vaccine is on the horizon or else research funding will quickly dry up." Indeed, the successful continuous in vitro culture of *P. falciparum* would lead to enhanced funding and, for some, fame and a fortune in research support, but for others it would lead to deceit and greed. It would also attract a new cadre of researchers who were idealists as well as those who fancied themselves as future Nobel laureates.

The successful culture of *P. falciparum* encouraged USAID (470) to advertise for contractors to improve on the cultivation of the erythrocytic stages of malaria with the goal of purifying the antigens produced (with special attention to removal of red cell membranes), culturing different strains, developing methods to produce gametocytes, and providing methods for harvesting parasite antigens and testing their immune potential. There were 25 responses, and "after intensive review" two additional projects were selected, bringing the "collaborating network on malaria immunity and vaccination" to 10 projects including Rockefeller University, University of Hawaii, WRAIR, GML, Parke-Davis and Company, University of New Mexico (to work on the antigenic nature of in vitro-grown parasites), and New York University (to conduct sporozoite immunity research).

Trager recruited a team to carry out vaccine research. Robert Reese, an immunologist, would study the isolated antigens. Susan Langreth, an electron microscopist, would monitor the purity of the isolated antigenic fractions. Harold Stanley, with expertise in tissue culture, would improve the in vitro growth conditions and synchrony. Araxie Kilejian, a biochemist, was to isolate and purify antigens. To support the USAID program, Trager was given a large budget and a number of laboratory rooms to accommodate the increased number of investigators. However, before vaccinations of *Aotus* were begun, it was learned that there were nine different kinds of chromosomal patterns (karyotypes) and that each had the potential to respond in an independent fashion to challenge with *P. falciparum* (387). Indeed, preliminary studies indicated that monkeys of the type VI karyotype (*A. boliviensis*) were more resistant than those of karyotype II or III (*A. griseimembra*). To minimize variability in immune responsiveness, *Aotus* monkeys of karyotype II were used. Schizonts and mature trophozoites, harvested from Trager-Jensen cultures of *P. falciparum* (FCR-3 strain), were used to prepare merozoites by saponin lysis, and this served as the vaccine. The first experiment employed six monkeys. Three received merozoites emulsified in FCA administered intramuscularly, and 3 weeks later a booster vaccination was given; this consisted of the same number of merozoites but this time emulsified in FIA. The controls received only an injection of the adjuvant. (A more suitable control would have been uninfected red cells; however, that was not done.) Three weeks after the second vaccination, the animals were challenged by intravenous inoculation of 1 million parasitized red cells of the same strain derived from an *Aotus* monkey. Within 2 to 3 weeks after being challenged, all of the monkeys were dead with blood infections ranging from 10 to 55% of the red blood cells parasitized.

Based on the assumption that the amount of antigen was more important than the manner of administration, the protocol was altered so that three intramuscular injections were given at 3-week intervals, with the first two containing merozoite antigen emulsified with MDP (muramyl dipeptide, a synthetic derivative of N-acetylmuramyl-L-alanyl-D-isoglutamine) and mineral oil (FIA) and the third containing solely the antigen in mineral oil (387). (The basis for using MDP was that Mitchell et al. [336, 338] found MDP to have some protective effect; i.e., two of six monkeys survived challenge when vaccinated with merozoite antigen emulsified in MDP and FIA, in contrast to the use of FCA, where five of six vaccinated animals survived challenge.) Although the number of merozoites in the vaccine given to each monkey had been increased and the challenge injection had been reduced to 500,000 infected red blood cells, two of three vaccinated monkeys died. One died a week later than the controls, a second died 2 weeks after the controls, and the third monkey had less than 0.5% of its red blood cells infected. All of the "protected" *Aotus* monkeys showed anemia, and in those that succumbed this was assumed to be the cause of death.

Reese et al. (387) claimed that the work was "significant for two major reasons." First, merozoites from *P. falciparum* cultured in vitro for over a year were still able to induce immunity; i.e., they had not lost their antigenic "punch"; and second, synthetic MDP could substitute for FCA when a sufficient amount of antigen was used. Subsequent studies at Rockefeller University reported on improved isolation of merozoites (476) and the inhibitory effects of immune serum on in vitro-grown *P. falciparum* (386); inhibition was claimed to be due to reaction with surface alterations (called knobs) on the *P. falciparum*-infected red cell, but not to merozoites!

Reese, smooth-talking, aggressive, and secretive, wanted to be independent of Trager, and, imbued with his own self-importance, felt that he alone was better prepared to develop a malaria vaccine. In 1980, he moved his laboratory and several of his Rockefeller University colleagues across the country to the Scripps Research Institute, La Jolla, CA, where immunology was at the forefront. At Scripps Reese was joined by Randall Howard and Harold Stanley, and together they studied the synthesis of merozoite proteins (239) and complement activation (477), developed an assay for measuring the effects of immune sera on in vitro parasite maturation (478), prepared monoclonal antibodies from *Aotus* monkeys and used them to identify a merozoite surface antigen (479), and constructed a cDNA library and screened it with serum from immune *Aotus* monkeys to identify three merozoite surface antigens (17, 18), one of which was related to so-called heat shock proteins. They suggested that the heat shock protein would be an excellent immunodiagnostic candidate and should be studied as a possible vaccine candidate. (The reasoning behind this was not revealed, however.) When Reese did not gain tenure at Scripps, the USAID project was moved to the Agouron Institute (also in La Jolla, CA). There the Reese laboratory carried out immunological modeling (397) and peptide mapping and antibody responses to the heat shock protein (432). They identified a high-molecular-weight phosphoprotein (238) and suggested that it was involved in parasite transport (475), and they also identified rhoptry antigens (237). However, for the most part these finds produced nothing of any importance for the development of a protective malaria vaccine. Indeed, in 1987, after a negative review of the project by a group of scientific consultants, the USAID contract to the Agouron Institute was terminated. Shortly thereafter, Reese left the field of malaria vaccine research and others in the field neglected the work done at Scripps and Agouron.

In 1968, Voller and Richards (533) prepared formalin-fixed, schizont-infected red cells by the method of Targett and Fulton (see p. 110), emulsified them with FCA, injected them intramuscularly into two monkeys, and gave the monkeys a second dose 5 weeks later. The vaccinated animals were challenged with 2,000 parasites obtained from an owl monkey infected with the Malayan Camp strain (which had been used to prepare the vaccine). All the animals died from fulminating infections. A year later, Sadun et al. (415) at WRAIR, following success in the vaccination of rhesus monkeys against the blood fluke disease parasite *Schistosoma*, used irradiated parasitized erythrocytes as the malaria vaccine. Infected blood obtained from an owl monkey was washed and irradiated (25,000 rads) using a ^{60}Co source, and then naive monkeys were given, at weekly intervals, four intravenous doses of the γ-irradiated parasitized red cells (Camp strain) without FCA. Upon challenge, two of the vaccinated animals did not develop a blood infection and five had lower levels of parasites in the blood; however, four of the seven vaccinated monkeys died within 5 weeks while three survived at least 150 days. Thirteen control animals, four of which received irradiated normal red cells, died in less than 8 days. Less protection was obtained with three vaccinations, and no protection was obtained with one immunization. (Ten years later, Wellde, Diggs, and Anderson [548], also working at WRAIR, found that only one of three monkeys that had received four immunizing doses of the irradiated parasitized red cells died after challenge and two of the surviving monkeys developed a low-grade infection.)

Graham Mitchell, having successfully vaccinated rhesus monkeys against *P. knowlesi* with merozoites as the antigen (see p. 123), used short-term in vitro cultivation of *P. falciparum*-infected blood taken from children living in The Gambia to obtain mature schizonts and isolated merozoites; he then emulsified them with FCA. One owl monkey was used as a control, and three were vaccinated three times at 2- to 3-week intervals; 3 to 6 weeks after the second vaccination, all the monkeys were challenged with a West African (Lagos) *P. falciparum* isolate from an infected owl monkey. All four monkeys, including the control, survived; however, in the control animal the infection appeared earlier and was more severe than in the vaccinated animals, where the appearance of parasites in the blood was delayed and infection was less severe (337). When these vaccinated monkeys were challenged with an East African strain (FUP), there was some evidence of resistance.

Beginning in 1975 the USAID research at the University of Hawaii (Principal Investigator, Wasim Siddiqui) was dedicated to developing a malaria vaccine and continuing to improve on methods for in vitro cultivation of *P. falciparum*. Siddiqui extended the earlier immunization studies of rhesus monkeys with *P. knowlesi* to *Aotus* by attempting to vaccinate them against *P. falciparum*. Taking clues from the *P. knowlesi* studies, Siddiqui and his colleagues (454) used blood stage antigens obtained from *Aotus* monkeys infected with *P. falciparum* (FUP) and later used a merozoite-enriched preparation from short-term in vitro cultures (456). In the latter experiment, two doses of vaccine (2.73 mg) emulsified with FCA were administered intramuscularly at a 3-week interval and, 3 weeks following the second vaccination, the animals were challenged by intravenous injection of 6.2×10^5 parasites. The two control animals died of fulminating infections within 2 weeks, whereas all three vaccinated monkeys survived; however, they all had low-grade blood infections. Subsequently, Siddiqui et al. (458) were able to show cross-resistance in monkeys vaccinated (using FCA) with FUP and challenged with FVO.

Siddiqui never questioned whether it would be practical to obtain sufficient numbers of merozoites from in vitro cultures to vaccinate millions of human subjects, nor did he ever consider the problems associated with the purity of his vaccine preparation or possible viral contaminants. He did recognize, however, that a barrier to the development of a protective malaria vaccine was the need for a suitable replacement for FCA. He wrote (455): "The ultimate objective of all malaria vaccine studies is to develop a vaccine that can be used to immunize and protect man, not monkeys or rodents. Therefore the development of an immunologically satisfactory and pharmacologically acceptable adjuvant is imperative in the development of a malaria vaccine acceptable for use in man." Siddiqui tried MDP, a substance that was reportedly able to replace whole tubercle bacteria in FCA and which had already been shown to enhance the immunological response of animals to an antigen when injected with FIA (mineral oil). However, vaccination trials involving *Aotus* monkeys vaccinated with merozoites in MDP in peanut oil or mineral oil were unsuccessful. Further, MDP or its derivatives could be used for immunization of humans only without the inclusion of mineral oil.

Despite these failures, the portly, pipe-smoking, jolly, and cocky Siddiqui heralded his own findings by writing, "the course of infection in immunized animals was definitely ameliorated as judged by: the lower level of parasites in the blood and a 7–12 day delay in time to death in the two immunized monkeys . . . and the immunity in the third immunized monkey was impressive, the parasite levels in the blood never exceeding 0.4%" (455). Siddiqui continued to pursue alternatives to FCA. He tried stearoyl-MDP adjuvant with carrier liposomes (cholesterol plus lecithin). The vaccine contained a crude antigen (50 to 60% schizonts with merozoites and the remainder consisting of other developmental stages) with 2.86 mg of protein emulsified in the adjuvant. Animals were vaccinated twice with a 4-week interval, and 17 days after the last dose the monkeys were challenged with 750,000 parasites (FUP) obtained from an *Aotus* monkey with an ongoing blood infection. Two of the controls died within 2 weeks of challenge; remarkably, the third control, despite having 25% of its red cells infected, survived. All four of the immunized monkeys survived challenge, with two developing low-grade infections that lasted a week and two developing infections ranging from 5 to 15%. A limited number of owl monkeys became negative for blood parasites after a month; however, since blood from these animals was not injected into naive monkeys, it was not known whether the animals still harbored small numbers of parasites undetectable by microscopy. The results, reported in a *Science* article (459) entitled "Vaccination of experimental monkeys against *Plasmodium falciparum*: a possible safe adjuvant," were considered to be important and significant (by Siddiqui) since the stearoyl-MDP adjuvant and liposomes did not produce an inflammatory reaction at the injection site and "an effective, safe malaria vaccine may be possible."

The vaccination trials with *P. falciparum* used a very limited number of owl monkeys. The pattern of protection against challenge was not unlike that seen earlier with *P. knowlesi* in rhesus monkeys: a delay in the onset of infection and a low-grade blood infection; in the few cases where protection was achieved, FCA (unsuitable for use in humans) was the most effective adjuvant. The level of protection provided a degree of protection in the *Aotus* monkeys; however, in humans a *P. falciparum* infection of similar magnitude (2 to 5% of red blood cells infected) would cause acute symptoms and pathologic damage. In spite of this, these findings were given great prominence, and there were claims that a protective vaccine for human malaria might be achieved. However, one skeptical critic, Carter Diggs, opined that it was also possible that survival of the monkeys was related to low host susceptibility and/or low virulence of the parasites used in the challenge (128). Further, and of greater concern, was that not only were the *P. falciparum* vaccine preparations exceedingly crude but also the nature of the specific antigens involved in protection had not been defined.

Siddiqui's finding that the vaccinated animals failed to develop sterile immunity after primary challenge was dismissed. He justified his failures (as well as those of many others) when he wrote: "The objective of vaccination against microbial infections is to prevent disease and mortality . . . populations immunized against polio continue to secrete viruses . . . is accepted . . . as indicative of successful vaccination . . . reduction in morbidity and mortality are the main criteria for assessing the potency of vaccines against tetanus, cholera, smallpox, rabies, pertussis and diphtheria. It is unnecessary to demonstrate that the protected host is completely free of the infecting agent i.e. that sterile immunity exists. The aim of human malaria vaccination at present is not to obtain sterile immunity but to lower the parasitemia and to modify the clinical course of the disease" (455). In other words, he contended that despite there being an absence of sterile immunity, vaccination with blood stage parasites might permit a malaria infection within clinically tolerable limits such that in an area of endemic infection the individual would survive the most dangerous period of the disease, the primary infection, while establishing acquired immunity.

Money and Malfeasance

The USAID Malaria Immunology and Vaccine Research (MIVR) program began in 1966 with a single project at the University of Illinois under the aegis of a project officer, Edgar A. Smith, a medical entomologist who not only was ignorant of the epidemiologic and clinical aspects of malaria but also knew nothing about immunology. In 1972, without input from scientific consultants, USAID carried out a self-evaluation and reached the conclusion: "the feasibility of vaccination was convincingly demonstrated" (470). Absence of peer review and evaluation by USAID had actually begun in 1965 when, without scientist administrators, the agency had sent the University of Illinois (Paul Silverman) proposal to an ad hoc panel of respected malaria researchers (127). They were unanimous in their judgment that, at the time, it was not feasible to develop a practical malaria vaccine for humans and that USAID should stay out of the research business and stick to its "handout" business of distributing food, insecticides, and advice. The recommendations of the malaria experts were ignored, and the University of Illinois was awarded $1 million to develop a malaria vaccine (see p. 125–126). Thus began a pattern of ignoring advice from the scientific community. Ultimately it would lead to a crisis in the USAID malaria network including criminal investigations, charges of sexual harassment, indictments and convictions for fraud and malfeasance, and a loss of confidence in those carrying out research on a malaria vaccine.

In the 1970s, several discoveries were made to indicate that the landmark investment by USAID in a malaria vaccine might pay off. The race to find a protective vaccine sharpened the rivalries between scientists and their sponsors, notably USAID, NIH, WRAIR, the Centers for Disease Control (CDC), NMRI, and the MRC. By 1974, flush with self-delusion, USAID sponsored a malaria workshop through the National Academy of Sciences to review the current status of research, to provide the agency with a more specific focus of approaches and priorities for development of a vaccine, and to suggest ways of stimulating increased interest and cooperation among scientists working in the area. In 1975 a panel of experts proposed that USAID expand its support to more than one laboratory. USAID did not engage in competitive bidding and simply chose four additional laboratories, one of which was the Trager laboratory at Rockefeller University. A year later, with the breakthroughs in the continuous culture of *P. falciparum* at Rockefeller and at WRAIR (213), USAID requested additional funding so that the number of contractors and projects could be expanded from 6 (University of New Mexico, Rush Memorial Institute, Parke-Davis and Company, New York University, University of Hawaii, and Rockefeller University) to 10 with inclusion of projects at WRAIR, GML, and the National Institute of Health in Bogota, Colombia (470).

In 1982, project officer Smith retired and was succeeded by James M. Erickson, an economic entomologist with a Ph.D. in population ecology from Cornell. Erickson was, like his predecessor, devoid of experience in vaccine development, yet both were enthusiastic boosters of the program. Under Erickson the MIVR, which had been funded at under $4 million per annum from 1966 to 1974 and included four laboratories, by 1985 grew to 11 laboratories funded at $13.466 million. But by 1989, the 41-year-old Erickson, a stubborn, abrasive, fast-talking, and irreverent malaria vaccine booster, had been removed from his position at USAID. On 29 November 1989, he was indicted by a federal grand jury that charged him with conflict of interest, conspiracy, illegally accepting gratuities, making false claims, and submitting false income tax returns. He was accused of making illegal profits through a tangled web of schemes from the vaccine program that he managed. Erickson never came to trial.

After protesting his innocence and filing countersuits against USAID, he eventually pleaded guilty to the charges of accepting an illegal gratuity, making false statements, and filing a false tax return (8). A lenient court sentenced him to 6 months, to be served in a halfway house, and fined him $20,000. He served his time and vanished.

By 1980 USAID was sponsoring a lavishly funded vaccine research program. More and more scientists, not wanting to lose out, began to enter the race. In 1982, Miodrag Ristic, a University of Illinois veterinarian and an expert on the cattle disease redwater fever, reported that when the responsible blood parasites (called *Babesia* and remotely related to malaria) were held in short-term cultures they secreted a proteinaceous substance into the medium and that this could be used as a vaccine to protect cattle. (This was similar to the work Paul Silverman did in preparing a vaccine against roundworm infections [see p. 125].) In 1983, Ristic submitted an unsolicited research proposal to USAID to continue his vaccine research for 3 years, suggesting that malaria parasites might also secrete such antigens and that these might serve as the basis for a protective vaccine. His budget was $2.38 million, and USAID sent out the proposal for peer review. Although the reviewers did not recommend funding because it was more like a preproposal, Erickson requested that it be funded and skirted the issue of scientific merit by stating that the expert panel "had endorsed the scientific methodology and the exceptional qualifications and experience of the researchers" (8). Ristic got his money, accomplished little toward a vaccine, and eventually was convicted of mail fraud. The contract was terminated by USAID in 1987.

This was not Ristic's only run in with the law (11). An audit by the University of Illinois found that he had improperly used over $30,000 in federal grant travel funds. He had purchased airline tickets which went unused, and then the refunds were credited to his personal account and an outside agency. Ristic was ordered to vacate his university office and research laboratory, but the university did not press criminal charges. The results of the audit were passed to the Illinois Attorney General's Office, which began its own investigation. In June 1991 Ristic was convicted on four counts of theft by deception and was given a 12-month sentence with conditional discharge and a $5,000 fine. In 1993 a U.S. District Court Judge in Minnesota sentenced Ristic, then 74 years of age and by that time an Emeritus Professor, to 5 years of probation, 300 h of community service, $90,000 in restitution, and a $50,000 fine (13). When Ristic pleaded guilty to accepting funds that were fraudulently obtained through a business partner, Gottfried Kellerman, he implicated Erickson in the scheme.

In 1984 Kellerman and Ristic submitted to USAID a research proposal requesting $750,000 for the development of a diagnostic field test for malaria on behalf of a nonprofit shell corporation, KT&R Laboratories Inc., formed by Ristic and Kellerman for the sole purpose of making the grant proposal. Erickson submitted the proposal to a panel of three malaria experts to review its technical merit, and despite their negative recommendation Erickson reported otherwise to his superiors. In February 1985, Erickson notified Kellerman that the grant had been awarded to KT&R Laboratories. However, when a preaward audit showed that KT&R had no assets, Kellerman told USAID that ProtaTek International Inc. (a shell company established by Ristic and Kellerman to manufacture vaccines and diagnostic tests) would furnish equipment, laboratory space, and support services to perform the work. More than $430,000 was disbursed under this award. Ristic collected more than $90,000 in consulting fees, and KT&R also paid about $170,000 to ProtaTek (8).

After Kellerman was notified of approval of the contract, he entered into an agreement to pay $4,000 per month to a Guatemalan company, International Insect Research and Development (IIRD), in which Erickson had a financial stake, and eventually all of the $88,000 paid to IIRD found its way into Erickson's personal bank account. USAID terminated the contract in 1987, and a criminal investigation ensued. A jury convicted Kellerman of defrauding the United States, making false statements to a governmental agency, and conspiracy to make false statements. He was sentenced to 8 months imprisonment with 5 months suspended, 3 years of probation, and $75,000 restitution. The 8th Circuit Court of Appeals rejected Kellerman's 1993 appeal of the decision (11). Kellerman disappeared without serving time or making restitution.

Erickson was also involved in other monkey business. In 1983 or 1984, USAID decided that a large number of candidate vaccines suitable for testing in humans would be available, and it was thought that before this could occur, the candidates would have to be tested in South American owl monkeys (*Aotus*) and squirrel monkeys (*Saimiri*) from Peru, Colombia, and Bolivia to measure the toxicity of the vaccine, possible harmful effects, and ability to protect against malaria. Between 1984 and 1988, approximately $6.7 million of USAID monies would be used to acquire, house, and maintain monkeys.

A General Accounting Office (GAO) investigation (8) found indications of fraud in a USAID-funded project at the National Institute of Health in Bogota, Colombia. In 1982 USAID awarded a 3-year subcontract to the Institute in Colombia with an estimated budget of $709,375, and 2 months later the American Institute of Biological Sciences (AIBS) (a respectable nonprofit group that had managed various biological and biomedical projects for the federal government) was asked by USAID to amend the contract and increase the budget to $1.53 million to cover AIBS overhead and research costs. The Principal Investigator for the Colombian program was Carlos Espinal. Although there were research components to the program, the primary motive for the contract was to obtain Colombian monkeys. By 1984 an AIBS audit found numerous billing and financial irregularities at the Institute's Malaria Unit. Three sets of books were found. Espinal was unable to account for all project funds, and the U.S. dollar checks mailed by AIBS to the malaria unit had been converted into local currency "through channels forbidden by national monetary authorities." Further, some 40 checks valued at ca. $150,000, made out to the Colombian Institute, had been deposited in Swiss banks and other accounts outside Colombia. AIBS alleged that Espinal had defrauded the U.S. government, and a U.S. District Court indicted him for the crime. The Institute suspended its malaria activities; no further work was performed during the remainder of the contract period, and nothing more was heard from Espinal. USAID did not seek to recover any expenses from AIBS or the Institute due to "sensitive relations" between the United States and Colombia (8).

USAID's first large procurement of monkeys began in 1984 with a purchase from the Peruvian Primatological Project (PPP). Through an agreement with NIH and the Pan American Health Organization, $1.1 million was provided for conservation activities and to identify surplus monkeys that could be exported. Eighty-three monkeys were sent in 1983, 100 in 1985, and 200 in 1986 and 1987. USAID did not specify the karyotype needed, so PPP sent monkeys that were more easily captured but had a less desirable karyotype. In 1986 Erickson requested that PPP provide 500 Peruvian *Aotus* and 500 Peruvian *Saimiri* monkeys at a cost of $532,190. At an August 1986 meeting of primate consultants (of whom I was one), the Peruvian *Saimiri* monkeys were ranked fourth among available species for testing blood stage vaccines, and when 30 were sent to CDC for evaluation they were deemed to be unsuitable. Nevertheless, USAID decided not to cancel the order because of prior commitments to PPP. USAID arranged for shipment of 150 *Saimiri* monkeys to other government agencies and accepted the remaining 350 at a cost of $183,000 (8, 127).

In April 1985, USAID decided to purchase up to 600 Bolivian monkeys with unexpended funds from the suspended AIBS subcontract with the

Colombian National Institute of Health. AIBS, the management arm of the vaccine program, was to arrange for the purchase of the monkeys, and a person named George Diaz would be given a contract to find and acquire the monkeys. Diaz and Erickson then contacted Worldwide Primates based in Miami to obtain 200 owl monkeys at $475 each and 400 squirrel monkeys at $375 each, for a total of $245,000. Diaz then told AIBS that the monkeys would be obtained from Gerrick International for $336,000—$630 for each *Aotus* monkey and $520 for each *Saimiri* monkey. It turned out that Gerrick International was a shell corporation for Diaz and Erickson. On 12 September 1985 Erickson had ordered stationery with the letterhead Gerrick International and Diaz had opened a bank account in the name of Gerrick International. AIBS issued a check for $168,000 to Gerrick International in partial payment for the monkeys, and Gerrick in turn paid Worldwide Primates $122,500. In January 1986 Diaz wrote a check for $8,500 to his brother-in-law, and this was then signed over to Erickson. In March $11,886 was withdrawn from the Gerrick account and used to purchase two cashier's checks made payable to J. Erickson. In October 1986 Gerrick was paid the remainder of the $168,000, and presumably Erickson and Diaz made a profit on this also. The grand jury used this to level charges against Erickson of conspiracy, conflict of interest, and accepting a gratuity. Since he paid no tax on the profits, Erickson was also indicted on three counts of submitting false tax returns. Diaz fled the country (8).

In January 1986, 341 Bolivian *Saimiri* and 20 Bolivian *Aotus* monkeys arrived in the United States. When the Bolivian press reported that the monkeys were being sent to zoos and pet stores, there was an adverse public reaction. The Bolivian government put an embargo on the export of owl monkeys, and Worldwide Primates' export waiver was rescinded before the remaining monkeys could be shipped. Colombia and Peru followed Bolivia in making it illegal to export *Aotus* monkeys, but by then more than 1,000 monkeys had been shipped. In March of 1989 USAID owned 1,452 monkeys including 508 Colombian *Aotus* and Bolivian *Saimiri* of suitable karyotype for vaccine research, 526 Peruvian *Aotus* with marginal suitability, and 237 unsuitable Peruvian *Saimiri*. The cost of housing these monkeys in the United States was reported to be $2 million (8).

Wasim Siddiqui at the University of Hawaii also became entangled in the web of deceit associated with the USAID vaccine program. Beginning in 1972, Siddiqui had set his sights on conducting vaccine trials with a human parasite, using *Aotus* monkeys as surrogates for a related primate, humans. As noted earlier, Siddiqui immunized *Aotus* monkeys with a crude vaccine made from cultured *P. falciparum* mixed with FCA, and although a few of the vaccinated monkeys were protected from challenge, all had small numbers of parasites in the blood. The vaccine could not be used with humans because it contained FCA; however, USAID publicized the findings as if a human vaccination trial was sure to occur in the near future. Press conferences were called, the State of Hawaii legislature expressed gratitude to him, and the University of Hawaii honored him with its most prestigious award for excellence in research, stating this was for his "finding the first promising candidate for a malaria vaccine." There were further exaggerated claims: he was the first to culture malaria parasites and the first to infect owl monkeys with human malaria. USAID was also wildly enthusiastic: "Technically speaking a vaccine may be available for human testing as early as 1985" (127). In August 1984, USAID announced a major breakthrough in the development of a vaccine . . . "and the vaccine should be ready for use around the world, especially in developing countries within 5 years." In the same year, Siddiqui submitted a proposal to USAID to extend his research for another 3 years at a cost of $1.65 million. USAID sent the proposal to two external reviewers. The first reviewer wrote, "The proposal is mediocre, overly ambitious and the budget is overwhelming and excessive," and the second reviewer said, "the proposal is unrealistic in terms of time, money and availability of material. The amount of money requested is outlandish and outrageous." Yet by 1986 Siddiqui would write "the asexual blood stage vaccine may soon be available for clinical trial" (127).

In 1988, acting on "information received," the Office of Inspector General of USAID began investigating Siddiqui's and the University of Hawaii's handling of research funds (8). The investigator reported that there was evidence to support the allegations that the Principal Investigator (Siddiqui) apparently diverted funds to his and his secretary's personal use, and that construction costs for refurbishing his offices at the University were charged to the grant as consultant payments. On 14 September 1989 the Grand Jury in Hawaii indicted Siddiqui and his administrative assistant Susan Lofton on charges of theft in the third degree and criminal conspiracy. The Deputy Attorney General of Hawaii charged that some of the USAID money was siphoned off between 1984 and 1987 through illicit accounting tricks and a kickback arrangement with a Honolulu travel agency that Siddiqui had engaged to run a 1985 Asia Pacific Conference on Malaria. The USAID conference check was deposited with the Research Corporation of the University of Hawaii and, at Siddiqui's direction, was to pay the bills for the conference. There was $100,000 as an advance payment for services to be used as a deposit to the Pacific Hotel, where the conference was to be held. In March 1985, a month before the conference, Siddiqui instructed the travel agency to begin direct payments of $1,260 per month to him and $1,000 per month to his secretary. These salary supplements were to continue for the next 2 years. Between December 1984 and April 1986 the travel agency paid Siddiqui $17,400 and Lofton $16,000. USAID was sent a bill for $35,425 for services and departmental rental related to the conference. USAID paid the bill, but the indictment disputes that the money was used for that purpose (316). This was only the beginning of USAID's troubles. Hawaii Senator Daniel K. Inouye, Chairman of the Senate Appropriations Committee on Foreign Operations, which has jurisdiction over the USAID budget, launched a GAO investigation of the entire $8.5 million malaria vaccine program budget. The 18 October 1989 edition of the *Washington Post* reported (49, 256, 320, 357) that the GAO had found irregularities at the University of Hawaii and that as a result Siddiqui and his Administrative Assistant were indicted on charges of stealing USAID research contract monies. By chance, the indictment came out on the same day that the University of Hawaii received word that USAID had renewed Siddiqui's contract for 3 years (worth $1.65 million). However, the agency insisted that Siddiqui be removed from the project until the charges were resolved. Legal maneuvering continued for years, but it finally ended in 1993 when Siddiqui pleaded no contest to embezzlement charges and was sentenced to 6 months of house detention (14). The University of Hawaii relieved him of teaching duties; however, he continued to report to the campus wearing a court-ordered radio bracelet to alert police to his whereabouts. The University Professors' Union successfully defended Siddiqui's assertion that he not be dismissed because there was no just cause, and he did not lose his retirement pension. On the day he retired in 1994, Siddiqui pleaded no contest to a civil suit brought against him for $250,000 (126, 572).

In spite of the troubles at the University of Hawaii, Siddiqui and his associates continued their vaccine-related studies. They used serum from vaccine-protected *Aotus* monkeys to identify 13 protein antigens (264), and then in a final vaccine trial they were able to protect *Aotus* monkeys against challenge using a single merozoite surface coat protein (Pf195 = MSP-1); protection against challenge was achieved only when the antigen was emulsified in FCA (457). During a decade of research (1987 to 1996), with generous support from a USAID contract, Siddiqui and his collaborators were able to show that serum from animals vaccinated with Pf195 inhibited the in vitro growth of *P. falciparum* (251) and that B30-MDP (a lipophilic MDP derivative) and LA-15-PH (a synthetic equivalent of monophosphoryl lipid A) could replace FCA in inducing growth-inhibitory antibodies to Pf195 in rabbits (252). A protective malaria vaccine, however, was far from being at hand.

The debacle of the USAID MIVR program has been ascribed by some to the rivalries between competing scientists in pursuit of limited funds. However, there was a more pernicious element—greed—and the center of it all was James Erickson, USAID's chief technical officer for the malaria program (8, 126, 127, 315). Coincident with his appointment in 1982, US-AID decided that it needed independent management to oversee the project. The management contract was awarded by noncompetitive selection to AIBS. AIBS assigned Dorothy Jordan to be the project manager, and in short order she and Erickson became more than good friends. In 1983, when I became a consultant to USAID, Erickson confided in me that he and Jordan were romantically involved; the plan was that after they divorced their respective spouses, they would be free to marry. Erickson held to his part of the bargain, but Jordan balked. When their affair ended in 1986 or 1987, Erickson claimed that AIBS' management was dishonest and incompetent. There were contract failures in not getting reports done on time, not carrying out scheduled site visits, and purchasing computers without formal approval. There ensued a bitter war of memos and affidavits. Erickson called for Jordan's removal. Jordan, in return, filed a charge against Erickson for sexual harassment. There were charges and countercharges, and by April 1987 Erickson's superiors had reassigned him pending allegations of mismanagement. To this was added allegations of sexual harassment and anonymously supplied information of illegal activities. In October 1987 Erickson was put on administrative leave, and he languished for a full year before launching an offensive claiming that he was the victim of a witch hunt. He sued AIBS through a whistleblower statute, and in July 1988 he legally enjoined USAID to demand that it come to a decision to reinstate him or fire him. The court agreed, and since nothing had been proven, USAID docked him a week's pay for bad judgment and personal misconduct. Dorothy Jordan refused to press a formal charge of sexual harassment. She resigned from AIBS. Erickson might have escaped without punishment had it not been for the continuing probe by the Inspector General and the GAO, whose investigations brought down the house of cards Erickson had built. As already described

(p. 147), in 1990 after pleading guilty Erickson was fined and received a short sentence to be served in a halfway house. He served his time and was not heard from again.

Along with Erickson, Ristic, Kellerman, Espinal, and Siddiqui, other reputations were damaged and funding of USAID-supported research programs was reduced, albeit temporarily. In 1987 to 1988, after USAID sent onsite review teams to the Agouron Institute (the Reese project) and to the Biomedical Research Institute in Rockville, MD, where Michael Holling-dale had a contract to develop culture systems for EE stages of *P. vivax* and *P. falciparum* and to characterize the liver receptor for the sporozoite and where Werner Zolg was to develop a DNA vaccine for blood stages, both contracts were terminated. At some institutions there were complaints of a lack of objectivity by the review team as well as incomplete disclosure of research findings lest data be given away to reviewers with a competing interest; one of the institutions reviewed stated that a member of the review team was prejudiced and had submitted a proposal for funding knowing that this would depend on availability of funds in a very restricted budget environment, and another member of the review team was accused of "pursuing precisely an identical line of research to ours." These objections were rejected by USAID. In 1988 the project to study antigen localization, assess the purity of blood stage antigen preparations, and determine the structure and function of knobs, headed by Susan Lan-greth of the Uniformed Services University of the Health Sciences in Bethesda, MD (rumored to be Erickson's paramour), was terminated without any review. Leveling a complaint, she claimed she had been threatened with repossession of an electron microscope that had been purchased under her USAID contract. The claim was denied.

To its credit, USAID was one of the first agencies to recognize the need for a malaria vaccine, and it funded such research when others had either ignored or abandoned the responsibility. However, those who were responsible for the USAID malaria vaccine program were neither patient nor realistic, and they did not want to face the fact that vaccine development can be a slow and deliberate process. Seeking to ensure that funds would continue to be provided, USAID and the network contractors did not produce a realistic timetable of 15 or 20 years or perhaps longer—something they feared the public and government would not like to hear—and so they promised a "magic bullet" in just a few years. The significance of the vaccination results was frequently exaggerated and given undue prominence. Within USAID there was a lack of critical self-assessment and accountability, a disregard for the advice of professionals, and an unbridled desire to be the first developer of a protective vaccine. The MIVR network had become an exclusive club whose members were its cheerleaders, particularly the project director. Dissent by members was not heard, since this would jeopardize both membership and funding. It was this combination that led a USAID technical officer, his cronies, and some investigators in the network to engage in sleaze and corruption.

Through manipulation and falsification of peer review evaluations, the MIVR not only was able to maintain itself but also was able to expand. Findings of doubtful significance continued to be heralded in the press, and despite some of MIVR's failures, the glamour of a malaria vaccine was undiminished and the funding by USAID (and other agencies) would continue. Indeed, in short order the crisis at USAID would be forgotten and researchers at WHO, WRAIR, NIMR, CDC, NIH, the Walter and Eliza Hall Institute (WEHI), and MRC would promote and prophesy that through biotechnology—recombinant DNA, monoclonal antibodies, gene cloning, peptide synthesis, and immunochemistry—not only was a vaccine possible but it was just around the corner.

The conclusion of the 1989 GAO report (8) was that in the MIVR there was evidence of fraud and waste of government funds, subprojects were selected under questionable circumstances, performance was not subject to adequate evaluation, and poor quality research may have been tolerated. To correct such deficiencies, USAID appointed a new management group (Atlantic Resources Inc.) with Carter Diggs, formerly director of the Army's malaria vaccine program, as a scientific consultant. Under his leadership there has been improvement in the evaluation of subprojects, using an expanded list of consultants to review proposals, and an insistence on high-quality proposals from institutions seeking funding (77).

The MIVR program illustrates the problem with so-called push programs (276), i.e., those that pay for research inputs rather than results and where funds are committed before a product is developed. In push programs, applicants tend to exaggerate the prospects that their approach will succeed, and once the project is funded researchers may divert the resources to other projects and to pursue avenues that will more rapidly advance their careers. Indeed, even when it was clear that there were problems in finding a protective malaria vaccine with monkey models, researchers kept requesting more funding and USAID administrators kept approving it. This resulted, in part, from researchers tending to look favorably on the promise of their own work (the Pygmalion effect) and administrators having an incentive to expand their own programs (the Emperor effect). In the development of a malaria vaccine, it is imperative that the perils inherent in push programs be recognized and avoided.

In 1992 USAID elected to shift emphasis from support of research to the development of promising vaccine candidates and to testing of investigational vaccines. USAID formed the Malaria Vaccine Development Program (MVDP), effectively replacing the MIVR, to build a pipeline from early preclinical vaccine development through the regulatory process and field-testing of vaccine candidates. The major target products of the MVDP are blood stage vaccines, particularly MSP-1 and AMA-1, for children living in areas where malaria is endemic. Recombinant protein vaccines are being developed in collaboration with WRAIR and are being tested in Kenya, and DNA recombinant vaccines are being developed in collaboration with the Naval Medical Research Center. The most advanced blood stage vaccine, the merozoite surface protein (FMP1/AS02A), developed on the initiative of USAID at WRAIR with GSK, is the culmination of work begun in 1995, and until 2000 the major operating expenses were borne by USAID. In 2006 the USAID budget for malaria vaccines was $6.2 million; however, this has been effectively leveraged through a pooling of resources by partners such as PATH (Program for Appropriate Technology in Health), MVI (Malaria Vaccine Initiative), GSK, WRAIR, and NMRI.

8

Hunting for Protective Blood Stage Antigens

The discovery of a simple means for continuously growing *P. falciparum* in laboratory flasks (see p. 137–139) seduced William Trager into believing that a protective vaccine not only was possible but also could be achieved in short order. Indeed, successful immunization experiments conducted by others using material from infected humans and *Aotus* monkeys (see p. 140–141) encouraged him to proceed. He recognized that infected *Aotus* monkeys could not provide sufficient amounts of infected red blood cells and suggested that the way around this impediment was to harvest abundant quantities of *P. falciparum*-infected erythrocytes from large-scale Trager-Jensen petri dish cultures, and when the parasite materials so obtained were freed from host red cell contaminants, this could serve as a vaccine. But on expressing this view, he was asked by critics, "Were these parasites antigenically similar to those obtained from *Aotus*?" Two years later he had an answer: *Aotus* monkeys vaccinated with laboratory-grown *P. falciparum* could be protected from challenge with parasites obtained from infected owl monkeys (515).

Trager, during the time when I was with him in his laboratory as a postdoctoral fellow, was without any airs of self-importance. He was a dedicated bench scientist, meticulous in both design and conduct of experiments. I remember seeing him day after day, peering through his wire-rimmed glasses at slides under the microscope or, with the sleeves of his shirt rolled up, bent over a Bunsen burner flame transferring cultures. He prepared his own media and recorded each day's results in a bound notebook in his own unmistakable and illegible handwriting. During a span of 40 years he devoted all of his efforts to studying the nutrition of malaria parasites with the goal of providing a list of their minimum daily requirements. Dedicated to a deeper understanding of the molecular basis for the intracellular life of *Plasmodium*, he had cared not one whit for vaccines. But by 1978 his attitude toward his favorite "animalcules" had changed. More and more (or so it appeared to me), his thoughts seemed to be centered on the recognition that he might receive. It was natural for Trager to consider his own work to parallel that of Weller, Robbins, and Enders, whose discovery enabled Jonas Salk and Albert Sabin to reorient and accelerate the development of a polio vaccine. Indeed, if a malaria vaccine were developed, it would have been entirely reasonable that Trager should receive similar recognition.

Trager, who had received a Ph.D. from Harvard University in 1933, was acutely aware of the seminal discoveries made in the 1940s at the Harvard Medical School and knew that this work had led to a shared 1954 Nobel Prize for John Enders, Frederick Robbins, and Thomas Weller (549). And Trager, who was a contemporary of the driven, curious, and dedicated microbe hunter Weller, knew him well. Thomas Weller (1915 to 2008), as an undergraduate at the University of Michigan, had spent two summers under the well-respected parasitologists L. J. Thomas and W. W. Cort at the University of Michigan Biological Station and had described a roundworm that parasitized perch. Upon completion of the A.B. degree, he intended to study parasites at the Harvard Medical School; however, after taking a microbiology course taught by Hans Zinsser (an expert on typhus and author of the classic *Rats, Lice and History*), his interest in microbes was piqued. He did not give up on parasites altogether, and one summer when he was a medical student he went to Florida to study malaria at a Rockefeller training center. During his time at Harvard, Weller visited the Rockefeller Institute in New York City and was exposed to the methods for growing viruses in chicken embryo tissue cultures. In his fourth year of medical school, Weller wanted to see whether the larval stages of the roundworm *Trichinella* found in muscle could mature into adults in the presence of living tissue cells, and so he began a tutorial project with Enders, an associate of Zinsser, who had earlier turned from the study of bacterial immunity to investigations on the growth of viruses in roller tube cultures. Shortly after Weller received his M.D. in 1940, his clinical training was interrupted by the outbreak of World War II and military service. The U.S. Army Medical Corps sent him to Puerto Rico, where he focused on the diagnosis and control of malaria. The U.S. Army Sanitary Corps posted Trager to New Guinea, where he worked on malaria control through atabrine treatment. At the end of the war, both returned to their respective academic institutions—Weller to Harvard and Trager to Rockefeller.

Weller had carried out a crucial experiment almost by chance (549). Using chopped animal tissues such as placenta, brain, and kidney, it was possible to grow the cells as a single layer in laboratory flasks. This technique was called tissue culture. After inoculating 16 tissue culture flasks with the throat washings from a child suffering from chicken pox, Weller had 4 flasks left over; he thought it would be interesting to inoculate these with a mouse brain suspension that contained poliovirus. The chicken pox virus did not grow, but the poliovirus did, and in the process the tissue culture cells were destroyed. Subsequently Weller and Robbins found that the fluids from the tissue-grown poliovirus would paralyze mice and monkeys. Mixing infected fluid with antibodies to the virus prevented growth of the virus in the tissue cultures, and the monkeys injected with virus and then antibody were protected from paralysis. In Weller's hands and with improvements in tissue culture methods, different antigenic types of the poliovirus were isolated from fecal samples. These could be grown in laboratory flasks and characterized. Finally, Enders, Robbins, and Weller were able to show—as had Louis Pasteur 75 years earlier—that multiple passages in tissue culture decreased the virulence of the poliovirus. This attenuation of the poliovirus would ultimately lead to a live protective vaccine against polio. Trager found reassurance in the words Weller had spoken in his Nobel lecture: "tissue culture in some form might eventually prove of value (in developing a prophylactic agent . . . (and) immunizing materials . . . through the use of in vitro techniques is already more than a theoretical possibility."

It was after his feat of "taming" *P. falciparum* had been achieved that Trager began to think more and more of Weller's serendipitous discovery and felt that his success with culture of the deadly *P. falciparum* was similar to Weller's poliovirus work. And if there was a malaria vaccine for which a Nobel Prize would be awarded, in the minds of many he would have been considered for such recognition. But before any dream of a Nobel Prize could be realized, there had to be a vaccine; that would require either collaborators or success by others. Early vaccination studies at Rockefeller University were carried out with material cultured by the Trager-Jensen method. Although there had been evidence for protection in initial experiments in 1978 and 1983 (387, 515), the "magic bullet" was clearly not near at hand. The "pure" merozoites were contaminated by red blood cell components, and protection required the use of FCA, an adjuvant unsuitable for use in humans. More seriously, by this time Trager had lost most of his "vaccine team." Reese and Stanley had moved across the country to The Research Institute of the Scripps Clinic (La Jolla, CA) under a USAID contract (see p. 142), and Langreth, provided with a new electron microscope and her own USAID contract, had joined the faculty of the Uniformed Armed Services Medical School (Washington, DC). Consequently, were Trager to be considered for a Nobel Prize, he would have to rely on the USAID-supported vaccine studies being conducted by his former postdoctoral student Wasim Siddiqui at the University of Hawaii (see p. 144). Trager, now a Member of the prestigious National Academy of Sciences, enthusiastically supported Siddiqui's vaccination studies and ensured that the findings were published by communicating them to the *Proceedings of the National Academy of Sciences*. Until the time of his death at age 93, Trager continued to support vaccine research by others, writing as late as 1995: "As one who, in 1976, with J. B. Jensen, provided methods for the continuous culture of *Plasmodium falciparum*, methods which have led to and greatly facilitated . . . work on an erythrocyte-stage vaccine for malaria . . ." (511).

However, by 1985 Trager realized that malaria parasites grown in vitro or in monkeys were structurally and antigenically more complex than poliovirus and that the immune responses to these two parasites were so vastly different that a malaria vaccine was not on the horizon. Trager, to his credit, clearly recognized that a chemically defined parasite antigen would never be obtained by fractionation of intact infected red blood cells by using his beloved cultures. Instead, he felt that their main use in relation to the development of malaria vaccines would come from the identification of target antigens for both the asexual erythrocytic stages and the sexual stages by using the advances being made in biotechnology (monoclonal antibodies and recombinant DNA), areas in which he had no expertise. The noble (or is it Nobel?) quest for a malaria vaccine would have to be pursued by others.

Monoclonal Antibody to the Rescue

Understanding immune protection began with the empirical vaccination studies by Edward Jenner and the attenuation of virulent microbes by Louis Pasteur and Emile Roux. Following this, Behring and Ehrlich discovered that immune serum contained a protective substance called antibody, and half a century later Kabat and Tiselius described antibody in molecular terms, gamma globulin (see chapter 3). By the 1950s it had become evident that a particular kind of lymphocyte, a B lymphocyte, played a key role in the antibody response to a pathogen. There are millions of B cells in the human body, and each has the capacity to make a particular antibody that is able to recognize a different antigenic determinant (epitope). It is this capacity of lymphocytes to react to each component of a pathogen that is the basis of immune protection. Under ordinary circumstances it is impossible to use immune serum to isolate a particular antigen because it contains a mixture of antibodies directed against a spectrum of antigens. However, were it possible to isolate a particular B lymphocyte from the others, place this in a test tube containing a nutrient medium allowing growth and division into identical progeny (clones), and allow it to secrete antibody against a single epitope, there would be a "molecular tweezer" able to pluck out a specific antigen. Although theoretically it might be possible to locate and isolate an individual B cell by using a microscope and pipette and then place this cell in culture, it would not serve to provide antibody in reasonable amounts because a B cell divides for only a finite length of time, the progeny secrete antibody, and then they die. Therefore, providing abundant amounts of an antigen-selecting antibody requires something unnatural—long-lived and dividing lymphocytes that continue to secrete antibody. In 1975, César Milstein (1926 to 2002) and Georges Köhler (1927 to 1995) were able to create this unnatural B lymphocyte.

Milstein, born in a provincial town in Argentina, studied chemistry at the University of Buenos Aires, where he wrote his first Ph.D. thesis on the active site of enzymes (332, 333). In 1958 he received an MRC scholarship to work in the MRC Laboratory of Molecular Biology at Cambridge, headed by Fred Sanger. In his small, windowless basement storeroom of a laboratory, he turned from enzymes to trying to understand how lymphocytes were able to produce millions of different antibodies. He believed he could solve the problem of antibody diversity by determining the amino acid composition of antibodies and began by using transformed B cells (myelomas) to obtain the amounts necessary for chemical characterization. Myelomas are malignant tumor cells of the immune system that can be grown indefinitely in a test tube and hence are considered "immortal." Although the myelomas in culture produced large amounts of antibody, the problem presented to Milstein was that the antibodies made were without specificity.

Köhler, a young, independent Ph.D. student at the Basel Institute of Immunology, had been struggling without success to find a way to obtain enough antibodies from mortal B lymphocytes in culture. After hearing a lecture by Milstein, he was convinced that if B lymphocytes making antibody against a known antigen were fused with Milstein's immortalized myeloma cell, they would form a hybrid cell (hybridoma) that could grow indefinitely in culture and secrete antibodies specific for that antigen—designer antibodies. The hybridoma could be maintained indefinitely in the laboratory, producing a large amount of identical antibody, called a monoclonal antibody, specific for a single antigenic determinant (epitope). In 1973, Köhler joined the Milstein laboratory as a postdoctoral fellow, and it was shortly thereafter that success was achieved (5). Once the right components were known, the procedure turned out to be absurdly simple. Mouse myeloma cells are mixed with spleen cells from a mouse immunized with a particular antigen in the presence of a fusion agent, polyethylene glycol. The desired hybridomas are then selected using a special growth medium, and they can be frozen for long-term storage. When a monoclonal antibody is needed, the hybridoma is thawed out and injected into mice that provided the original cells for fusion. The mice develop tumors that secrete the specific monoclonal antibody, and the serum of such mice also contains a high concentration of antibody (>10 mg/ml of serum).

In 1975, Köhler and Milstein submitted their work for publication. Although the referee reports were positive, the editors of *Nature* did not consider it of sufficient general interest to publish it as a full article and so the original text was severely pruned to fit the length of a letter. The final sentence in their three-page paper is prescient for understatement: "Such cultures could be valuable for medical and industrial use." However, the scientific community at large paid little notice. By the time Köhler and Milstein received the 1984 Nobel Prize for Physiology or Medicine "for the discovery of the production of monoclonal antibodies," Köhler had returned to Germany, where he held a position at the Max Planck Institute in Freiburg until his sudden death in 1995, and Milstein continued at the MRC Laboratory in Cambridge until his death in 2002 from an inherited vascular disorder.

Today, thousands of monoclonal antibodies are being mass-produced in hundreds of laboratories and biotech companies. This has enabled unparalleled developments in biomedical research. Monoclonal antibodies have been used for diagnosis as well as for therapy for cancer, infectious diseases, complications of transplantation, allergy, asthma, and some autoimmune diseases. In 2005 it was estimated that the monoclonal antibody market was worth $9 billion; however, in the British tradition Milstein and

Köhler did not patent their invention. At the time of their discovery, neither employees of nor the MRC itself was allowed to register patents, so they sent the unpublished paper to a patent officer of the National Research and Development Corporation (NRDC) to consider its patenting potential. The NRDC expressed the view that the work as published was not patentable. Since developing the invention to a state where it would have been adequately protected would have required further work under conditions of secrecy, Milstein considered himself extremely lucky that he was never asked to do this or to refrain from sending the myeloma partner (which had actually been obtained from Michael Potter of the NIH) to colleagues around the world. Others, however, did not share their attitude to sacrifice of personal gain and quickly took out patents, in 1979 for monoclonal antibodies against tumor cells and in 1980 for antibodies to viral antigens.

In addition to the therapeutic advantage of hybridomas (due to the high specificity with which they bind to their target antigen with limited side effects), a principal value of hybridoma technology has been the ability to derive an antibody to a single component of a crude mixture of antigens; this opens the way to purification of a single antigen and makes possible a dissection of a mixture of completely unknown substances into its components. When wed to molecular biology and recombinant DNA, monoclonal antibodies have also been of particular significance in the development of vaccines and, significantly for vaccinologists, in the quest to discover and purify malaria parasite antigens that could be protective.

Rescued by Recombinant DNA

In the early 1970s, recombinant DNA technology made possible the isolation of genes. The method (as originally developed by Herbert Boyer and Stanley Cohen) used plasmids—small circles of naked DNA. Both strands of a circular plasmid double helix could be severed at a specific location with molecular scissors known as restriction enzymes. When mixed in the same test tube with a second DNA (such as *Plasmodium* DNA) that had been similarly cleaved, and in the presence of another enzyme (called ligase), the snipped ends could be "glued" together; the end result was the creation of a hybrid or recombinant plasmid. These ligation mixtures could then be used to infect bacteria. After growth of the bacteria in a petri dish containing antibiotics, only the recombinant plasmids would enable the bacteria to survive. In some instances the bacteria containing the recombinant plasmid would express the protein encoded by the foreign DNA. Although plasmid-based vectors were used early on, it was later found that engineered bacteriophage (especially λgt11) were more efficient vectors and more clones were generated than with plasmids. In addition, handling of large numbers of phage and screening of recombinant clones was far easier.

Expression libraries can be constructed using genomic DNA, that is, the DNA in the chromosomes. Genomic expression libraries are often constructed from sheared DNA or DNA partially digested with nuclease, including restriction enzymes, which is able to chop up the chromosomes into small pieces of DNA. To be able to handle these pieces of DNA, it is necessary to copy and store them, just as a book needs to be printed and bound. In the copying process, each fragment of DNA is attached to the DNA of a bacteriophage. After the chromosomes of *Plasmodium* have been chopped into small pieces of DNA and attached to phage DNA, the phage can be used to infect the bacterium *Escherichia coli*. Each phage carries its own DNA as well as fragments of malaria parasite DNA. When the surface of a petri dish is covered with a lawn of *E. coli* infected with such a phage, clear spots appear on the bacterial lawn where the viruses have killed (lysed) the bacteria. These spots, called plaques, contain millions of virus particles with millions of copies of the original pieces of malarial DNA. Theoretically, genomic libraries have all the DNA sequences present at equal frequency.

Another kind of library, a cDNA expression library, is somewhat more difficult to prepare because it first requires the isolation of the transitory and unstable mRNA; however, in this library the sequences are present in proportion to their abundance as mRNA molecules and thus represent differentially expressed genes. In other words, the mRNA is a concise working copy of the *Plasmodium* DNA code. The mRNA can be faithfully copied into a stable and complementary form, called cDNA (complementary DNA), by using the enzyme reverse transcriptase. As with the method for producing a genomic DNA library, when the cDNA is inserted into a plasmid or a phage and *E. coli* is infected, the malaria parasite DNA is copied.

Using filter paper, it is possible to remove DNA from petri dishes containing *E. coli* with either genomic or cDNA libraries. The filter paper is soaked overnight in a solution of the DNA probe for the plasmodial DNA. To determine whether it has bound to the cDNA in the library, the probe is labeled (i.e., made radioactive). The filters are then washed to remove any of the radioactive probe not attached to DNA fragments, dried, and placed on an X-ray film for a few days. Positive colonies or plaques—areas where the radioactive probe has bound on to the target DNA fragment—are revealed on the X-ray film as black spots. Once that colony has been identified and sufficient DNA is produced, it is possible to "read" the message of the DNA by working out the order of the four nucleic acid bases (CGAT) that form the "rungs" of the DNA double helix. The reading of the bases is called sequencing, and Frederick Sanger at the MRC Laboratory of Molecular Biology in Cambridge, England, first described the basic method. In May 1975 Sanger reported on the first complete sequence of a phage, consisting of 5,375 bases, and he later sequenced the 17,000 bases in the human mitochondrion (the powerhouse in our cells). In 1980 Sanger received the Nobel Prize for his achievements in sequencing methods.

John Scaife (1934 to 1991) was the first to produce a *P. falciparum* cDNA library. In the late 1970s he and other bacterial geneticists sought new challenges, and he recognized the potential of applying the emerging techniques of recombinant DNA technology to understanding the nature of malaria parasites. Influenced by his Edinburgh University colleagues, the *Paramecium* geneticist Geoffrey Beale and the *Plasmodium* geneticist David Walliker, he made the bold decision to set up a group to study the molecular biology of *Plasmodium*. His major goal was to search for antigens suitable for the development of vaccines. Awarded an MRC grant in 1980, Scaife and his group involved themselves in cloning the genes for a number of parasite proteins; by combining the use of monoclonal antibodies with molecular genetic techniques, they were able to characterize several blood stage antigens.

The Australian group at the Walter and Eliza Hall Institute (WEHI) followed the accomplishment of the Scaife laboratory in Edinburgh. WEHI was established in 1915, and its principal focus has been immunology, immunopathology, transplantation biology, and virology. Malaria research at WEHI began at a time when advances in recombinant DNA technology made possible its application to parasites and funding agencies began to support the study of neglected tropical diseases. In the mid-1970s, Graham Mitchell, an immunologist who had turned his attention to the study of host-parasite interactions, established an Immunoparasitology Unit at WEHI. A meeting with Michael Alpers of the Papua New Guinea Institute of Medical Research and funding from the Rockefeller Foundation precipitated the move to studying human malaria. Recruitment of Robin Anders and Graham Brown led to the establishment of the

Trager-Jensen culture system for *P. falciparum*, methods of immunochemical analysis, access to serum from residents of countries where malaria is endemic, and methods for preparation of affinity-purified antisera. By 1980, investigations on the molecular biology of malaria under the direction of David Kemp (born in 1945) had begun. In 1973 Kemp obtained a Ph.D. in biochemistry from the University of Adelaide. In a time before recombinant DNA techniques had been fully developed, he purified feather keratin mRNA by using nucleic acid hybridization techniques. In 1976 Kemp went to Stanford University to study with David S. Hogness, a leading expert in the study of gene regulation in bacteriophage who had recently started using this expertise to study *Drosophila*. Kemp's project was to isolate developmentally regulated genes, and this involved construction and analysis of cDNA libraries. In 1978 Kemp returned to Australia and joined WEHI, starting as a Research Fellow working on immunoglobulin genes. In 1979 he joined the Parasitology Program at WEHI. He and his colleagues spent the next 2 years developing a system for cloning malaria antigens. The goal of cloning malaria genes was to try to make a vaccine, and the approach used was to identify *P. falciparum* genes critical to immunity. Work on *P. falciparum* was hampered by the small scale and low yield of parasites from the Trager-Jensen cultures and the inefficiency of enzymes such as reverse transcriptase and terminal transferase, incredibly valuable and rare reagents that at that time could be obtained from only one laboratory in the world. Malaria parasites had to be cultured continuously for months before enough of them could be harvested to allow a single cDNA synthesis to be performed. Although a malaria cDNA library was available by 1982, progress was slow since the vector used, pBR322, expressed antigens in very small amounts. Indeed, the only way to detect expression was to use the incredible sensitivity of the mouse immune system: random lysates of clones from the library were injected, and production of antibody to malaria antigens was assessed. These studies led to the identification of a schizont antigen; however, such methods did not provide a great enough throughput for a meaningful attack on the repertoire of important protective antigens.

Kemp's colleague at WEHI was Ross Coppel (born in 1952), who graduated in 1976 from the University of Melbourne with a medical degree, and interned at the Royal Melbourne Hospital. During his medical course, he had undertaken a year of research for the B.M. degree, researching the function of radioresistant macrophages with Jacques Miller and Graham Mitchell at WEHI. He went on to do postgraduate studies in tropical medicine at the London School of Tropical Medicine and Hygiene (1976 to 1979), and in 1979 he returned to Australia to join Graham Mitchell at WEHI in the recently established Immunoparasitology Unit. After an unsuccessful attempt to study the surface antigens of microfilaria, he was given the project of applying recently developed techniques in molecular biology to the study of the malaria parasite. He extended many of the still rudimentary approaches to mRNA extraction and cDNA synthesis and expression screening to successfully work with the small amounts of material available in the malaria system. In collaboration with coworkers, Coppel was able to clone and characterize a number of blood stage antigens including a number of merozoite surface proteins. Although he constructed the first cDNA expression libraries for the asexual stage of the parasite, the existing expression methods were incapable of expressing sufficient amounts of protein for most purposes. The turning point was the introduction into the laboratory of the bacteriophage vector λgt11 by Robert Saint, a postdoctoral fellow who joined the WEHI group following a stint at Stanford University, where the vector had been developed in the laboratory of Ronald Davis. With all the steps required for construction and screening of cDNA libraries already in place, the group was quickly able to take advantage of the new vector. Expression levels were at least 1,000-fold higher than pBR322 could provide, and direct screening of libraries became possible. Improvements to the screening procedure, including the preparation of affinity-purified pooled human serum, provided sufficient sensitivity. Hundreds of phage plaques expressing epitopes of proteins recognized as immunogenic by the sera from residents of areas of endemic infection were found, many of which remain uncharacterized even today. The clones could then be used to raise monospecific sera either by immunization of laboratory animals or by affinity purification on expressing clones. In turn, the monospecific antisera could be used to define properties of the corresponding malaria protein. The requirement for affinity-purified serum was circumvented by the development of methods for screening with individual sera. This allowed comparisons between exposed individuals who differed in their level of immunity to malaria. Clones expressing many defined malaria antigens were identified by these studies, and reports characterizing their role and utility in inducing host protective immunity were published over succeeding years.

Initially the WEHI group functioned in an integrated fashion, and a significant infusion of Australian government money supporting the development of a malaria vaccine led to a large increase in the number of workers and in efforts to define host-protective malaria proteins. In time, however, the group became riven with dissent, and many researchers departed to work elsewhere. Despite this, many members of the original team continue to be productive in malaria research, and a malaria research group remains at WEHI.

In retrospect, it became clear that the screening approach used, although particularly good at identifying repetitive antigens and proteins containing linear epitopes, was much less efficient at identifying the important subgroup of cysteine-rich proteins. Successful at identifying merozoite antigens such as apical membrane antigen (AMA-1) and merozoite surface proteins (MSP), the approach used proved incapable of identifying the surface molecule responsible for antigenic variation, PfEMP1 (*P. falciparum* erythrocyte membrane protein 1), the family of merozoite reticulocyte-binding proteins, and the erythrocyte-binding antigens. Working as they were at the beginning of the molecular biology/biotechnology revolution, the WEHI group was one of the first to encounter the complexities involved in commercialization of basic research and struggled with the resulting problems of direction and management that come from large-scale funding of big science projects, a failure to understand the regulatory requirements, and the complexities of bringing a recombinant protein formulation to market. Nonetheless, the technological contribution of the WEHI group was enormous, and by building an atmosphere of excitement and optimism about malaria research the institution attracted a new generation of talented scientists.

Powerful as the recombinant methods are, there is a significant limitation: they require growth of large volumes of bacteria to amplify the DNA of interest as well as a considerable investment in time and research funds to identify recombinant clones of interest. By 1985 they had largely been replaced by PCR, a technique called "molecular photocopying." In brief, the technique works as follows. Two short stretches of single-stranded DNA (called primers), corresponding in nucleic acid base sequence to the regions bracketing a DNA expanse of interest such as a specific gene, are synthesized. The primers are added to the DNA template, i.e., total genomic DNA or a cDNA of interest, and the DNA double helix is "melted" by heating to 90 to 95°C to separate the strands. Upon cooling, the primer can bind to its complementary stretch of single-stranded template DNA. Also present in the test tube is a DNA polymerase and all four nucleic acid bases. The polymerase begins incorporating bases only where the DNA is already double stranded, and so it begins adding nucleotides at the end of the primer and synthesizes the DNA region that follows. By using a thermal cycler, one that heats ("melts") and cools ("anneals"), the process can be repeated every 5 min and the stretch of DNA of interest will be copied again and again; in 2 h the amount of DNA of interest will be increased about 34 million-fold. This amplified DNA can then be sequenced by a variety of methods.

In the Sanger sequencing method, the strands of DNA are duplicated by means of DNA polymerase in the presence of a mixture of the normal nucleic acid bases A, T, G, and C plus some dideoxyadenosine (ddA), ddT, ddG, or ddC with a radioactive label. If the polymerase incorporates the normal base the DNA chain grows, but when it encounters a dideoxy nucleotide it stops lengthening. The procedure results in four different samples, each containing a series of DNA chains of different lengths depending on where in the growing chain the different dideoxynucleotides were incorporated opposite the complementary T, A, C, or G template bases during the replication process. The result is a mixture of DNA fragments of different lengths, each ending with a radioactively labeled C, G, A, or T depending on which base contained the radioactive base. Each of the four samples is placed on a gel, and the fragments are separated in an electric field according to size: short chains move faster and longer chains move more slowly. When the gel is dried and placed on X-ray film, four parallel tracks of black bands appear. The positions can be read off such that the shortest fragment will contain the first base, the next larger will contain the second, and so on. For example, if the first, smallest piece of DNA is in the C track, then C is the first letter. If the next black mark is in the A track, then A follows C, and so on. In this way it is possible to read the sequence of bases. This method was tedious, and often things went wrong: the lanes would not run parallel to one another, one of the reactions failed and there were no black marks in a lane, or the gels cracked on drying. Later, this four-lane method of manual sequencing would be replaced by one that was automated and used a single lane. A different colored dye for each type of chain-terminating dideoxy base was added to the polymerase mix and incubated. By subjecting the single-lane sample to an electric field, all the DNA pieces can be sorted according to size. Under UV light illumination, each fragment fluoresces differently depending on its terminal dideoxy base; if the fluorescent pattern is scanned and fed into a computer, a sequence can be printed out.

When such an automated system is used, half a million bases can be sequenced in a day. In 1996 an international consortium of scientists from more than a dozen institutions set out to determine the 23 million bases in the nuclear DNA of *P. falciparum*. The 14 chromosomes were physically separated by pulsed-field gel electrophoresis. Then the DNA was mechanically sheared into random fragments; after the fragments were cloned in bacteria, they were sequenced using high-throughput, automated dideoxy sequencing and the order of the bases was determined for individual chromosomes by assembling overlapping sequences on a computer. The results of the *P. falciparum* genome project were published in 2002: 5,279 genes were postulated, with fully 60% being of unknown function as they match no other gene sequences in the international data banks (181).

Two Blood Stage Vaccine Candidates

The concept of a blood stage vaccine is based on a fundamental principle: mimic the immunity that occurs in nature. In areas where malaria is endemic, natural acquired immunity is induced by continued exposure to the bite of infected anophelines, and although blood parasites persist, successive bouts of clinical disease are diminished due to reduced parasite multiplication and growth. The assumption made by proponents of a blood stage vaccine is that it will accelerate the acquisition of acquired immunity, inhibit parasite invasion into red blood cells, and affect parasite multiplication to reduce the parasite burden, thereby limiting disease severity and death. This would "offer enormous benefit to the public health, particularly infants and children living in endemic areas who suffer the most . . . due to malaria" (132).

The majority of blood stage vaccines are clustered around two hypothesized mechanisms of protection: block red blood cell entry by the merozoite and/or attack the development of the malaria parasite within the red blood cell, directly or indirectly. Leading blood stage vaccines include the merozoite surface protein 1 (MSP-1) and apical membrane antigen 1 (AMA-1).

The road to MSP-1

Invasion of the red blood cell by *Plasmodium* merozoites is a complex multistep process (27). Initially, any part of the merozoite surface may make contact with the surface of the red cell. This binding, which is reversible, is followed by reorientation of the merozoite so that its pointed (apical) end becomes attached via an irreversible junction; once attached, the junction moves to surround the merozoite, enclosing it within a vacuole; finally, its entry is completed by resealing the red blood cell and vacuolar membranes. Several studies suggested that immunization with merozoites could induce some protection in *Aotus* monkeys and that antibodies directed against the merozoite can inhibit invasion and confer protection against a blood stage infection (see p. 123). Sera from immune subjects are able to agglutinate or opsonize merozoites as they emerge from the schizont and hence inhibit invasion, and it was suspected that antibodies capable of binding to the merozoite surface were blocking attachment and thereby preventing invasion. Before 1980, the structural and antigenic complexity of the merozoite made it impossible to identify and characterize the merozoite antigens responsible for protection. In 1980 the problem of characterizing the merozoite antigens to which the protective antibody was directed was solved through the use of monoclonal antibodies.

Under the leadership of George Cross, the Wellcome Research Laboratories embarked on a malaria vaccine program in which the convenient malaria model *P. yoelii* in mice and *P. falciparum* grown in Trager-Jensen cultures were used. The study of *P. yoelii* produced a critical discovery: monoclonal antibodies specific for the antigens exclusive to the merozoite were protective upon passive transfer. The researchers went on to suggest that the protective monoclonal antibodies would be useful in the isolation of the important antigens of this parasite.

George Cross, born in 1942 in Cheshire, England, received his undergraduate degree in 1964 and his Ph.D. in biochemistry in 1968 from the University of Cambridge. From 1970 to 1977 he worked in the Biochemical Parasitology Unit of the MRC at the Molteno Institute in Cambridge, studying the variant surface glycoproteins of African trypanosomes. At Cambridge he attended a lecture by the immunochemist Rodney Porter, who, just before he ended the lecture, made a short and memorable statement to the effect that he had some exciting news from his friend Professor Sydney Cohen: they were in line to have a malaria vaccine in the next 3 years! It was not this remark alone, however, but the fascinating biology of malaria and its importance that were determining factors in his decision to establish a research program on malaria vaccines. Cross set out with what was a very naive combination of information and ideas (a naiveté shared by many investigators at that time and for some time thereafter). The long-standing studies in Sydney Cohen's laboratory at Guy's Hospital in London had shown that antimerozoite antibodies could control an infection with the simian malaria parasite *P. knowlesi*; there were rodent malarias that could serve as laboratory models; Trager and Jensen had recently succeeded in growing *P. falciparum* in cell culture; Miller's laboratory at the NIH had started to study the merozoite invasion process, using *P. knowlesi*; and, finally, Cross made the assessment that the interaction between the simplest and best understood mammalian cell, the red blood cell, and the invading merozoite stage of the malaria parasite, which specifically invaded only the red blood cell and had host-species specificity, would provide an amenable experimental system in which the parasite molecules that guided invasion could be identified and possibly serve as targets for a vaccine.

In 1977, the recently appointed head of the Division of Immunobiology at the Wellcome Laboratories, the immunologist James Howard, was intent on rebuilding and reorienting the Department of Immunochemistry to better reflect emerging molecular approaches to microbiology and parasitology. Sir Henry Wellcome founded the Wellcome Bureau of Scientific Research in London in 1913 to carry out research in tropical medicine. Although funding was derived from the commercial activities of the Wellcome Foundation Ltd (later Wellcome PLC), including Burroughs Wellcome Inc., its wholly owned American subsidiary, the scientific staff was relatively free to establish its own lines of investigation. Henry Wellcome's tenet for his research laboratories had always been "Freedom of research—Liberty to publish," and this attracted some of the most talented scientists of the day to work there. In 1934, the Bureau changed its name to the Wellcome Laboratories of Tropical Medicine, and in 1965, these laboratories moved to Beckenham to be merged with the existing Wellcome Research Laboratories, which had been there since 1922. In the 1970s there was a view at Wellcome that Brazil and South East Asia represented growing economies that could become substantial future markets; Wellcome already had subsidiary companies in these territories and, in Latin America, was a major player in the foot-and-mouth disease virus vaccine business. Wellcome's corporate decision to move into the new areas, using emerging technologies such as in vitro culture, monoclonal antibodies, and the first steps in genetic engineering, created a favorable climate for research on a malaria vaccine. Cross first met Howard at a meeting of the Royal Society of Tropical Medicine and Hygiene early in 1977, and after successive meetings he was offered a job at Wellcome to build a group that would include a prominent commitment to malaria immunology and vaccinology. Although he had no formal training with malaria, he had a free hand to learn "on the job."

Cross began to hire smart, experienced, and keen young scientists to spearhead malaria research. The first was a talented young immunologist, Robert (Robbie) Freeman, who had recently finished his Ph.D. on the immunology of *P. yoelii* in Ian Clark's laboratory at the Australian National University in Canberra. Freeman arrived in 1978 and persuaded Cross that *P. yoelii* was a better model for human malaria than was *P. berghei*—existing as it did in virulent and less virulent clones—and was capable of being scaled up for biochemical studies. With an excellent supporting staff, especially Avril Trejdosiewicz, who had been hired to run the monoclonal antibody facility, the first monoclonal antibodies to the asexual blood stages of rodent and human malaria parasites were made. One of Freeman's first contributions was to show that the best way to identify useful relevant monoclonal antibodies was to screen, from the beginning, by the time-consuming but informative method of immunofluorescence microscopy. The next step was to see if any of the monoclonal antibodies would suppress a mouse malaria infection. The first experiment involved two monoclonal antibodies: one that clearly reacted with the merozoite surface, which could be recognizing a protective antigen, and the other, which recognized a merozoite internal structure and was unlikely to be protective. In fact, both were strongly protective in "passive-immunization" experiments (172).

From this point, the Immunochemistry group moved rapidly to purify the target antigen of the surface-staining monoclonal antibody, which was done mainly by affinity purification, for which the monoclonal antibody worked well. It was at this point that Cross persuaded Anthony Holder, who had been hired to work on the variant surface glycoproteins of African trypanosomes, to get involved with the malaria project and contribute his skills in protein purification and analysis. In one experiment, using parasites harvested from 300 mice, they were able to purify at least 1 mg of protein. This enabled true vaccination experiments to be done (229).

Cross left Wellcome for his present position as a Professor at the Rockefeller University in 1982, but was retained as a consultant from 1982 to 1987, and Robbie Freeman left Wellcome in 1985 to pursue a different career and lifestyle, and died tragically by his own hand 8 years later. In 1988, Holder moved to head the Division of Parasitology at the NIMR, but the Wellcome malaria program continued until 1990, when the program was terminated because Wellcome had moved from being owned entirely by a charitable foundation to being partly a public company. This triggered a review of the company's position and future direction, which resulted in the sale of its vaccine business to Medeva and a subsequent reorganization of its molecular biology resource that terminated all work on vaccines and other biological therapies at the British research site.

Cross was an early member of the Steering Committee of the Scientific Working Group on the Immunology of Malaria of the Special Programme for Research and Training in Tropical Diseases of the WHO (1983 to 1987), and this provided a way to keep in touch with and to foster promising developments at the leading edge of malaria vaccine research. He was not always popular for framing the vaccine quest as something that should not be expected until the next century, rather than the rolling 3-year time frame many others were promoting. At Rockefeller, Cross had difficulties in obtaining the long-term commitments of extramural funding from NIH that were needed to do meaningful malaria research and to build a stable foundation of knowledge and technical resources (rather than just cloning genes). This was the major factor in his decision to drop the malaria project and return to work with trypanosomes.

A year after Holder joined the Wellcome group, the monoclonal antibodies were used to identify a *P. yoelii* merozoite-specific protein with a molecular mass of 235,000 (229). The protein (named Py235) could be purified and prepared in large amounts by using an affinity column: the monoclonal antibody was bound to a column of Sepharose beads, a merozoite extract was poured over the column, and the protein that bound to the column had the same molecular mass. Mice immunized with this protein in FCA survived challenge, whereas controls died. But, Holder and Freeman wondered, was this of any relevance to *P. falciparum*? They answered their own question upon finding a cross-reaction between the *P. yoelii* antigen and a merozoite antigen from *P. falciparum,* and they suggested (228) "this 195,000 mol wt *P. falciparum* protein, and its form on the merozoite surface, are . . . of considerable interest for malaria vaccine development." They had discovered the first and most abundant merozoite surface protein, MSP-1. (MSP-1 has also been called at various times PMMSA [precursor to major merozoite surface antigens], Pf195 [*P. falciparum* 195,000-molecular-weight antigen], and FMP-1 [falciparum merozoite protein 1]).

Anthony Holder (born in 1951) completed his undergraduate studies in biological sciences, graduating in biochemistry at the University of East Anglia in Norwich in 1972, and for his Ph.D. studies (1972 to 1975) he worked with the geneticist John Fincham at the University of Leeds, using protein-sequencing methods to obtain amino acid sequences of an enzyme (NADP-dependent glutamate dehydrogenase) from the mold *Neurospora crassa*. At the time, DNA sequencing was not routinely available, and the dansyl-Edman procedure for peptide sequencing was very slow and extremely laborious; several person-years were devoted to obtaining the complete 452-amino-acid sequence of this protein. Wellcome Trust and Royal Society Fellowships allowed for a postdoctoral stint (1975 to 1978) at the newly rebuilt Carlsberg Laboratory in Copenhagen, doing studies on the evening primrose, barley, and potato-tomato hybrids; 3 years later, a serendipitous flick through the back pages of *Nature* revealed an advertisement for a job to work on African trypanosomes with Cross at Wellcome. In this group the variant surface glycoprotein was the molecule of choice and molecular biology was starting to have its pervasive influence. However, the variant surface glycoprotein and the nature of the so-called "cross-reacting determinant" was the main area of research, leading to one of the first studies of what subsequently turned out to be a ubiquitous membrane anchor of proteins, the glycosylphosphatidylinositol (GPI) moiety. Holder later joined Freeman and Cross in their specific remit to establish malaria research. By 1981 Holder and coworkers had been able to purify two parasite proteins and showed that they could be used individually to immunize mice against subsequent challenge infection with *P. yoelii*—the first time that this had been done, establishing the principle of a subunit vaccine against malaria.

Holder also studied MSP-1, a leading blood stage vaccine candidate (231). The initial studies of MSP-1 in *P. yoelii* were rapidly expanded to other *Plasmodium* spp. including sequencing of the gene in *P. falciparum*. How could they be sure it was the right gene? Interestingly, it was a protein sequence from a fragment of MSP-1 released into culture supernatants that provided the definitive proof. Studies of the processing of MSP-1 followed, particularly with M. J. (Mike) Blackman, and the discovery that a monoclonal antibody isolated by Jana McBride in Edinburgh directed against one end of the molecule inhibited merozoite invasion of red blood cells focused attention on this part of the molecule. Human studies done by Eleanor Riley in The Gambia and Roseangela Nwuba in Nigeria supported the view that this part of the molecule was important as a target of protective immunity and a vaccine candidate. Holder, Irene Ling, and Sola Ogun, in parallel to work being carried out by Carole Long, showed that a recombinant protein ~100-amino-acid fragment (MSP-1$_{19}$) of *P. yoelii* MSP-1 was effective in protecting mice against a parasite challenge infection.

MSP-1 is found in both *P. falciparum* and *P. vivax* (22). It is encoded by a single gene on chromosome 9 in *P. falciparum* and is synthesized late in the blood stage developmental cycle. It has not been possible to "knock out" the gene for MSP-1, suggesting that it is essential for invasion and parasite survival. Sequence analysis shows that MSP-1 contains conserved regions interspersed with polymorphic regions. It undergoes extensive processing during merozoite formation, and soon after merozoite release, the resulting four fragments remain as a complex on the merozoite surface with the 42,000-molecular-weight protein (MSP-1$_{42}$) being attached via the GPI anchor. A final processing step mediated during invasion cleaves this fragment into 33,000- and 19,000-molecular-weight fragments (MSP-1$_{19}$); the larger fragment is shed from the merozoite surface along with the remainder of the MSP-1 complex, whereas the smaller MSP-1$_{19}$ fragment remains on the merozoite surface during invasion.

Full-length MSP-1 has been suggested to bind to red cells in a sialic acid-dependent manner, or it may form a complex that binds to band 3 protein (see p. 316). Antibodies directed against MSP-1$_{19}$ block invasion and prevent processing of MSP-1 (47). Since immunization of *Aotus* with recombinant MSP-1$_{42}$ in the presence of FCA partially protected against challenge and appeared to be as effective as the use of whole merozoites, MSP-1 was suggested to be a blood stage vaccine candidate. It was clear, however, that the immunogenicity of the recombinant protein needed to be improved (230). Antibodies directed against MSP-1$_{19}$ in mice and monkeys elicited protection against challenge, and similar results were reported with antibodies raised against MSP-1$_{42}$ (465). Immunization of *Aotus* monkeys with the MSP-1$_{42}$ recombinant protein produced in *E. coli* resulted in high-titer antibodies and led to protection against an ordinarily lethal challenge with *P. falciparum* (465). However, with recombinant MSP-1$_{42}$ and MSP-1$_{19}$, protection of *Aotus* was seen only when high antibody titers were obtained using formulations with FCA; only one of seven monkeys required treatment for uncontrolled blood infections (486).

So what were the results of a Phase 1 (safety) vaccination trial with adult human volunteers in Kenya, using MSP-1$_{42}$ (also called FMP-1) in combination with AS02A (an oil-in-water emulsion containing 3-deacylated monophosphoryl lipid A and QS21, a saponin-like adjuvant) or Imovax rabies vaccine? Slightly more recipients of the three intramuscular monthly vaccinations of FMP-1/AS02A experienced more pain than did recipients of the Imovax group, but otherwise the two groups were equal. The Kenyan volunteers already had high preexisting antibody titers to

FMP-1, and the titers increased after each immunizing dose (as was the case with malaria-naive subjects in the United States) (483). The vaccine appeared to be well tolerated and immunogenic, but its efficacy in adults and in children was disappointing and further work on it has been abandoned.

Carole Long (born in 1944) grew up in Syracuse, NY, and received her B.S. degree from Cornell University in 1965. Given her background—she worked on basic immunochemistry studying antibody affinity and valence while a graduate student at the University of Pennsylvania under the direction of Fred Karush, a physical chemist who had moved into studies of antibody structure and function—it seemed unlikely that she would work on malaria. Indeed, after receiving her Ph.D. in 1970, she did postdoctoral training, also at the University of Pennsylvania, in T-cell immunology, studies having little to do with malaria. However, as a young Assistant Professor at Hahnemann University School of Medicine (now Drexel University), she met up with William Weidanz, a Professor in the Department of Microbiology and Immunology who had worked on mouse and bird malarias by using various immunodeficiency models to dissect the roles of antibodies and lymphoid cells in protective immune responses to malaria infection. Weidanz asked for assistance with some immunochemical experiments that were running in Long's laboratory. Soon Long and Weidanz were carrying out analyses of parasite proteins to identify the numerous red blood cell stage targets of antibodies produced by mice that were protected against challenge infection.

Prompted by the observation that sera from mice immunized by infection could passively protect naive mice against a virulent challenge infection, Long and Weidanz started producing monoclonal antibodies from the spleens of mice protected by repeated infection. At the time, other laboratories were taking a similar approach, notably that of Anthony Holder at the NIMR. Will Majarian in Long's laboratory was fortunate enough to find one monoclonal antibody that provided significant although partial protective activity when administered to naive mice even after infection with an ordinarily lethal strain of *P. yoelii*. This observation led to a major effort to identify the antigen that was the target of this protective monoclonal antibody. Accomplishing this required the addition of new molecular biological techniques to the laboratory, which were performed by Jim Burns and Akhil Vaidya. After learning to construct and screen phage libraries of parasite DNA, Long and coworkers eventually identified the target of the monoclonal antibody as the major merozoite surface protein, MSP-1, particularly

MSP-1_{19}. Long was running neck and neck with Holder's laboratory, which was also focused on this protein.

Now the challenge for Long remained to produce a recombinant protein that mimicked this portion of MSP-1. At that time (1992) there was little experience with expression of correctly folded recombinant proteins. Initial attempts to produce MSP-1_{19} of *P. yoelii* directly in *E. coli* were unsuccessful; however, eventually a method was worked out for producing it as a fusion protein with glutathione *S*-transferase that appeared to be correctly folded. When Tom Daly in Long's laboratory came to her with the immunization challenge data which indicated that the mice had been partially protected against an otherwise lethal infection, tears came to her eyes because this was the first time that active protection from a blood stage malaria infection had been elicited by a recombinant protein. Shortly thereafter, a small workshop was held at NIH to discuss the status of research on MSP-1. When Holder presented his data on MSP-1, it was clear that he was pursuing the same approach as that used by Long; however, his laboratory was not as far along with the immunizations. Patrick Farley, a student from Long's laboratory and also an attendee, looked at her after Holder's presentation; when their preliminary results were presented after his, the tension in the room was palpable. Long's results with MSP-1 were expanded over the years by using various rodent models, and others at NIH and elsewhere attempted to extend them to nonhuman primates infected with *P. falciparum*. Subsequently her studies moved to production of MSP-1 protein suitable for human clinical trials, and currently a number of laboratories are working on MSP-1 as a candidate blood stage vaccine for human malarias. Unfortunately, the first Phase 2 trial of this antigen in children in Kenya was unsuccessful, and analysis of that trial is still under way.

In 1999 Long moved to NIH to participate in the new efforts there to actually produce and test malaria vaccine candidates in humans. The efforts at NIH, sparked by NIH Director Harold Varmus and National Institute of Allergy and Infectious Diseases (NIAID) Chief Anthony Fauci, sought to fill the gap between vaccine candidates discovered in the laboratory and process development/clinical testing. In contrast to the situation with human immunodeficiency virus (HIV), there appeared to be reasons for optimism about a malaria vaccine, given the resistance seen in persons living in areas of endemic infection and the success in protecting laboratory animals by immunization. However, making the transition from a University academic laboratory to a quasi-biotechnology system was not easy. A whole new vocabulary had to be learned, and then laboratory procedures had to be modified so that the data would be suitable for submission to the U.S. Food and Drug Administration as part of Investigational New Drug applications. The learning curve to transform one's laboratory procedures and to complete the massive record keeping and paperwork involved was very steep. Eventually, there was success. MSP-1 and AMA-1 blood stage vaccine candidates, as well as some sexual stage proteins, were produced in the MVDB (Malaria Vaccine Development Branch) laboratories and tested in animals and in the clinic. Evaluation of all these products is still under way, and elsewhere other groups are testing similar candidates. Regrettably, the limited efficacy trials conducted with blood stage vaccines have not shown protection to date.

Long says, "Perhaps we were somewhat naive in thinking that a vaccine using single recombinant proteins directed to such sophisticated and complex organisms would elicit protective responses. In the case of the blood stage infections, it is likely that we will have to elicit a complex combination of immune responses to achieve significant protection in children by vaccination. While most existing vaccines are directed to production of specific antibody, the remaining vaccine development challenges such as malaria or tuberculosis or HIV will likely also require cellular responses as well." While the possibility of recombinant proteins representing blood stage antigens has by no means been exhausted, Long has taken the approach of stepping back to put vaccination efforts in the context of what happens to those who live in areas where malaria is endemic, asking: What can be learned about the process by which infants and children develop resistance to malaria? A deeper understanding of the relevant immune mechanisms that evolve with time in children and young adults, coupled with elucidation of the strategies that the parasite employs to manipulate the host immune system, may illuminate new directions in the search for vaccines for malaria.

During her professional life, Carole Long has been a witness to some of the best and the worst in human behavior. She has also met interesting people from around the world, been to villages in Africa that have been little affected by time, and been privileged to participate in an important global quest. Currently Chief of the Malaria Immunology Section at NIAID, Long continues to work on innate and adaptive immune responses to rodent and human malarias and is evaluating CD4 T-cell responses to specific malaria antigens in mice and humans.

In 2005, according to the WHO, there were 93 candidate malaria vaccines in development. Of these, 26 were based principally on MSP-1. By 2006, the total number of candidate vaccines in development was 23, of which 6 were based on MSP-1. And by November 2007, as a result of failures during safety and immunogenicity trials in volunteers in a nonendemic setting, the number of vaccines had been reduced and three of the remaining five MSP-1 projects had been terminated. Clearly, MSP on its own as a subunit vaccine is ineffective in protection.

The road to AMA-1

The demonstration that antibody from rhesus monkeys immune to *P. knowlesi* inhibited the multiplication of the parasites in vitro (100) and that passive transfer of immune monkey serum also confers protection provided a rationale for using sera from immune hosts to characterize antigens and identify those with putative protective activity. Since it had already been shown that vaccination with merozoites with FCA induced some degree of protection (see p. 123), a search for the protective antigens of the merozoite began. Of the 11 antigens recognized by immune serum, 2 were found to be from merozoites (120). But were these protective, and how could sufficient amounts be obtained for vaccination purposes? Sydney Cohen's group at Guy's Hospital, now including Judith Deans and Alan Thomas, elected to apply the recently developed hybridoma technology to an antigenic analysis of *P. knowlesi* merozoites to identify putative protective antigens and produce them for vaccination.

In 1982, supported by funds from the MRC and the malaria component of the UNDP/World Bank/WHO Special Programme for Research and Training in Tropical Diseases, Deans and Thomas prepared two rat monoclonal antibodies that inhibited the in vitro multiplication of *P. knowlesi*; both of them reacted specifically with a single protein of molecular weight 66,000 (121). The antigen was processed at the time of red cell rupture with merozoite release to give rise to two smaller molecules. Fab fragments of the monoclonal antibody inhibited merozoite invasion (504), and it was postulated that this antigen played a role in merozoite invasion and not in parasite proliferation. Later, it was found, using an antiserum prepared against the 66,000-molecular-weight antigen, that the antigen did not prevent merozoites from attaching to erythrocytes but did inhibit their orientation (339). Since saponin had been shown to be effective with a vaccine containing whole merozoites (see p. 238), rhesus monkeys were immunized with the purified 66,000-molecular-weight antigen with saponin as adjuvant; monkeys produced antibody that inhibited merozoite invasion of red cells, and four of six monkeys vaccinated with the purified antigen were protected (122). It was assumed that where immunization failed to protect, it was due to a waning of antibody titer.

Alan W. Thomas (born in 1954) grew up in London and obtained a B.Sc. (honours) in bacteriology and virology from Manchester University in 1976; he then obtained a Ph.D. (in 1985) under Cohen's guidance at Guy's Hospital. Days were spent looking down a microscope trying to divine the smallest hints about the location of molecules identified by the monoclonal antibodies. In 1978 to 1979, the chief technician, Dave Dennis, a creative engineer, developed an automated machine (called Philomena for reasons that are now obscure), to culture parasites in glass receptacles on an automated cycle of tipping to remove medium, replacing medium, and shaking at intervals. In the early days Dennis and Thomas used material from these cultures and a cell sieve apparatus in attempts to obtain merozoites. Hours full of anticipation were spent watching the proportion of schizonts increase in the cultures as the sieve removed red blood cells, and in the evening the two waited for the merozoite explosion to occur. Sadly, although the apparatus worked well in obtaining merozoites for *P. knowlesi*, it was unable to replicate the feat for *P. falciparum*. Nevertheless, the *P. knowlesi* merozoites obtained were used for many purposes, including much of the work on monoclonal antibody production.

Using the phage technology of the time, Andy Waters, who also joined the group at Guy's to do his first postdoctoral research, had been given the task of finding the *P. knowlesi* AMA-1 gene, PkAMA-1. The monoclonal antibodies generated by Thomas and Deans gave him the tool to eventually do so, although it was certainly not straightforward (545).

After obtaining his Ph.D., Thomas spent a while visiting Louis Miller's laboratory at NIH before taking an NRC fellowship at WRAIR, splitting time between working with Jeffrey Lyon on a program to identify the *P. falciparum* equivalent of PkAMA-1. For the AMA-1 of *P. falciparum*, Thomas had some leads, had part of the gene in hand, and had presented the preliminary data at a meeting in Annecy, France. However, Robin Anders' group at WEHI had already published on the molecule after identifying it by a different route. Around that time Thomas was asked to set up a research group at the University of Maryland, secured a large grant from USAID to work on AMA-1, and in 3 years (1989 to 1992) made considerable progress in characterizing PfAMA-1.

In 1992, an opportunity arose to establish a new malaria research programme at a nonhuman primate research center in Rijswijk, The Netherlands, now known as BPRC (Biomedical Primate Research Centre). Thomas was struck by the benefits that nonhuman primate models had to offer to several critical aspects of malaria research, as well as the prospects for his growth, so he accepted the offer and never looked back. At BPRC he established a Department of Parasitology staffed by a dedicated team of 25 people; this has become the largest research activity at BPRC. During this period they have, among other things, pioneered transfection techniques for malarias of nonhuman primates, been intimately involved in malaria genome-sequencing efforts, established models for assessment of antimalarial drugs and malaria and tuberculosis vaccines, and made considerable progress in developing AMA-1 as a vaccine candidate. The group has been keen to play a role in translating ideas into testable products and in accelerating the pathway to the clinic.

Since it has been impossible to delete (knock out) the gene for AMA-1, it is presumed to be critical for parasite survival. The complete amino acid sequence of AMA-1 from several strains of *P. falciparum* has been determined. Levels of antibody to AMA-1 are higher in populations exposed to natural malaria infections than are levels of antibody to most other antigens including MSP-1, and the prevalence of such antibodies increases with age. AMA-1 has been considered to have potential as a vaccine candidate. However, the finding (in 1990) that the AMA-1 gene is polymorphic, i.e., that different strains may have up to 100 different amino acid sequences and therefore may produce differently shaped or sized forms of AMA-1, "has tempered expectations in terms of vaccine-induced broad protective immunity" (388). A further complication for the use of AMA-1 as a vaccine is the finding that it has to fold correctly to induce parasite-inhibitory responses (417).

There have been several published reports on Phase 1 (safety, immunogenicity, and efficacy in a nonendemic setting) vaccine trials with AMA-1 as an *E. coli*-produced protein with the adjuvant Montanide ISA 720 (containing squalene, a metabolizable oil). There were no adverse effects with increasing doses of 5, 20, and 80 µg given intramuscularly at 0, 3, and 6 months; however, AMA-1 showed immunogenicity in only 6 of 29 subjects, suggesting a loss of potency (420). In another study with "Combination 1" containing the AMA-1 protein produced in the yeast *Pichia pastoris* and adsorbed to Alhydrogel, and with a similar dose and schedule, it was also well tolerated, but again immunogenicity occurred in only 4 of 22 subjects

(309). When AMA-1 was packaged in a poxvirus and delivered as NYVAC-Pf7, antibody production was poor and the protective effects could not be ascribed to any one component in the vaccine (354). In the most recent trial, the malaria vaccine FMP2.1/ASO2A, consisting of a recombinant protein based on AMA-1 with the adjuvant ASO2 (an oil-in-water emulsion with deacylated monophosphoryl lipid A and QS21, a saponin agent derived from the soap bark tree *Quillaja saponaria* and manufactured by GSK), was found to be safe and immunogenic when tested in the Phase 1 and Phase 2 trial (safety, immunogenicity, and efficacy determination of dosage in an endemic setting in children in Mali, West Africa, with doses of 25 and 50 μg given at 0, 1, and 2 months). The sera inhibited parasite growth in vitro (503). As of this time there is no surrogate measure for protection other than testing in human subjects; this continues to hamper evaluation of vaccine responses. The authors of the Phase 1 and Phase 2 trial sagely concluded: "Although these results [with AMA-1] are promising until a blood stage vaccine demonstrates clinical efficacy, immune correlates of vaccine induced protection and the choice of immunogenicity endpoints for clinical development decisions will remain a matter of reasoned conjecture" (503).

It is likely that the efficacy of an AMA-1 vaccine will be restricted due to its many forms; however, some vaccinologists are optimistic that this will not preclude protective effects since "crucial conserved epitopes exist, the function of which can be blocked by antibody" (388). Indeed, it has recently been possible to design three diversity-covering sequences of PfAMA-1 that incorporate 97% of the variability observed in several isolates (389). When rabbits were immunized with this preparation, they produced antibodies that were able to inhibit the in vitro growth of three strains of *P. falciparum* (FVO, 3D7, and HB3) from different geographic areas. This suggests that the diversity of AMA-1 was adequately covered, and there is the possibility such a preparation could serve as a protective vaccine. Whether it will pass muster in monkey and human trials remains undetermined. There are two other caveats in the use of PfAMA-1 as a vaccine: vaccination with AMA-1 could induce the parasite to switch its antigens, and a more effective adjuvant will be required before it can serve as a practical protective vaccine.

A Failed Promise, SPf66

In 1987, with his announcement that he had developed the first synthetic malaria vaccine (369) the Colombian physician Manuel Elkin Patarroyo stunned the scientific community. He said he had protected monkeys from malaria and now was beginning to test his vaccine on thousands of his countrymen. Patarroyo could have tried several approaches in the development of a protective vaccine, i.e., identify the proteins recognized by immune sera, measure the T-cell responses in infected populations to antigens identified by other means (such as antibody), and identify the proteins on the surface of the merozoite or the infected red blood cell. Patarroyo used none of these approaches but instead directly identified antigens of asexual-stage parasites that would protect monkeys against a fatal *P. falciparum* infection and then synthesized a 45-amino-acid peptide.

Working in laboratories housed in three Victorian-era hospital buildings in Bogotá and financed by a series of Colombian presidents, the strong-willed and flamboyant Patarroyo seemed to have done what no one else had been able to do: make a polymer consisting of three peptides from the merozoite surface, including MSP-1, and then with a deft stroke of legerdemain link them by using a peptide from the sporozoite surface (55). Patarroyo, after a short period of training at the Rockefeller University with Bruce Merrifield (who in 1984 received the Nobel prize for his "development of methodology for chemical synthesis [of peptides] on a solid matrix"), was leading a team of 60 scientists—chemists, molecular biologists, and computer scientists—in the preparation of a multiantigenic, multistage synthetic peptide vaccine, named SPf66. Initially, SPf66 was reported to be able to protect four of eight *Aotus* monkeys. After repeating the monkey work several times with larger groups of animals (368), he was planning to go forward with tests of humans. This leap forward to human trials was undertaken despite the fact that Socrates Herrera, a former student of Patarroyo, was unable to replicate the work and the CDC, at the request of WHO, had tested two preparations in *Aotus* monkeys and found an inconsistent immune response and a lack of protection against challenge. Patarroyo, in response, claimed that the CDC had failed to couple the peptides properly!

In 1988, in a preliminary trial with human volunteers, Patarroyo claimed that two of the five immunized volunteers were protected. In a field trial involving 1,548 human volunteers in Colombia, the overall protection rate was reported to be 39%; in adults it was 67%, with immunity lasting 1 to 3 years (318, 522). Patarroyo was not interested in why the vaccine worked: he simply held news conferences. Like Rumpelstiltskin, the little man in the Grimm brothers fairy tale who could spin straw into gold, Patarroyo, with a twinkle in his eye and a smile of satisfaction on his lips, showed the assembled reporters a jar of the new vaccine and boasted that the "three series dose costs thirty cents and that's less than the price of Coca-Cola. It's a chemical product, it's completely reproducible, it's completely pure. There is no possibility of mutation" (55). Some were enthusiastic, others were skeptical, and yet others were jealous (63).

Patarroyo was hailed as a modern Louis Pasteur and was received as if he were a rock star. By 1993 SPf66 had been tested on 41,000 people in Brazil, Colombia, Ecuador, and Venezuela, where it was claimed that the vaccine had 40 to 60% efficacy. Critics charged that these field trials had been faulty in design (317). The trials in South America had not included a placebo control, the identities of those immunized with SPf66 were not hidden from the vaccinators, and those who had received the vaccine knew they were not getting a placebo. Louis Miller of the NIH said the trials were unethical since volunteers were placed at an excessive risk of a dangerously high parasite load. A further concern was that no one outside of Patarroyo's group had examined the data in his laboratory notebooks. Yet another criticism was the vaccine had not been produced by good manufacturing practices. Undeterred, Patarroyo donated a license for SPf66 to WHO, and WHO responded by backing field trials.

In a 1994 field trial, backed by the Swiss Tropical Institute, in Tanzania with 586 children vaccinated, the incidence of disease was reduced by 31%, falling far short of earlier claims of much higher efficacy (6, 194). In a 1993 to 1995 U.S. Army-sponsored trial among 1,200 children in Thailand, said to be the most expensive ($1.5 million) and the best designed to date, there was no efficacy whatever. In another trial supported by the MRC and held in The Gambia, 630 infants (6 to 11 months of age) were immunized with three injections but the vaccine had little effect on protection against first or second attacks of malaria; delay or prevention of malaria was seen in only 8% of those vaccinated. During the second year of follow-up (50) there was no significant protection. In 1999 Acosta et al. wrote: "Given the modest protection previously documented in older age groups and the lack of efficacy in younger infants this vaccine in its current alum-based formulation does not appear to have a role in malaria control in sub-Saharan Africa" (2).

In 1984, Patarroyo, in a fit of hubris, said, "Why should we wait so long? The technology is now ready to be used, why not use it right now? And that's what happened—we made it." At the time of Patarroyo's announcement there was optimism, but by 1996 and after exhaustive field trials, Ripley Ballou of WRAIR (who was now involved in developing the competing vaccine RTS,S) concluded that the SPf66 vaccine "does not protect against clinical falciparum malaria and the . . . results are so disappointing that further efficacy trials are not warranted" (317). Nick White of the Oxford Tropical Medicine Research Programme of Wellcome Mahidol University in Thailand put it more gently: "We have not reached the end of the journey towards an ideal vaccine against falciparum malaria . . . but we are still on track" (553).

9

Blocking Transmission

The oldest and most effective public health measures have been interventions that prevented or reduced the transmission of malaria. In Italy in the early 19th century—years before Ronald Ross had implicated the mosquito in malaria transmission—legislation required that irrigated land be situated at least 500 m from rural habitations and at least 8 km from towns (212). In time, however, this approach to reduction of contact between humans and mosquitoes by ensuring that the distance between the two was beyond the flight range of the breeding sites of the mosquitoes would be replaced by other methods designed to eliminate the mosquitoes themselves.

In February 1899, when Ross left India, he was disgruntled that he had not had the chance to complete his research on the transmission of human malaria. He took some comfort, however, in the fact that his discovery could have practical implications for the eradication of the disease. In April, shortly after assuming the position of Lecturer at the Liverpool School of Tropical Medicine, he set sail for Sierra Leone to "extirpate malaria at once" (404). Sierra Leone, consisting of some satellite villages and its capital Freetown, was the British West African headquarters for trading and settlements on the Gold Coast and at Lagos and by the end of the 19th century was a protectorate of Great Britain. It was a black settlement firmly governed by whites, and it prospered by trade with the interior. It was also disease ridden and remained the "White man's grave." The goal of Ross' project was to make it safe for Europeans by improving the sanitation. By August 1899 he had arrived in Sierra Leone with his "army of mosquito extermination" and began to identify the *Anopheles* mosquitoes that were the main transmitters; he then set about to eliminate their breeding sites. Ross' cheap and simple method, i.e., draining the *Anopheles*-containing puddles, turned out to be a failure since the mosquitoes survived the dry season by using small pools in the streambeds, as well as using wastewater and spring runoff. There was, however, another approach. A team appointed by the Royal Society's Malaria Commission noted that the greatest danger to the health of Europeans lay in their spending the night near the native huts. Therefore, the government decided on a "different" solution from that of Ross: remove the 200 or so Europeans in Freetown to the hills, where they would be protected from the parasites living in the blood of the natives. Ross understandably was irritated by the government's approach to malaria control, i.e., avoidance, and discounted his own failure by stating that the attacks on the mosquitoes were never more than half-hearted and the project had been abandoned too quickly (212).

By 1902, Sir Rickard Christophers and Captain Sydney Price James of the Indian Medical Service set out to eradicate malaria from a British troop contingent living in Lahore, India. The plan was simple: depress the mosquito population by clearing and oiling the irrigation ditches, remove infected individuals from the vicinity of the mosquito breeding grounds, and administer quinine to those who were infected. This, too, failed. Ross thought it was due to mosquitoes hatching in the canals between applications of oil, but killing mosquito larvae (larviciding) was not the main problem; instead, it was a constant migration of mosquitoes, both larvae and adults, by water and air, respectively. The attack on the mosquitoes in Lahore continued for another 8 years, yet malaria persisted. A bitter Ross wrote, "it (larviciding) proved nothing at all and its only effect was to retard antimalaria work in that and in other countries for years" (404). Ross was wrong in concluding that nothing had been learned. The studies in India had proven that an attack on *Anopheles* was not cheap and easy, that the mosquito was adaptable and able to fly considerable distances, and that mosquitoes could breed in a variety of water sources.

At War with Mosquitoes

The success in building the Panama Canal can be directly attributed to the control of disease through an attack on mosquitoes (442). In 1880, the French (Panama Canal Company with Ferdinand de Lesseps as President) acquired the land rights to build the Panama Canal from Colombia and gained title to the Panama Railroad (which had been built in 1855 by a group of New York executives). They began digging in 1882, but by 1889, after removing 76 million cubic yards of earth, they gave up due to mismanagement, poor skills, theft, and disease. More than 22,000 workers died during that period. The California gold rush had stirred U.S. interests in a canal to connect the Atlantic and the Pacific as early as 1849, and this prospect became even more enticing during the Spanish-American War of 1898 because of the difficulty of sending ships from San Francisco to Cuba to reinforce the Atlantic fleet. (The distance from New York to San Francisco with the canal is 5,200 miles; without it, it is 13,000 miles because ships have to go around South America.) In 1899, Congress authorized a commission to negotiate a land deal, and the French sold their rights and the railroad. Colombia, however, refused to sign a treaty in 1903 to agree to construction of the canal. A group of Panamanians feared that Panama would lose the commercial benefits of a canal, and the French worried about losing its sale to the United States, so with the backing of the French, the Panamanians, encouraged by the United States, began a revolution against Colombia. The United States, in accord with its treaty of 1864 with Colombia to protect the Panama Railroad, sent in troops. Marines landed at Colón and prevented the Colombian troops from marching to Panama City, the center of the revolution. The United States recognized the Republic of Panama, and the Hay-Bunau-Varilla Treaty was signed 2 weeks later, in November 1903. It gave the United States permanent and exclusive use of the Canal Zone (10 miles wide, 50 miles long), and the United States guaranteed Panama's independence. Major William C. Gorgas was appointed the Sanitary Engineer on the Canal Project, and he applied his energies to the single-minded goal of eliminating the mosquito. Swamps and other breeding sites were drained and sprayed with kerosene, weeds were cleared, bed nets were used, and workers were housed in copper-screened houses; when there was no way to keep mosquitoes out, they were swatted by hand. Construction of the Panama Canal began in 1907 and was completed in 1914. The magnitude of the accomplishment can be best appreciated by the number of deaths. During the French period of canal construction, the death rate was 176 per 1,000 workers. When the canal was completed by the United States, the death rate from all causes was 6 per 1,000 workers. The success in Panama provided convincing evidence to all who would listen that mosquito-transmitted diseases could be controlled if the effort had enough money, commitment, and leadership behind it.

At the end of the 19th century, John D. Rockefeller, the oil billionaire, decided to alter his public image and become a philanthropist after a life as a robber baron. In 1901 he set up the Rockefeller Institute for Medical Research, and 2 years later he launched the General Education Board, whose work was directed to the states in the American South character-ized by one writer as "a region full of little else but lynchings, shootings, chain gangs and poor, lazy white trash" (442). It was not long before the Board had identified the "germ of laziness" as the bloodsucking hook-worm. Accordingly, in 1909 the Rockefeller Sanitary Commission was founded, and it mounted a successful campaign to eradicate the worm. Four years later J. D. Rockefeller set up the Rockefeller Foundation to pro-mote "the well-being of mankind," and encouraged by the successes with hookworm diseases it decided to widen the war to include other diseases. When the Foundation's war on malaria was authorized, its goal (which it thought was achievable) was an application of existing scientific knowl-edge to public health through relatively cheap and simple techniques. In 1915, the Foundation announced that it was "prepared to give aid in the eradication of this disease [malaria] in those areas where the infection is endemic and where conditions would seem to invite cooperation for its control." From 1910 to 1950 the International Health Division (IHD) of the Rockefeller Foundation employed over 400 scientists who operated glob-ally. The IHD scientists and sanitarians had as their goal the extirpation of malaria.

In 1923, IHD's Lewis H. Hackett (1884 to 1962) began work in Italy and identified the six species of mosquitoes that were vectors of disease and delineated their different blood-feeding habits; some feed primarily on animals, and others feed primarily on humans. He established the doc-trine of anopheline species eradication and controlled malaria by dusting the breeding sites of the responsible species with insecticide. By 1929 Hackett was able to banish malaria from several towns in Italy and boasted: "the discovery of Paris green as a larvicide is the most important addition to our knowledge of malaria control in a decade." Paris green (copper acetoarsenite) was cheaper than oil, did not injure the spreaders, did not hurt cattle or fish, and cost only a little more than 9 c a person to re-duce the parasites in the blood of 18,000 people to insignificant levels. The IHD work in Italy demonstrated a profound and often ignored truism: the epidemiology of malaria varies with each locale, and only a study of that locale can provide the basis for the successful control of malaria. In addi-tion, by this time it had become appreciated that the essential mechanisms of transmission of malaria could be expressed by mathematical formulas first illuminated by Ronald Ross beginning in 1904. However, Ross's pio-neering efforts were almost wholly ignored until the 1930s, when George MacDonald at the Ross Institute in Ceylon, in the aftermath of the Great

Ceylon Malaria Epidemic of 1934 to 1935, provided several elegant models (303, 304).

MacDonald was able to show that "since the number of carriers and the number of infected mosquitoes increase as functions of each other the multiplication of a few all-but-invisible parasites can—and does—occur with a speed that will always seem miraculous however mathematically routine." MacDonald pointed out that, in the simplest form, for an infection to persist in a population, each infected individual on average must transmit the infection to at least one other individual. The number of individuals infected by transmission from each infected person at the beginning of an outbreak is given by R_0; this is the basic reproductive ratio of the disease or, more simply, the multiplier of the disease. The multiplier helps to predict how fast a disease will spread through the population. If the value of R_0 is larger than 1, then the "seeds" of the infection (i.e., the transmission stages) will lead to an ever-expanding spread of the disease. However, if R_0 is less than 1, each infection produces fewer than one transmission stage and the parasite cannot establish itself.

The value of R_0 can be visualized (442) by considering the children's playground game of "touch tag." In this game, one person is chosen to be "it" and the objective of the game is for that player to touch another, who in turn also becomes "it." From then on, each person touched helps to tag others. If no other player is tagged, the game is over, but if more than one other player becomes "it," the number of "touch taggers" multiplies. Thus, if the infected individual ("it") successfully touches another (transmits), the number of diseased individuals (touch taggers) multiplies. In this example the value of R_0 is the number of touch taggers that result from being in contact with "it."

The longer a person is infectious and the greater the number of contacts that the infectious individual (or in the case of malaria the infected mosquito) has with those who are susceptible, the greater the value of R_0 and the faster the disease will spread. Perhaps paradoxically, an increase in the human population in an area will, per se, lead to a reduction in R_0 (because R_0 is proportional to the number of female *Anopheles* mosquitoes per person, and this ratio falls if the number of people is increased). This fact and consequence, which at first sight appears strange, is probably why, in general, malaria transmission is much lower in urban areas, where human population densities are high, than in rural areas, where they are relatively low. On the other hand, an increase in the number of female *Anopheles* mosquitoes clearly increases R_0 and the rate of spread of malaria while an increase in mosquito mortality will reduce the spread of disease in a population (R. Carter, personal communication).

The essential message of the mathematical models of Ross and Mac-Donald is this: killing all the vector mosquitoes is not by any means necessary for the reduction and, indeed, the elimination of malaria. It is only necessary for the number of malaria-infected mosquitoes to fall below a certain number for malaria itself to decline toward zero; i.e., R_0 is proportional to the number of female *Anopheles* mosquitoes per human in a locality, and if R_0 is less than 1, malaria cannot be sustained and must sooner or later die out.

The R_0 equation can be taken a step further to take account of several other key parameters in malaria transmission. One is the probability that a female mosquito takes a human blood meal. This recognizes the importance of the "human-biting habit" to malaria transmission. Oddly, neither Ross nor MacDonald considered including in their equations the distance that mosquitoes must fly to take a blood meal. What happens in reality, however, is that at distances of more than about 1 km, and often much less, R_0 falls below 1. In other words, putting distance between humans and mosquito breeding sites eventually cuts out the possibility of malaria transmission (Carter, personal communication).

There is another important part of the Ross-MacDonald malaria equation that needs to be put in place: the life expectancy of a female mosquito after a blood meal and how long it takes the malaria parasites to reach infectivity (maturation) in the mosquito. The shorter the life expectancy of a mosquito after a blood meal, and the longer the maturation period of the parasites in the mosquitoes, the lower will be the value of R_0. Either reducing life expectancy or increasing the maturation period will pack a powerful and more than linear punch in reducing R_0. Still other factors affect malaria transmission. MacDonald thought about the susceptibility of a particular mosquito, which obviously matters since some are much more susceptible than others; this would include a factor for the infectiousness of gametocytes in the blood, in other words a factor that can take account (among other things) of "transmission blocking anti-gamete antibodies" (Carter, personal communication). What one learns from this is that transmission-blocking vaccines and antimalarial drugs and, indeed, antiblood stage and preblood stage vaccines each reduce R_0 in a simple linear fashion. With this notion in hand, it is possible to think more or less rationally about the various aspects of malaria transmission and its reduction.

At the turn of the 20th century it had already been possible to kill adult mosquitoes by fumigating houses with the smoke of burning sulfur or tobacco; however, the problem was that the insects might not be killed outright. Furthermore, the use of smoke required that rooms be sealed for hours. It was unpleasant, inconvenient, expensive, and damaging to the household contents. Recognizing this, Gustav Giemsa (1867 to 1948)—famous for the staining method he described in 1904—in 1910 began thinking of substituting a fine mist for the fumigating smoke (212). He suspended pyrethrum powder (an insecticide) in glycerin, green soap, and water and devised a suitable pump and spray nozzle. He tried out his spray on hibernating *Culex* mosquitoes living in the cellars of houses in Hamburg. It worked well, and in the laboratory it killed *Anopheles*. Giemsa reported his invention in 1913; however, since it seemed unimportant scientifically it was neglected in most places. Nevertheless, within a decade in the United States pyrethrum sprays came into use against nuisance insects—"Quick Henry the Flit!" In 1928, a Brazilian doctor reinvented Giemsa's pump, modeling it on a paint sprayer then in use, and was able to kill *Aedes* mosquitoes. In the 1930s, this spray gun filled with pyrethrum was used in a limited trial on Brazilian estates and in the barracks of plantations during a yellow fever outbreak. It worked.

In South America, IHD's Fred L. Soper, "a dark, handsome Kansan with a thin black moustache, glittering eye, and fierce, yet good-humored, with a determination to command every sanitation" (554), was intent on ridding places of disease-carrying mosquitoes. Soper was a blunt and difficult man who, though medically trained, always thought of himself as an administrator and sanitarian. He, like Ross and Gorgas, was an enthusiast for direct action, had formidable talents for organization and command, and was capable of browbeating man or mosquito. Initially working in Brazil, he targeted the *Aedes* mosquitoes that transmitted yellow fever and was determined to exterminate them. Following the path taken earlier by Gorgas, he eliminated their breeding places by emptying, covering, and oiling domestic water containers. He established an "eradication army" with uniformed mosquito inspectors, forbidding them to take bribes on penalty of immediate dismissal, and set over them district inspectors who were specifically trained and better paid. Inspectors were given the right to enter houses, and penalties were levied for obstructing their work. Maintaining a breeding place on one's property was a punishable offense. The strategy worked, and by 1934 the domesticated *Aedes aegypti* had been eradicated in Brazil. The eradication work was then extended to Bolivia, Paraguay, Peru, and Colombia. The Brazil campaign was so successful in reducing the number of yellow fever deaths that the Rockefeller Foundation elected to carry out a large-scale malaria eradication experiment in India. In 1941 Sir Gordon Covell, head of the Malaria Institute, wrote: "The spray killing of adult mosquitoes is now recognized as a major control measure and . . . represents the most important advance which has been made in malaria prevention in years" (212). Covell exaggerated: in areas of hyperendemic infection spraying did not work since it was overwhelmed by the sheer numbers of mosquitoes, and with mosquitoes that normally stayed in the house long enough to bite and then rest, the spray worked only when used daily and in large amounts.

By the mid-1930s it was clear that the mosquito transmitter of human malaria, *Anopheles gambiae*, had been imported into Brazil from Africa. Soper, having succeeded with the yellow fever mosquito, set his sights on eliminating the malaria-transmitting *A. gambiae*. By mid-1939, the Brazilian government along with the IHD funded the Malaria Service of the Northeast. Soper began to use Paris green in place of oil for killing larvae, and the adult mosquitoes were attacked with pyrethrum sprays by using the same system of hunt and destroy that was used effectively with yellow fever. Two years later, Soper and his men could celebrate their victory: in the districts under their control 185,572 cases had been reduced to only 6,124. However, Soper could not settle for a reduction in the number of vectors. He thought it to be "a good philosophy for the defeated" (212). There were still *Anopheles* mosquitoes in existence, and although their numbers had been severely reduced, they would shortly recover. Soper's view echoed that of Ronald Ross: the malaria-carrying mosquitoes could have been extinguished if only enough "men, money and necessary materials" (404) had been provided. Because the battle in Brazil in the 1930s was the biggest so far in which spray-killing played a major role, it was concluded by some health authorities that spraying (and not larviciding) had made the critical difference. Although this may not have actually been true, the Brazil campaign did show that control measures require a well-trained, well-disciplined organization under decisive leadership with full government backing. A less well appreciated lesson was that the prevalence of disease is influenced not so much by the abundance of carriers as by the numbers and efficiency of the mosquito vector. Indeed, even many human carriers in Brazil could not keep malaria at a high rate when the vector transmitted poorly, as had been predicted by MacDonald's mathematical model (see p. 192).

In 1874, Othmar Zeidler (1859 to 1911), an obscure German doctoral student studying chemistry with Adolf von Baeyer at Strasbourg, synthesized dichloro-diphenyl-trichloroethane from chloral hydrate with chlorobenzene in the presence of sulfuric acid. Although he made note of the new chemical compound in the *Berichte der deutschen chemischen Gesellschaft*, he did not describe its use, and it received little attention until 1941, when Paul Müller (1899 to 1965), a chemist at Ciba-Geigy in Switzerland, synthesized the same compound as one in a series of related compounds in the search for new insecticides. He called it DDT and found it to be a safe and efficient killer of the clothes moth. DDT killed clothes moths as well as flies, mosquitoes, beetles, and a host of other insects. Through the U.S. military attaché in Berne, the British and Americans procured small amounts of DDT and began performing experiments in 1942 (212). They confirmed the Swiss results, and during World War II Britain gave the manufacture of DDT its highest priority along with radar and penicillin.

In 1943, DDT production began first in Britain and then in the United States. In 1942, Soper had left Brazil and ended up as a civilian on the U.S. Typhus Commission working in Cairo, Egypt. There he met up with his old enemy *A. gambiae*. Following the model used in Brazil, *A. gambiae* was wiped out of Egypt in 6 months. After the U.S. occupation of Italy in 1944, Soper was put in charge of the typhus eradication campaign in Naples. Although it is generally believed that DDT served as the delousing agent in this campaign, pyrethrum spraying was actually the effective agent. In 1944, the Allied Control Commission asked Soper and his Rockefeller team to see what they could do about ridding the area of malaria by controlling *Anopheles* since this was impeding the Allied advance. It was found that a single spraying of DDT to coat interior walls where feeding mosquitoes rested halted the transmission of malaria for an entire season. In addition, the method was simple and cheap. By 1945, the United Nations Relief and Rehabilitation Administration (UNRRA) decided to embark on a campaign to eradicate malaria from Italy. It provided the DDT and money for the spray teams. In 1946 the Italian government took over the project, and within 3 years the number of deaths due to malaria in Italy had been reduced to zero. The lesson learned in Italy was that the value of DDT was not that it was more efficient than Paris green or pyrethrum in killing mosquitoes but that as a house spray it stood guard over an enemy that might recur in an area. It was a defensive weapon whose action resembled that of a vaccine. DDT, by making the environment unsuitable for the transmission of the parasite, held out (for some) the possibility for the eradication of malaria.

The UNRRA's involvement in Italy set a pattern for international cooperation in the war against malaria, and by 1946 its assets had been provided to UNICEF, which decided that malaria was so great a killer of children that a continuing investment in malaria control was warranted. Although UNICEF was an important supplier of pesticides and other materials to developing countries, the principal agency involved in the war against malaria was the WHO. (After World War II the Rockefeller trustees phased out the IHD because they felt that the WHO would mount comparable long-term studies on tropical diseases.) The WHO absorbed the Health Organization of the League of Nations, and in 1947 it established an expert committee on malaria to formulate strategies, review problems, and generally function as a board of directors. The committee's great hope was the eradication of the disease—and even *Anopheles*—from entire countries. Initially, the committee refrained from recommending an immediate worldwide campaign, but by 1949 the experts at WHO had become impatient. An eradication campaign was sold to the nations of the world based on the near-miraculous blow that DDT had dealt to malaria. By1950 many nations had antimalaria programs dependent on routine annual spraying of DDT. Because eradication was successful in some places, public health authorities hoped that continued spraying would be unnecessary. Although the number of *Anopheles* species resistant to insecticides had risen from 5 in 1956 to 38 in 1968 (212), the WHO went so far as to claim that mosquito resistance had been exaggerated and it was a local phenomenon.

Another approach to blocking transmission is to reduce the infectivity of the human population itself by widespread distribution and use of antimalarial drugs. In 1955 the World Health Assembly endorsed an aggressive campaign to eradicate malaria by overwhelming force: interrupt transmission by insecticides and treat the infected people with drugs. The WHO provided technical aid, and UNICEF and USAID provided the funding. In 1960, 52 nations that attempted eradication had campaigns completed or under way. Funding for the campaign peaked in 1965 and dwindled thereafter. By 1966, the World Health Assembly reviewed progress and noted that malaria had been eradicated from areas that were inhabited by more than 600 million people and had once harbored endemic infections, while the disease burden in another 334 million people had been reduced. Ten of the countries had achieved eradication, while 11 others had banished the disease from some part of their territory. Of the 10 countries where eradication had been achieved, 4 were in Europe and the other 6 were in the Americas. Nevertheless, 638 million people still lived in areas where malaria was actively transmitted, and there it remained a major cause of morbidity and mortality. In Africa alone there were 360 million people living in malarious areas, and in half of these areas there was no effort to fight the disease. Not a single victory had been achieved in any major tropical area. There was a resurgence of malaria in places where it had been under control, and the tide of disease had turned around. International organizations saw no end to the demands on their funds, and with only diminishing returns from them they became fed up. More and more, the all-or-nothing campaign came to be regarded as a failure. By 1969, the World Health Assembly demanded that the malaria eradicators change course and adopt a more flexible approach. Where eradication was judged impractical, countries might revert to measures for control. The controllers would essentially do the same thing the eradicators had done but with less pressure and no time limit. One of the lessons learned from the eradication era is that striving for perfection must not go on too long. Not only do scientists and funding agencies get bored, but also there is a natural impatience with a prolonged attack. An attack must succeed quickly, or it will not succeed at all (212).

The WHO gave in slowly, but it was not easy for those who had embraced eradication to abandon the old ways and adopt new ones. The eradication campaign had turned the subtle and vital science of malariology dedicated to managing a complicated natural system—mosquitoes, malarial parasites, and people—into a war waged by spray guns and antimalarial tablets. For the eradicators, the fight against malaria had meant spraying walls with DDT, collecting and reading blood films, and administering drugs in doses sufficient to eliminate residual parasites. The older tactics that had been superseded by the use of these "miraculous pesticides" would have to be resurrected and changed: drainage, environmental manipulation such as filling and drainage, larvicide, colonization of mosquito breeding places with predatory fish, house screening, bed nets, and repellents.

"The use of antimalarial drugs for transmission control is, however, by and large not very effective compared with those that directly affect the human-mosquito interaction. The reason is that it is virtually impossible, using drugs alone, to ensure that a malaria-infected individual has received the necessary treatment at the time when he or she is infectious to mosquitoes. In endemic areas drug cover can only be intermittent . . . unless permanent drug prophylaxis is employed across an entire endemic human population . . . and in most situations, would be neither practical or affordable" (70). An alternative to the use of drugs to suppress human infectivity for mosquitoes, some have suggested, is to develop and deploy vaccines designed to limit the transmission of human malaria infections to mosquitoes—a transmission-blocking vaccine.

Mutualistic or Cooperative Transmission-Blocking Vaccines

During the 24 h of development within the mosquito midgut, malaria parasites are outside of cells and are bathed with the constituents of the blood meal. Thus, the midgut stages are accessible to the immune factors of the vertebrate host, i.e., primarily antibody with or without activation of the complement cascade. This situation has been exploited by inducing antibodies in the vertebrate host against the pre-oocyst stages (gametes and ookinetes) in the midgut stages. (Because of the difficulty of intact antibody crossing the mosquito midgut, the oocyst and sporozoite are unlikely to be affected.) These antibodies, when ingested by the female mosquito as a part of the blood meal, may lead to sterilization of the infection in the mosquito. Since the antibodies have the potential to suppress the transfer of malaria parasites from human to mosquito, this suggested the possibility of development of a mosquito stage vaccine that would block transmission by preventing infection in the vector. A malaria transmission-blocking vaccine (TBV) would not prevent disease in newly infected individuals but would contribute to herd immunity: although it would not directly influence the course of infection in the vaccinated individual, it would affect the potential for infection in others (71, 267, 418, 419). Thus, TBVs are sometimes called altruistic vaccines. This, however, is misleading. Malaria is transmitted by mosquitoes over small distances, generally less (and often much less) than 1 km. Indeed, a considerable proportion of malaria transmission occurs between members of the same household. Thus, any measure that prevents an individual from infecting mosquitoes, such as a TBV, will significantly reduce the chance that others in a household will get malaria and ultimately will reduce the chance that the vaccinated individual will be reinfected. In short, with malaria, as in other walks of life, what goes around comes around. Therefore, a house, a small community, or a village whose members elect to receive a malaria TBV are not acting altruistically but are acting to their collective mutual and, indeed, individual personal benefit. Vaccination with an anti-mosquito-stage malaria TBV is more correctly referred to as "mutually beneficial, or co-operative vaccination" (Carter, personal communication). Indeed, an effective TBV might also effect a reduction in the spread of parasites resistant to drugs and/or those vaccinated against blood or liver stages, which is likewise a matter of mutual and communal benefit.

The path to a TBV began with Richard Carter (born in 1945), who always wanted a career in biology. Late in 1967, after graduating with a B.Sc. in biochemistry, Carter joined the Institute of Animal Genetics (which doubled as the Genetics Department of the University of Edinburgh) and the laboratory of Geoffrey Beale. Beale was one of the first to pick up on the contributions that biochemical analysis, in the form of protein (and more precisely enzyme) electrophoresis, could make to the investigation of genetic differences between organisms. Carter's task was to conduct starch gel electrophoresis analysis of "enzyme variation" in malaria parasites of rodents with the objective of obtaining "genetic markers" for use in crossing experiments between the parasites. These rodent malaria experiments were successful, and after completing his Ph.D. studies Carter joined the NIH laboratory of Louis Miller, intending to do a genetic cross with *P. falciparum* by using *Aotus* monkeys. When he arrived at the NIH, he realized that this was a very tall order. Apart from anything else, the chances of having two monkeys infected with two different strains of *P. falciparum* carrying gametocytes, let alone infectious gametocytes, on the same day was extremely low. The only alternative at that time was to try to deep-freeze gametocytes from two infections and then later bring them back to life together on the same day for feeding through an artificial membrane. However, to be able to work out the principles of how to deep-freeze and recover gametocytes alive and infectious to mosquitoes, a working model that was more practical than monkeys with malaria was needed. As Ross had done 75 years previously, Carter began to use an avian malaria, *P. gallinaceum,* the best available. He never looked back or considered doing a genetic cross with *P. falciparum* again until almost 10 years later, after infectious gametocytes had become available from Trager-Jensen cultures.

The first requirement, Carter realized, was to keep the gametocytes calm and quiet during the test tube manipulations that would have to be done, probably over many hours, while he worked on the gametocytes following their removal from an infected chicken. The problem was that gametocytes are highly excitable and almost always transform spontaneously into gametes within a few minutes after blood is removed from an infected animal, whether by a mosquito or by a working scientist. Using *P. gallinaceum* gametocytes and following a tip from Miller, Carter "invented" gametocyte "suspended animation medium," which was just Tris-buffered physiological saline with glucose (Carter's addition) and without bicarbonate (Miller's tip); this medium kept them alive and quiescent for at least an entire working day, but they were still ready to begin exflagellation immediately when the right stimulus was given. In collaboration with Mary Nijhout, Carter found that almost instant induction of exflagellation occurred when bicarbonate ion was added (Ann Bishop and colleagues at the Molteno Institute had already elegantly demonstrated the requirement for bicarbonate in the 1950s) and that this occurred in a strictly pH-dependent manner. Thus, in the presence of bicarbonate ion, the amount of exflagellation increased to a maximum at pH 8.0 and thereafter declined rapidly again; it did not take place at all above pH 8.5 or below pH 7.7, and the CO_2 gas tension was quite irrelevant (347).

It was while this work was under way that Carter became exposed to ideas that led to the development of the concept of malaria transmission-blocking immunity. It was late 1974, and talk of a malaria vaccine or vaccines was thick in the air. Elsewhere, work was beginning on attempts to immunize against the blood stages of malaria parasites (see chapter 8), and so prevent or reduce the clinical effects of malaria, or to immunize against the sporozoites and prevent infection altogether. What struck Carter, in the midst of this talk, was that unless either type of vaccine was 100% effective, neither would prevent malaria transmission in a community where the infection was endemic. His experience as a malaria geneticist had already taught him that even a tiny number of gametocytes, such as might escape either form of "asexual" vaccination, could very effectively infect mosquitoes. In Carter's mind, vaccination served one or both of two purposes: to reduce the risk of disease in a vaccinated individual and equally, if not much more important, to reduce its spread to others. He might have mused: Isn't that what the triumph of vaccination against polio was all about? Wasn't that what smallpox eradication from the world by vaccination was all about? Surely this must also be a major objective in the struggle to produce vaccines against malaria.

However, none of these thoughts were any part of Carter's thinking until the day in 1974 when a conversation took place between himself, Miller, and the entomologist Robert Gwadz. Carter learned of Gwadz's interest in the possibility that infectivity of malaria parasites to mosquitoes could be modulated, in part, by host immunity. This was an idea first investigated more than two decades earlier by Don E. Eyles of the U.S. Public Health Service. Eyles had shown, using *P. gallinaceum*, that serum associated with a peak in the level of blood parasites was detrimental to the development of gametocytes in the mosquito (166, 167). He suspended gametocyte-carrying chicken red cells in a variety of chicken sera and fed these to mosquitoes through an artificial membrane. After some time he carried out Ross-type mosquito dissections and counted the oocysts on the stomach wall. Compared to sera from normal birds, that taken from chickens 1 day after the peak of a blood-induced infection caused a reduction in oocyst numbers, indicating suppression of infectivity of the gametocytes to mosquitoes. On the other hand, normal serum as well as serum taken from chickens at the peak of blood infections that had been sporozoite induced had no effect. Then in 1958, Clay Huff and colleagues at the Naval Medical Research Institute (NMRI) found that immunization of turkeys with five successive intravenous inoculations of formalin-killed *P. fallax* gametocytes led to a subsequent reduced infectivity to mosquitoes (250). Neither the numbers of asexual stages nor the numbers of gametocytes in the blood differed from those in the controls. However, the potential relevance of antigametocyte immunity to malaria transmission and public health was overlooked by Huff et al. and Eyles.

Before the time of this conversation, Carter had no idea that his own work and thinking would turn in this direction—he was still thinking about how to conduct genetic crosses with *P. falciparum*. Likewise, neither was Gwadz thinking primarily about gametocyte vaccines since he was preoccupied with studying the factors involved in the infectiveness of *P. cynomolgi* blood infections in rhesus monkeys to mosquitoes.

From that day onward, however, as Carter became progressively involved in studying exflagellation, originally for reasons of genetics and then for its own sake, his mind began to revolve around the idea of the possibility of vaccinating against the sexual stages of malaria with a view to preventing their transmission to mosquitoes. In his mind this was now the self-evident major goal of vaccination against malaria as a public health measure to reduce or eliminate malaria from human communities, just as vaccination had done in some communities for polio. The idea then occurred to Carter that immunizing a host with the gametes of a malaria parasite (sexual forms which are normally to be found only in the mosquito midgut but would now be available in the laboratory via the methods that he had been working upon) could be used to induce antibodies against their surface antigens by immunization of a host. When ingested together with gametocytes in a mosquito blood meal, such antibodies would possibly do great harm to the gametes as soon as they had left the protective enclosure of the host red blood cell. The idea appeared to him to be so obvious that it seemed surprising that it had not already been proferred by Ronald Ross himself, a man with an exceedingly fertile and energetic mind. Nevertheless, as ideas began to flow, the talk in the laboratory about immunity that might prevent a malaria-infected individual being infective for mosquitoes was always about antigametocyte immunity and what might be going on in the blood of the vertebrate host to harm the gametocytes as they circulated in the bloodstream. There was no discussion whatever concerning the effects of an immunity that originated in the vertebrate host but exerted its effects on the gametes of the parasites as they were released into the mosquito midgut during a blood meal. Even later, when this phenomenon had been conclusively demonstrated, the whole idea of antibodies, or other kinds of immunity of vertebrate host origin, acting in the mosquito vector tended to be disregarded, especially by professional immunologists, some in high places, who more or less considered the notion to be nonsense. It was therefore with not a little trepidation, and a few days after the idea had occurred to him, that Carter put the concept, somewhat hesitantly, to Miller.

Carter remembers receiving a rather blank look—as happens when something almost ridiculously obvious has been said. But the implications in both their minds must have been jumping out at them. There was also the matter, and this was of concern to both Miller and Carter, that this was entering a research territory running very close to the interests being actively followed by Gwadz. Although neither Carter nor Gwadz had a stated goal at that point of working on anything directly to do with immunization for the purposes of suppressing the infectivity of malaria parasites for mosquitoes, let alone making a vaccine, each was moving toward it from his own particular line of investigation. What Carter proposed to Miller was the injection into chickens of extracellular gametes of *P. gallinaceum*, in whose preparation he was now developing specific expertise, to see whether there was any effect upon the infectivity of subsequent blood infections of *P. gallinaceum* to mosquitoes through the effects of antibodies on the gametes as they emerged into the immune blood meal.

Miller was supportive of the idea, but it was clear to both of them that Carter would also have to discuss it with Gwadz and hopefully not find that there was an unworkable conflict of interest. Carter went down the corridor to Gwadz's office and told him he was interested in testing the idea that extracellular gametes of a malaria parasite could induce antibodies that would interact not with the gametocytes in the bloodstream of the chicken but with the extracellular gametes themselves in the mosquito midgut. This did not directly conflict with Gwadz's immediate research interest in the factors affecting infectivity to mosquitoes of *P. cynomolgi* infections in rhesus monkeys. It did follow, however, from Carter's developing expertise in the preparation and handling of the extracellular gametes of malaria parasites. Thus, in the late winter or spring of 1975, the two agreed that Carter would conduct the experiments.

As often happens in science, as in life, a plan and the course of events do not always coincide. It had happened to Ronald Ross, when, on the point of completing the demonstration of the transmission of human malaria through mosquitoes and back to humans, he was called away on medical duty (he was after all an Army medical officer) to other things (see p. 24). "Other things" now intervened, delaying the start of Carter's proposed experiments by almost half a year—a set of experimental investigations had already been promised to a senior colleague, and these now took precedence. Then Carter left the NIH for a summer break at home in the United Kingdom, where he discussed the idea for gamete immunization with Robert Killick-Kendrick, Robert Sinden, and the doyen of malaria, P. C. C. (Cyril) Garnham.

By the time Carter returned to the NIH that autumn, he learned Gwadz had beaten him to the punch: he had immunized chickens with formalin-killed gametocytes of *P. gallinaceum*, following the protocol Huff had used. Gwadz obtained equivalent results with *P. gallinaceum* in chickens to those obtained by Huff with *P. fallax* in turkeys. But he now went further. Using the membrane feeding apparatus, he showed that gametocytes from the immunized chickens entirely recovered their infectivity when they were washed free of their own plasma and resuspended in normal chicken serum whereas gametocytes from nonimmunized chickens were almost completely unable to infect mosquitoes when resuspended in serum from the immune chickens. Then, finally, he took the gametocyte-infected blood from the immune chickens and examined it for exflagellation under the microscope. There, for the first time, he observed the "microgamete immobilization reaction" (204).

The results were almost exactly what Carter had anticipated. What he had not envisioned was that this kind of immunity could have been induced by the gametocytes themselves, or at least not very effectively. This was because part of his thinking had been that such antibodies would be induced mainly by the stages that emerged in the mosquito midgut, i.e., the gametes. He had anticipated that within the midgut of a blood-fed mosquito, a set of antigens would be expressed de novo on the gametes and that these should be the targets of the immunity that would suppress their infectivity to the mosquitoes. In this way gamete vaccination would be a way of creeping up on the parasites unsuspected, without any immune evasion mechanisms having been evolved.

Gwadz's experiment was not the experiment that Carter had proposed and was, in fact, based upon the study done previously by Huff and colleagues in 1958. However, it moved well beyond Huff's work by revealing the site and effectively the nature of a form of immunity that suppressed the infectivity of gametocytes to mosquitoes. After Gwadz's work had been done, there could be little doubt of the reality of antigamete infectivity suppressing immunity, essentially as Carter had thought.

Carter, nevertheless, decided to go ahead with his original plans in collaboration with David Chen, who had been hired earlier by Gwadz to purify sporozoites. By this time, however, Chen had been redirected to making gametes, so it seemed preordained that he and Carter would work together. They made semipurified preparations of male and female gametes of *P. gallinaceum* by using the methods and principles that Carter had developed earlier (they contained a high preponderance of male and female extracellular gametes and also a fair amount of asexual stages and general debris), X-irradiated them to eliminate their infectivity to the chickens, and inoculated them directly, without adjuvant, into the chickens by the intravenous route. There were six immunized chickens and one control. After three such weekly injections, all the chickens were infected with live blood stages of *P. gallinaceum*. Normal, or almost normal, blood infections ensued in all the immunized chickens and in the control. *A. aegypti* mosquitoes were fed upon the chickens daily throughout their infections, and the oocysts in the infected mosquitoes were counted. The normal chicken produced the expected hundreds of oocysts throughout the infection. The immunized chickens produced either none at all at any time or a very occasional one. The overall reduction in oocysts in the immunized chickens over the full course of their infections was 99.99 to 100%. Of course, their sera agglutinated and immobilized male gametes very effectively (72).

When Carter went to Gwadz's office and queried him about the successful immunization against the sexual stages of *P. gallinaceum* without any involvement on his part, Gwadz rolled back in his chair and simply replied, "Ideas are cheap." Carter was taken aback; however, he would have been less so had he heard the words of the biochemist Nathan Kaplan: "good ideas are a dime a dozen for a smart person, and the only thing that alternately distinguishes good from great is in how an idea is executed—how it becomes reality" (528). Great ideas can be simultaneously conceived, and scientific history is littered with stories of one person having an idea but not following through on it, only to see another person have a similar inspiration and then prove it to be valid. There is no reason to suppose that Gwadz and Carter had not approached their insights partially through their own independent thinking. They were certainly each traveling along apparently unconnected tracks that were nevertheless headed toward the same destination. Yet without the other, would either have taken the steps that led to the experiments that opened the field of malaria transmission-blocking immunity? Perhaps not; however, within the year Gwadz alone was awarded an NIH medal for the work.

The Nature of Transmission-Blocking Immunity

After Carter began to explore more closely the mechanisms involved in this immunity, he coined the term "transmission-blocking immunity" (73, 74). The first of two studies was an attempt to see whether extracellular gamete preparations were, in fact, any better, pound for pound, than gametocytes. The result showed that exflagellated gametocytes did in fact induce better transmission-blocking immunity than did the exact same material that had been prevented from exflagellating before being used for immunization. Although the results were pleasing, Carter still cannot explain why it should be the case. As discussed below, the major known immunogens and target antigens of antigamete immunity are present in mature gametocytes and are more or less (but not quite) in the same form as they are on the gametes themselves. Perhaps, as he says, it has something to do with antigen presentation.

In the second study, using an in vitro system for fertilization developed from the experience obtained with his experiments manipulating gametocytes and also using an ookinete culture system based on work by David Chen, Carter explored when and how the transmission-blocking antibodies exerted their effect. First, he investigated exactly when fertilization took place in vitro by using a high-dilution-of-gametocytes method to prevent gametes from being able to fertilize each other after exflagellation. When gametocyte-infected blood in glucose-saline (pH 7.4) was diluted 200-fold within 10 min or sooner after induction of exflagellation (gamete formation), there was no subsequent formation of ookinetes (the products of gamete fertilization). However, when dilution was done more than 20 min after induction, the numbers of ookinetes that subsequently formed in vitro was the same as that in never-diluted controls. The conclusion was that fertilization must begin and be completed between 10 and 20 min after the induction of exflagellation. In similar fashion, the transmission-blocking serum completely prevented subsequent ookinete formation if it was added earlier than 10 min after the moment of induction of exflagellation but had no effect on subsequent ookinete formation when applied later than 20 min after the time of induction, i.e., when it was applied after fertilization had already taken place. In other words, the antibodies were, indeed, fertilization-preventing antibodies and had no detectable downstream effect (73, 74).

Carter went on to explore how, and for how long, the immunized chickens "remembered" the transmission-blocking immunity that had been induced by immunization with the gametes of the parasites (73, 74). He found that up to at least 6 months after immunization and at a time when antigamete antibodies had ceased to be detectable in the circulation of an immunized chicken, a new blood infection dramatically boosts these antibodies back to effective transmission-blocking levels even before the first blood stage parasites are detectable in the blood. In short, the chickens "remembered" almost instantly how to activate transmission-blocking immunity as soon as it was confronted with a new malarial blood infection.

Discovering the Antigens for a TBV

The next series of studies that Carter undertook were devoted to characterizing the protein antigens on the surface of male and female gametes and zygotes of *P. gallinaceum* and their developmental successors in the mosquito midgut, the parasite ookinetes. Carter taught himself to prepare these stages to high levels of purity, enabling the identification and characterization of the surface antigens of each stage. The method involved using immune sera raised specifically against these stages of the parasites in chickens and rabbits and also using monoclonal antibodies from gamete-immunized mice through the newly developed technology of Köhler and Milstein (see p. 161). Labeling the surface of gametes with radioactive iodine and precipitating the now radioactive proteins by using transmission-blocking monoclonal antibodies that reacted uniquely with either of these two specific protein antigens of the gametes led to the identification of two antigens. Such radioactively labeled proteins can be readily identified on a photographic film laid over a slab of gel in which they have been separated from each other, usually by application of an electric field. As stated above, monoclonal antibodies react with a single unique antigen, or actually a unique bit of an antigen called an epitope, and thereby identify, as accurately as a DNA fingerprint, the proteins they react with. This, together with the fact that these particular monoclonal antibodies also efficiently blocked gamete fertilization and transmission of the parasites to mosquitoes, was effective proof that the antigens they identified were the actual target antigens of the transmission-blocking immunity.

It was by these means, therefore, that the protein antigens that could be used for a malaria TBV were identified—at least for chicken malaria. The equivalent antigens of *P. falciparum* were soon to be identified by exactly equivalent methods (530). These *P. falciparum* proteins were named Pfs230 and Pfs48/45 based on their apparent molecular weight from migration through a gel under an electric field. As was to become evident, these two antigens were expressed on the surface membrane of gametocytes as they developed inside the human bloodstream. Once the parasites were taken up in the blood meal by a mosquito, they both remained associated with the surface of the emerged gamete for several hours. The presence of these antigens in the gametocytes is the probable reason why these stages induce transmission-blocking immunity almost as effectively as do the gametes themselves. The antigamete activity of certain antibodies against these antigens, and especially against Pfs230, is dramatically effective (556). When the antibodies are of the complement-fixing type, they mediate their transmission-blocking activity by the lysis of male and female gametes and early zygotes within seconds to minutes; the destruction of these stages is effectively total. There are no survivors and no onward transmission of the parasites through the mosquitoes.

Meanwhile, still working with the *P. gallinaceum* system in chickens, Carter and his colleagues identified two additional antigens, Pgs25 and Pgs28, on the developing zygotes of the parasites appearing respectively at around 5 and 12 h after emergence (379). Thereafter, the antigens continued to be present on the surface of the zygote throughout its development as an ookinete. Thus, unlike Pfs230 and Pfs48/45, these antigens are expressed only during the mosquito phase of the life cycle and are not found at all in the gametocyte stages in the blood circulation. These proteins, too, were shown to be targets of very effective transmission-blocking antibodies. The mechanism of the effect of these antibodies remains unclear to this day; one thing is certain, however: they are not involved in preventing fertilization, as this event is long past by the time their target antigens appear. As had been the case with the gametocyte and gamete surface antigens, homologous protein antigens were soon identified in *P. falciparum* (531) and were named Pfs28/Pfs25.

To Make a Vaccine

Plasmodium spp. have 14 chromosomes, and the gene for Pfs230 has been located on chromosome 2. When this gene is deleted, the number of oocysts is reduced but not abolished (556). At least one of the functions of Pfs230 is probably in the fertilization mechanism itself, performing some essential action on the male gamete, whereas the Pfs48/45 class of proteins plays a role on the female gamete. It may also be that Pfs230 is involved in protection of the parasites from the digestive enzymes or nitric oxide or from antimicrobial peptides within the blood meal. Pfs230 was originally cloned from a *P. falciparum* cDNA library. Efforts to make synthetic forms of the Pfs230 and Pfs48/45 proteins that could be used for a vaccine have encountered significant problems. These have to do largely with the fact that each of these proteins is folded together by a large number of molecular bonds called disulfide bonds. Disulfide bonds turn out to be extraordinarily difficult to reproduce accurately by synthetic means. So far, only a small region of Pfs230 has been expressed in *E. coli* and is able to induce antibodies with some transmission-blocking effect (556).

Pfs48/45, of course, is the other identified target antigen of transmission-blocking antibodies that have their effect at and around the time of fertilization of the gametes of a malaria parasite in a mosquito midgut. Recently there has been significant success in expressing Pfs48/45 in an immunogenic form, and this appears to have solved the riddle of how to re-create successfully the elusive disulfide bonds (361).

One of the great potential strengths of Pfs48/45 and Pfs230 of antigens for a malaria TBV is that these antigens are expressed in the human host so that there can be a boosting effect. This was already anticipated in the study of gamete immunization against *P. gallinaceum* in chickens (73) and subsequently in a powerful study with *P. knowlesi* in rhesus monkeys (205). This work showed that monkeys vaccinated with material containing gametes of a malaria parasite induced an immunity that had a very effective memory for transmission-blocking immunity that continued for at least 3 years, with no suggestion that it was waning. We do not know for how long such immune memory may last, but it could continue for many years. If so, it would be of tremendous significance to the deployment of gametocyte/gamete antigen-based TBVs for the containment or even elimination of malaria transmission in regions of endemic infection. Thus, vaccinated populations could have their transmission-blocking antibodies boosted by stray incoming blood stage infections even many years after most malaria transmission had ceased, thereby locking out the onward transmission of infection and sustaining the low-malaria environment.

The likely reality of human immune memory for the gametocyte/gamete surface antigens has been shown in studies of populations from regions where malaria is endemic. Patients who experienced more than one *P. vivax* attack within 4 months had considerably less infectivity for mosquitoes than did those experiencing their first attack, suggesting the induction of transmission-blocking immunity by a prior infection (383). Moreover, antibodies to both Pfs230 and Pfs48/45 have been demonstrated in the sera of humans living in areas where *P. falciparum* is endemic, and the presence of these antibodies correlates with transmission-blocking activity in membrane feeding assays (180, 196).

In spite of the manifest potential of the gamete surface antigens of the Pfs230 and Pfs48/45 type as TBVs, the most advanced of the TBV candidates are the postfertilization antigens. Identification of the target antigens of antigamete immunity was pursued by using monoclonal antibodies developed against purified *P. gallinaceum* ookinetes as immunogens. These monoclonal antibodies acted to block transmission (198) and behaved as would be expected, i.e., acting well after the fertilization event. The effect was mediated by antibodies against a protein with a molecular weight of 26,000 (named Pgs26), one of two de novo ookinete surface proteins—the other being a protein with a molecular weight of 28,000 (75, 281). These proteins are now known respectively as the Pfs25 and Pfs28 classes of ookinete antigens; they have been found in all species of malaria parasites investigated. The Pfs25 protein, the *P. falciparum* equivalent of the Pgs26 ookinete protein, has become the basis of the only advanced human malaria TBV. In 2005 the WHO portfolio of candidate vaccines listed five vaccines in trial. The December 2006 WHO portfolio lists two TBVs: Pvs25 and Pfs25, both made in yeast, with the former designed to block transmission of *P. vivax*. In 2007, the WHO portfolio of vaccine candidates lists only Pvs28 as a DNA construct for an early Phase 1 clinical trial.

David C. Kaslow (born in 1958) grew up in Solvang, CA, and received a B.S. in biochemistry from the University of California, Davis, in 1979 and an M.D. from the University of California, San Francisco, in 1983. This was followed by a fellowship in human genetics in the Division of Pediatric Genetics at the Johns Hopkins School of Medicine. He joined the Laboratory of Parasitic Diseases at the NIH in 1986, where he worked on (among other things) the identification and characterization of transmission-blocking antigens (266–268). In 1986, a year after the transmission-blocking antigen Pfs25 was identified, Kaslow set his sights on producing a recombinant protein. Since expression in *E. coli* had failed, he used conventional biochemical methods to isolate the protein and then used this protein to obtain three fragments. In 1988 he isolated the first gene encoding a target antigen of malaria transmission-blocking immunity, the one for the *P. falciparum* zygote/ookinete surface protein, Pfs25. He isolated the gene and sequenced it by using degenerate nucleotides (based on the amino acid sequence of the three fragments) to screen genomic libraries. However, after failing to create a recombinant plasmid that encoded full-length Pfs25, he settled for a truncated version. Rabbits and mice vaccinated with the fusion protein produced an antiserum that recognized the native Pfs25; however, none blocked transmission. Undeterred, he made a recombinant antigen encoding the full-length Pfs25 by using the WR strain of vaccinia virus, vSIDK, and produced an antiserum. Transmission-blocking antibodies recognized the surface of the vSIDK-infected cells expressing Pfs25; however, production of potent transmission-blocking antibodies took three vaccinations of mice. Although Kaslow had demonstrated a "proof of principle" that a TBV was possible by piggybacking a malaria gene packaged in a virus, the project was abandoned because the virulent WR strain cannot be used in immunocompromised humans and some of the target population for malaria TBV have a high prevalence of HIV positivity.

Much of Kaslow's work with TBVs was done in collaboration with private industry. The success with the WR strain of vaccinia virus prompted an effort to develop an attenuated strain of recombinant vaccinia virus (NYVAC-Pf7). Although Pfs25 was expressed, the antibodies produced did not reproducibly block infectivity to mosquitoes. Kaslow also pursued the expression of recombinant Pfs25 in *S. cerevisiae*, baculovirus-infected insect cells, and mammalian cells; however, only in yeast were sufficient quantities of purified protein (Pfs25-B) made. Promising studies with primates, using the adjuvant alum (suitable for use in humans), were not vigorously pursued after a mutual decision by the biotechnology company and the funding agencies to move the product to Good Manufacturing Process (GMP) production, and further development of Pfs25-B was eventually abandoned entirely. Redesign of the product, however, produced a new product, TBV25H, which has been used in two clinical trials. In one, the Pfs25 was immunogenic but transmission-blocking activity was not observed. In the other, TBV25H was used to vaccinate human volunteers three times and was found to be immunogenic; however, because some of the volunteers had adverse reactions, its safety remains to be determined. The most promising TBV has involved vaccination of animals three times with NYVAC-Pf7 followed by further boosting with TBV25H plus alum. This regimen was tested in humans. A second-generation yeast-produced vaccine, TBV25-28, consisting of a fusion protein of the two sexual stage antigens, Pfs25 and Pfs28, appears promising in mouse studies: fewer doses, less antigen, and a longer immune response. Additionally, work on the analogous protein targets from *P. vivax* has led to the production of Pvs25 and Pvs28.

In 1999, Kaslow left the NIH for Merck, where he was Senior Director of Vaccine Research and then Head of the Department of Vaccine Research and Technology. In 2001 he joined Vical Inc.; in 2006, however, he returned to Merck as Vice President of the Division of Infectious Diseases, where he continues to direct efforts on development of vaccines and medicines to prevent and treat infectious disease. Kaslow is the author or coauthor of over 150 scientific publications and over 25 review articles and book chapters, and he holds or coholds more than 15 patents or patent applications.

When it was discovered that 99% of TBV25H did not have the proper shape and hence had reduced antigenicity, a new and improved version of TBV25H, now called Pfs25H, was prepared using the yeast *Pichia pastoris* (573). GMP at WRAIR manufactured 0.5 g of Pfs25H-A, and safety trials were carried out in 2003. The poor immunogenic properties of Pfs25, even with adjuvant, prompted another approach. By linking 2 to 10 molecules of Pfs25 and then adsorbing the conjugate onto alum, long-lasting transmission-blocking activity of antibodies was achieved (279). Indeed, mice immunized with this conjugate had antibodies 3 to 7 months after the second and third injections.

Rabbit and monkey sera against Pfs25 and Pvs25 emulsified in alum or Montanide ISA 720 induced transmission-blocking activity, and the degree of blocking was related to the antibody titer (340); however, a Phase 1 (safety) trial in which the vaccine was formulated with Montanide ISA

151 produced an unexpectedly severe local reaction, making such a formulation unacceptable for human use (561).

Mice immunized with Pfs25 as a nasal preparation that included the mucosal adjuvant cholera toxin produced a robust antibody response, and the antibodies recognized ookinetes and interfered with oocyst development in mosquitoes (16). Although these studies show that nasal vaccination may be possible, cholera toxin cannot be used clinically due to its toxicity and potential hazardous effect on olfactory nerves; therefore, although cholera toxin serves as a model for the effectiveness of mucosal vaccines, an alternative adjuvant will have to be developed before such a strategy can be employed with humans. A DNA construct encoding Pfs25 was used to vaccinate mice and also produced potent antibodies capable of reducing oocyst numbers by 97% and decreased the number of infected mosquitoes by 75% (295); however, no safety trials have been conducted with human volunteers.

TBV or No TBV, That Is the Question

Those working on a TBV cite as the major strengths of the lead candidates Pfs25 and Pfs28 to be the following: (i) they are present in both *P. falciparum* and *P. vivax*, (ii) they act synergistically in inducing transmission-blocking immunity, and (iii) vaccination of humans induces the correct immune response. However, as David Kaslow observes (268), the development of an effective TBV still has to surmount several obstacles including vaccine composition, formulation, administration, and deployment as well as commitment. How predictive is the membrane feeding assay of in vivo transmission-blocking activity, and, even more important, what is the association between reduction in morbidity and mortality as a result of blocking transmission as measured by the standard and cumbersome membrane feeding assay? Can methods be developed that will faithfully re-create the natural conformation of the antigen in bulk amounts? Can a stable TBV vaccine be produced? What adjuvants can be used for producing a robust immune response? What should the route of vaccination be, and what should the vaccination schedule be? Do these TBVs have to first be tested in monkeys before being used in human trials? There is a need to retest vaccine formulations that appear to be "safe" in malaria-naive populations in Phase 1 trials involving previously infected volunteers before there is a final decision to proceed with Phase 2 studies. The decision on site selection for Phase 2b studies can be complex. The site must allow for an accurate assessment of transmissibility of the parasite at the site, including measures of the rate of infection, rates of individuals with gametocytes, prevalence of preexisting transmission-blocking immunity in the target population, and determination of a suitable sample size.

In sub-Saharan Africa and in other tropical places, drainage, brush and grass cutting, and oiling and spraying of Paris green to kill larvae cannot be a practical tactic to block transmission of malaria by mosquitoes. Another strategy for blocking transmission before there is a TBV has been the use of bed nets (mosquito nets). The impetus for using bed nets stemmed from a desire to have uninterrupted sleep rather than as an attempt at preventive medicine (294). The earliest recorded use of bed nets was in the sixth century BC in the Middle East; the Assyrian General Holofernes was reputed to have used a purple bed net interwoven with gold, emeralds, and precious stones, and Cleopatra had one made of gold cloth. The Sepik peoples of New Guinea slept in large woven baskets to avoid mosquitoes. According to the historian Herodotus, Egyptian fishermen slept swathed in their nets, and so did Punjabi fishermen according to Ronald Ross. The use of bed nets in the Mediterranean spread from Egypt to the mosquito-ridden ancient civilizations of Rome and Greece. In Greece, bed nets were appropriately known as canopies from the Greek word "kunopestis" meaning "venomous insect," and in modern Greek, "kanops" means "gnat" or "mosquito." The British living in the tropics were passionate users of bed nets; however, their use predates by more than 70 years Patrick Manson's theory that mosquitoes transmit malaria. Despite their popularity, bed nets can present problems: they must be properly closed and must not have holes.

The use of insecticide-impregnated bed nets arose independently in many places: in the 1930s, they were used in Russia and by American and German forces during World War II. Insecticide impregnation greatly enhances the protective efficacy of bed nets; and the insecticide has repellent properties that reduce the number of mosquitoes entering the house. If high community coverage is achieved, the numbers and longevity of the mosquitoes are reduced. When this occurs, all members of the community benefit; however, for this to occur there must be high community coverage. Recently, there has been revived interest in the use of insecticide-treated bed nets as a strategy for controlling malaria by reducing the contact between humans and mosquitoes.

Beginning in 1980, bed nets impregnated with the insecticide permethrin were distributed in some African villages with populations of more than 400 through the Primary Health Care Programme. This has significantly reduced mortality and morbidity in Gambian children. Following this success, a National Impregnated Bednet Programme was initiated in 1992 with the objective of distributing bed nets to all villages in The Gambia over a 2-year period and establishing a system of cost recovery. However, once charges were introduced, usage dropped from 80% to 14% (343). Although impregnated bed nets are able to prevent a larger absolute number of malaria deaths than would a vaccine that had an efficacy of 48%, they do this at a considerable cost.

In areas where R_0 is less than 10, TBV coverage of approximately 90% would be needed to completely break the cycle of transmission. When R_0 exceeds 10, the TBV coverage required rapidly approaches 100%. Achieving a TBV coverage of 90% is possible; however, in areas where malaria is endemic, the requirement for 100% coverage is more realistic. Nevertheless, breaking the cycle of transmission in one small focus at a time, whether through TBV or impregnated bed nets, would still contribute to overall case reduction via herd immunity (485). What about costs? The estimates for costs per death averted by using bed nets averaged ~$700 compared to ~$250 for a TBV (195, 291). Clearly, a TBV would be of considerable benefit to those suffering with malaria. For the present, however, it remains an elusive goal.

10

Battling Blood Stages

Rome, Italy, 29 September 1890. The patient, a farmer, lived in the Roman Campagna; when brought to the Santo Spirito Hospital, he was feverish and near death. At the beginning of the fever paroxysm, a drop of blood was taken from the tip of the finger, smeared out on a microscope slide, and stained. When the smear was examined, it showed crescents and small ring stage parasites. Two days or so after the ring forms had been seen, they began to grow, and then these developing forms disappeared from the circulation. Where did they go? Observation of samples of tissues showed adult forms of the parasite, forms that radically altered the red blood cells. These altered cells, trapped in the small blood vessels deep in the body, were clumped together, sticking to the walls of the vessels, blocking and slowing the movement of the blood. All this was occurring while, inside the red blood cell, the malarial parasite was doing what parasites do so well—growing and reproducing. The disappearance of the more mature parasites from the bloodstream (and thus their absence from a drop of blood taken from the finger) was given the name "sequestration" by the attending physicians, Amico Bignami (1862 to 1919) and Giuseppe Bastianelli (1862 to 1959); they remarked that the phenomenon was a signal characteristic of *P. falciparum* infections (44).

What these Italian clinicians discovered more than a century ago was that the red blood cells containing the ever-enlarging parasites could clog the blood vessels, depriving that part of the body of oxygen, and, as was the case with their patient, when the congestion was massive enough in the brain it led to coma and death. It is this special propensity of *P. falciparum*-infected red cells to develop and deprive the inner organs of oxygen that makes it one of the world's greatest killers.

Although this observation by Bignami and Bastianelli explained why the large, growing asexual stages were absent from a drop of blood taken from the fingertip whereas the young, nonpigmented ring stage parasites circulated freely in the bloodstream, it did not reveal the mechanisms for such preferential binding. Almost a century later, when more powerful microscopes able to magnify objects 100,000-fold became available, the differences on the surface of the *P. falciparum*-infected red blood cell would be seen, providing clues to the reason for their stickiness. Examination under the electron microscope showed that red blood cells with mature asexual stage parasites lose their normal biconcave disc shape, become distorted, and are covered with minute pimples called knobs (4, 199) whereas red cells infected with ring stage parasites retain a smooth, pimple-free surface (i.e., they are knobless).

Knobs, first detected when using blood samples from patients from an area where malaria was endemic (516), were also found in *P. falciparum*-infected *Aotus* monkeys (327), where the parasite attached to the cells lining the small blood vessels via the knobs. Additionally, when *P. falciparum*-infected red blood cells were exposed to immune serum and only the knobs were recognized (287), it became clear that the knobs were "nonself"; that is, they were antigenically distinct. These early "picture studies" led to a search for the nonself antigens that were the "glue" of the knob. Moreover, there was a stated practical goal: when the glue was identified and characterized, it could be the basis for a vaccine that would prevent sequestration.

Finding the Glue of Sequestration

Identification of the surface antigens of *P. falciparum* can be traced to studies carried out by Monroe Eaton in the 1930s. Eaton showed that serum from monkeys infected with *P. knowlesi* could agglutinate schizont-infected red cells (see p. 115). This schizont-infected cell agglutination (SICA) reaction was a clear indication that the surface of the malaria-infected red cell had been altered. Since the SICA reaction was species and stage specific, it implied (but did not prove) that the reaction was due to the presence of a parasite antigen on the surface of the red blood cell. Eaton went on to show that over time there was a sequential expression of SICA, with new variants each having a different antigen. Thirty years later, when Neil and Ivor Brown at NIMR explored this phenomenon of antigenic variation (see p. 116), they were able to show that relapse parasites from a single isolate differed in their SICA antigens (60). Since Eaton and the Browns had worked with a mixed population of parasites, it was conceivable that these findings were the result of immune selection of a distinct antigenic subpopulation. However, when these experiments were repeated (31) with a purebred line of *P. knowlesi* (derived from a single infected red cell and hence a clone), the results proved that the parasite had the genetic capacity to express all the antigenic variants. Biochemical and immunological analyses of the SICA antigens showed they were polymorphic (literally, "many forms") parasite proteins that differed in size and shape. The SICA antigens could be "stripped" from the infected red blood cell's outer surface by treatment with the protein-digesting enzyme trypsin and could be surface tagged with radioactive iodine (240–242). Serial passage of the cloned parasites in a monkey without a spleen gave rise to lines that no longer expressed SICA; however, on transfer back into a spleen-intact animal, the antigen reappeared (32).

It was only natural after the identification of the *P. knowlesi* SICA antigens to see whether similar antigens were present on the surface of *P. falciparum*-infected red blood cells and whether they played a role in sequestration and immune evasion. By tagging the red blood cell surface with radioactive iodine and precipitating with isolate-specific immune sera, a consistent protein named PfEMP1 (*Plasmodium falciparum* erythrocyte membrane protein 1) was identified (290). PfEMP1, like SICA, is an antigen that can be removed from the surface of *P. falciparum*-infected red cells by exposure to low concentrations of trypsin, and after the stripping process the infected red cells lose their stickiness. That is, trypsin removed the "glue." Sera taken from people infected with one malaria strain precipitated PfEMP1 from the same strain but not from others. These sera also blocked binding to cells from blood vessels in a strain-specific way. The conclusion was inescapable: the parasite made a molecule, PfEMP1, responsible for both attachment to blood vessels and antigenic variation; i.e., it was the strain-specific glue. Significantly, PfEMP1 allows the "leopard to change its spots" and permits the malaria parasite to evade attack by the immune system.

For almost a decade, the PfEMP1 protein itself could not be isolated (presumably because of its low abundance). Indeed, because of this I boldly (and unwisely) questioned the very existence of PfEMP1 (447); however, in 1995 the gene for PfEMP1, named *var* (for "variant"), was cloned and sequenced independently in three laboratories (36, 471, 488). Today, we know the PfEMP1 proteins are coded for by ~60 *var* (variant) genes (282). The *var* genes are localized to the ends of all 14 chromosomes of *P. falciparum*, are extremely divergent in their DNA sequence, and theoretically could code for very large proteins (~200,000 to 500,000 molecular weight).

Through switching to different forms of PfEMP1, i.e., "changing its spots," *P. falciparum* is able to evade the host immune system, and because PfEMP1 acts as the glue of sequestration, passage through the spleen—the graveyard of the red blood cell—is avoided. It has been claimed that antibodies to PfEMP1 are responsible for regulating the parasite density in the blood such that there is a concomitant decrease in the severity of disease, leading to chronic infection and nonsterilizing immunity. Immunity to PfEMP1 may have other effects: (i) it can influence mosquito transmission by regulating the numbers of asexual blood stages with the potential to become transmission stages (gametocytes) and (ii) by directly targeting early gametocytes it could prevent their maturation.

Sequestration and immune evasion are linked to the finding that PfEMP1 has three different adhesive domains: DBL (Duffy ligand binding), a CIDR (cysteine-rich interdomain region), and an acidic region, called the ATS (acidic terminal sequence), that is presumed to anchor the molecule to the red cell surface. The DBLs are grouped into five different varieties, α, β, γ, δ, and ϵ, and the CIDR region can be divided into three different types, α, β, and γ. Although there is less than 50% identity in amino acid composition between the various DBL domains, the DBLα domain has the highest degree of sequence conservation among all domains. It has therefore been suggested as a possible vaccine candidate to act against adhesion-related processes of rosetting, pregnancy malaria (see p. 221), and sequestration in the brain, leading to cerebral malaria.

The association of acquired immunity with immunological recognition of PfEMP1 by people living in areas where malaria is endemic has provoked some investigators to argue for its inclusion in a malaria vaccine. The objective of a PfEMP1-based vaccine would be to not eliminate the parasite entirely but instead to produce a low-grade and asymptomatic blood infection that is able to boost broad protection. However, since PfEMP1 is involved in the change from one variant to another, this challenges the developer of a vaccine to make the appropriate choice of a conserved antigen. Several proof-of-principle studies with animal models have provided a modicum of encouragement that such an approach may be feasible. The first immunization study utilized the conserved CIDRα domain responsible for binding to the molecule CD36, which is present on the surface of cells that line the small blood vessels (34). *Aotus nancymai* monkeys were vaccinated with a protein fragment that contained 179 amino acids and was produced in yeast; for the vaccination the protein was emulsified in FCA or MF59, an adjuvant used for influenza vaccination.

Monkeys vaccinated with the protein in FCA "demonstrated a very high level of protection" (i.e., the blood infections were not severe enough to require chloroquine treatment). Two monkeys never developed any blood infection, and four others had fewer parasites in the blood than did the controls. Those vaccinated with the protein in MF59 had variable protection, with more severe blood infections than the FCA-immunized monkeys. It was suggested that this vaccine resulted in antibodies that blocked adhesion, agglutination, and/or opsonization of infected red blood cells and that this led to a low-grade chronic infection; there was also some evidence for variant-transcending immunity with the same strain but not with a different strain expressing another variant.

In follow-up studies, when plasmid DNAs encoding several CIDRα variants (35) or DNA vaccination followed by a boost with three CIDRα recombinant proteins (193) or chimeric proteins produced by shuffling of PfEMP1 genes (296) were used to vaccinate mice, cross-reactive antibodies developed. This again suggested the possibility for developing a vaccine able to induce variant-transcending immunity; however, whether this regimen will actually protect *Aotus* (or humans) against challenge remains unknown. Further, it must be borne in mind that because the human responses to DNA vaccines are so different from those of mice, the findings may have little practical relevance for developing a PfEMP1-based vaccine.

Since its discovery, PfEMP1 has been considered an attractive vaccine candidate. Indeed, it has been an especially tantalizing target because it is exposed on the surface of the infected red blood cell for ~20 h and correlations have been found between antibody recognition of it and clinical protection against severe malaria. Based on this, it has been hypothesized that vaccination with PfEMP1 will promote an early buildup of immunity to the most virulent parasites in children and significantly reduce malaria-related deaths. However, based on the lack of protection by recombinant proteins and DNAs, the persistent argument for inclusion of PfEMP1 as a component of a protective vaccine rings hollow. The selection of the "correct" antigen that mediates sequestration in cerebral malaria has not been identified despite decades of intensive effort. In consequence, the focus of most recent research on PfEMP1 has been on the mechanisms of switching of antigens (422). This would allow a better appreciation of the means of immune evasion as well as the manner by which it is trafficked from its place of synthesis within the parasite to the outer surface of the infected red cell (308).

Pregnancy Malaria

Each year over 50 million women are exposed to the risk of malaria during pregnancy. Pregnancy malaria (PM) results in substantial maternal and especially fetal and infant morbidity and mortality, causing 75,000 to 200,000 infant deaths annually. Pregnant women are more susceptible to malaria than are nonpregnant women, and this susceptibility is greatest during the first and second pregnancies. The reduced susceptibility to PM in women over several pregnancies suggests that they are able to develop immunity to PM after infection (178). Further, women develop broadly reactive antibodies to infected red blood cells isolated from the placenta (150, 151).

Michal Fried (born in 1960) grew up in Israel, received her B.Sc. and M.Sc. from Ben Gurion University of the Negev (in 1984 and 1987, respectively), and then studied the sexual stage antigens of the coccidian parasite *Eimeria maxima* for her Ph.D. at the Hebrew University in Jerusalem, which she received in 1991. Upon completion of her doctorate, she wanted to continue studying parasitic diseases that impact human health, particularly malaria. After spending 3 years at NIH studying malaria (with David Kaslow, and where she met Patrick Duffy) she had the opportunity to visit and carry out research in Kenya (dreams do come true!). In the process of designing a study on malaria pathogenesis, she read articles on the impact of PM on both the mother and her baby, and around that time she heard a lecture by Robert Desowitz describing the outcomes associated with PM. Despite the large number of publications describing the outcomes associated with this syndrome, it remained unclear why pregnant women were more susceptible to malaria. In the belief that a better understanding of the disease might help in the development of interventions, she set her sights on uncovering the molecular basis of the syndrome.

Parasite sequestration in the placenta had been described by a number of studies more than 100 years ago; however, only in 1958 did Cannon note the association between parity and the frequency of malaria infection during pregnancy. In succeeding years, other studies confirmed these observations. It was assumed that the increased susceptibility to PM was due to pregnancy-related immunosuppression; however, this did not explain the differences seen between the first and later pregnancies. An alternative hypothesis to this model was developed as a result of a study begun by Fried in 1995 with her colleague Patrick Duffy; i.e., the placenta presents new receptors for parasite adhesion. This landmark study described the unique binding of placental parasites to chondroitin sulfate A (CSA), a molecule expressed on the surface of placental cells. Because CSA is not readily accessible for parasite adhesion in the nonpregnant female, primigravid women are naive to this parasite subpopulation. In subsequent years, this work was extended to show that with successive pregnancies, women develop immunity to placental parasites.

The possibility of a specific vaccine for pregnant women was unthinkable prior to the discovery of humoral immunity to placental parasites able to block parasite adhesion to CSA, leading to reduced infection rates and lower parasite densities. However, over the past decade all that has changed. There is now not only increased funding (especially by the Bill & Melinda Gates Foundation) but also an increased awareness of the problem by governmental organizations.

Fried thinks that the key ingredients to developing a PM vaccine are a better understanding of disease pathogenesis and identification of the protective immune responses rather than a simple association between levels of antibody to specific antigens and a reduction in disease severity. The latter may indicate only exposure or experience, not necessarily protective immune responses. Guided by these factors, together with the availability of advanced technologies as well as funding, she is optimistic that a PM vaccine will be developed. Currently Fried and Duffy, as well as several other groups, are working on developing such a vaccine.

As Fried and Duffy have shown, sequestration of *P. falciparum*-infected erythrocytes in the placenta is central to the pathogenesis of PM (150, 151). Unlike sequestration in the deep tissues of the body, the parasitized erythrocytes from the placenta bind predominantly to CSA whereas infected red cells from the peripheral blood of pregnant and nonpregnant women bind to other receptors, primarily the molecule CD36. Such findings suggest either that the placenta secretes a factor or that serum-specific factors are present only during pregnancy and "turn on" parasite genes that allow binding to the placenta. The placenta, in turn, has a unique set of high-affinity adhesion receptors for binding the rare CSA-binding malaria-infected red blood cells (349), and these preferentially transcribe the *var2csa* gene (307, 416), encoding a member of the PfEMP1 family (see above). The *var2csa* gene encodes DBL domains that bind CSA in vitro, and in Malawian women *var2csa* predominates (148, 149). An interesting observation is that CSA-binding lines are phagocytized at significantly lower levels than are CD36-binding lines and when the *var2csa* gene is knocked out the red cells have impaired placental adhesion (430, 532).

The distinct pattern of surface protein expression on placental *P. falciparum*-infected erythrocytes (177) and the fact that women can develop immunity to PM after infection continue to fuel optimism that a CSA-binding ligand might be a prime vaccine candidate for therapeutic and preventive interventions (21, 254). However, tempering this enthusiasm for a vaccine is the finding that in fresh placental isolates there may be binding to several receptors (IgM, IgG, and hyaluronic acid) in addition to CSA (385). Since there is so much diversity of DBL sequences, even in individuals living in a single isolated locale, it may be difficult to produce a protective vaccine by using a single *var2csa* DBL sequence as the immunogen (114).

Still, all may not be lost for a vaccine to prevent PM. There are approximately 5,300 *P. falciparum* genes that theoretically could code for a like number of proteins. However, 2,938 proteins were detected in placental isolates and 15 did not bind to CSA; 7 proteins were exclusively associated with placental parasites (177). The functional role of these proteins on the surface of placental infected erythrocytes is unknown; however, some may act in concert with VAR2CSA to promote sequestration and, once identified, may serve as vaccine candidates.

Once such vaccine candidates are in hand, PM could be amenable to vaccination by specifically targeting the parasitized red cells sequestered in the placenta (401). Such a vaccine could be given to young women before their first pregnancy and would be aimed at inducing high levels of the PM-specific antigens generally seen in women who have become resistant to PM after successive pregnancies. Ideally, such a vaccine would result in protection against maternal anemia, preterm delivery, and fetal growth restriction but would leave unaffected the parasites that do not express the non-pregnancy-associated malaria PfEMP1 and thus would be of little benefit to the child.

What is the cause of PM? A recent study found that B-cell-related genes were among the most active in the placenta (341). In addition, a B-cell chemoattractant and a B-cell-activating factor were detected in the placental macrophages. Heavy deposits of IgG and IgM were also observed in the placenta. Birth weight was inversely correlated with increased amounts of these immunoglobulins. Taken together, these findings suggest that B cells and macrophages contribute to chronic PM, and it may be possible to abrogate PM with a vaccine or with drugs.

Rosetting

In the 1980s it was discovered that *P. falciparum*-infected red cells were able to bind not only to the cells lining the small blood vessels but also to uninfected red cells, forming clumps of cells called rosettes (538). Rosetting has also been observed in *P. vivax*, *P. ovale*, *P. malariae*, *P. coatneyi* in the rhesus monkey, and *P. fragile* in the toque monkey. Rosetting in some of these hosts may be benign, but in others, i.e., humans infected with *P. falciparum*, it may contribute to pathogenesis leading to severe malaria (405). However, even with this malaria there may be differences: rosetting was associated with severe malaria in some parts of Africa but not in Papua New Guinea (93).

A rosetting model in the rodent malaria parasite *P. chabaudi* (306) showed that rosetting increased with parasite maturity and could be reversed by exposure to trypsin, suggesting that rosetting was probably due to a surface protein made by the parasite. Although rosetting in *P. chabaudi* may seem to be analogous to that in *P. falciparum*, no *var* genes encoding the PfEMP1 family of proteins have been detected in any rodent malaria parasites (260). Therefore, it is highly unlikely that this rosetting protein is identical to PfEMP1 (171), and this provides another example of the differences between mouse and human malarias.

The substances to which the rosetting red blood cells bind are heparan sulfate, blood group antigens, and complement receptor 1 (405). Stable rosettes require serum whose essential components are mimicked by complement factor D, albumin, and anti-band 3 IgG antibodies (300). The A and B blood group antigens are thought to be responsible for rosetting on uninfected red cells, and although *P. falciparum* rosettes can form in group O cells, they tend to be smaller; therefore, it has been hypothesized that there is a protective effect of blood group O and that this operates through reduced rosetting (406). It has been proposed that "rosette-inhibiting drugs or vaccines may have the potential to save many patients' lives" (405). Presently, such drugs and a vaccine are not available.

Studies with a rosetting line of *P. falciparum* suggested that "CIDR1α has an affinity for a structure present in members of the IgG family i.e. IgM but [there is also] a separate binding site for CD36" (82). Immunization with DBL1α generated functional antibodies able to disrupt rosettes and to block adhesion in a rat model (83), and immunization with either one or three DBL1α domains showed a reduction in adhesion in four of five monkeys. Could a vaccine incorporating DBL1α block rosetting and reduce pathogenesis? We simply don't know.

The past 25 years has witnessed an explosion of new research in identifying the erythrocyte surface proteins of *P. falciparum*-infected red cells as well as in defining their role in antigenic variation, adherence, and pathogenesis. This has been achieved largely due to the availability of entire DNA sequences and the development of gene cloning, recombinant DNA, monoclonal antibodies, and other sophisticated molecular tools able to knock out specific genes. Yet there is much work ahead before it is possible to prevent or reverse sequestration/rosetting and to circumvent pathogenesis. Understanding how signaling between host and parasite is accomplished is still beyond our ken, as are the molecular and biochemical mechanisms whereby variant-antigen switching occurs. Finally, it should be emphasized that until we have a clearer understanding of the manner by which immune adults manage to control their infections, there will be little prospect for developing a practical protective vaccine, and as a consequence we may have to rely on chemotherapeutic interventions for many years to come.

Blocking Merozoite Invasion

Entry into a red blood cell is a necessary and specific requisite for *Plasmodium*. As such, the molecular mechanisms of invasion have been considered the Achilles' heel that might be exploited for the development of new therapies or vaccines for malaria. Despite a half century of "invasion research," a practical and effective means for interrupting the process of entry into red blood cells has not been achieved. However, hope for novel interventions remains.

The electron microscope has revealed a complex of organelles at one end of the invasive stages of malaria parasites. In sporozoites these included "conoid bodies and paired apical bodies." Similar structures were seen under the electron microscope with erythrocytic merozoites (407). Although the invasion of red blood cells by merozoites was described in studies carried out at WRAIR in 1969, it took six more years and technical improvements in electron microscopy to provide a detailed description of the process by using *P. knowlesi* (28). Since other electron microscopy studies have shown similar structural characteristics for erythrocytic merozoites, it has been assumed that the steps in invasion are similar for all species of malaria parasites including *P. falciparum* and *P. vivax*.

At the pointed tip of the *Plasmodium* merozoite and within its cytoplasm is a pair of membrane-bound pear-shaped and ducted structures called rhoptries; more numerous (up to 40) fusiform bodies that converge around the ends of the rhoptry ducts, called micronemes; and spheroidal membranous vesicles, the dense granules. The surface of the merozoite is covered with an electron-dense fibrillar coat. Primary attachment of the merozoite occurs at any point on the merozoite surface, and then reorientation occurs so that the tip is juxtaposed to the red blood cell surface. At the point of contact with the red cell, an electron-dense area is seen beneath the red cell membrane. The electron-dense area, termed a tight junction, forms a circumferential ring around the merozoite. (The components and constituents of the junction are not known.) During invasion the rhoptries discharge their contents, the ends of the ducts fuse with each other and the merozoite plasma membrane, and then they collapse. The micronemes also release their contents. Concomitant with rhoptry-microneme discharge, the tight junction moves from the front end of the merozoite to the back end, powered by its actin-myosin motor, and in the process the merozoite's fuzzy coat is shed (26, 211). As the tight junction moves backward, the molecules mediating invasion (see below) are removed by one or more enzymes (111). With the discharge of the contents of the rhoptries, the red cell membrane begins to expand, allowing the merozoite to induce an invagination; when this is sealed, it forms an enclosing membrane separating the parasite from the red cell cytoplasm. (Think of this as a finger being poked into a semi-inflated balloon—the balloon being the red cell, the finger being the merozoite, and the rubber of the balloon around the finger being the membrane.) The entire process of merozoite entry into a red blood cell is quick: it takes less than 30 s.

Electron microscopy studies suggested that there is a merozoite "glue" that binds to something on the surface of a red cell, that is, an attachment site or receptor. When I arrived in William Trager's Rockefeller University laboratory, I was told of the discriminating taste of P. lophurae merozoites for duckling red cells, as discovered by his former associate Barclay McGhee. However, since the identity of the duckling red blood cell receptor was still unknown, I was encouraged to identify it. I used a glass flask culture system that had been developed by Trager. Following McGhee's lead, attempts were made to determine whether enzyme treatment of duckling red cells could remove specific surface receptors and thus prevent P. lophurae invasion (437). None of the enzymes (i.e., trypsin, chymotrypsin, neuraminidase) I tried had any effect. This lack of success led to my abandoning further research on merozoite invasion. However, I learned an important lesson from this futile experience: success can depend on the choice of a particular experimental system. Fortunately, there were other investigators who were not so faint of heart and who chose more wisely.

In 1973, Geoffrey Butcher, Graham Mitchell, and Sydney Cohen (with support from the MRC and WHO) used an in vitro short-term culture system to study invasion and found that red cells from Old World monkeys (kra and rhesus monkeys) were susceptible to *P. knowlesi* merozoites whereas New World monkeys (*Aotus*) and human red cells were less susceptible and nonprimates were completely resistant (67). Although these workers were convinced that these results were due to differences in erythrocyte susceptibility and the basis of host cell specificity, they did not identify the red cell receptor. However, in a series of what are now considered to be classic studies, Louis Miller and coworkers (330, 331) used enzyme "stripping" of red cells (the method of McGhee and Sherman) to define the chemical nature of an erythrocyte membrane receptor. When the surface of human red cells was treated with the protein-digesting enzyme chymotrypsin or pronase, invasion of *P. knowlesi* merozoites was blocked, whereas other enzymes (neuraminidase and trypsin) were ineffective. However, rhesus red cells treated with chymotrypsin remained susceptible. It was suggested that the differences in invasion between the two kinds of red cells were due to the "high affinity between *P. knowlesi* merozoites and [rhesus] monkey erythrocytes [which] may require greater alteration of the receptor to inhibit invasion" (329). This interpretation would subsequently turn out to be incorrect; however, the study did allow the formulation of a hypothesis for invasion (328) and eventually led to the identification of a human red cell receptor for the merozoite (330).

Because human red cells are only one-fourth as susceptible to *P. knowlesi* invasion as are rhesus red cells, enzyme stripping was effective in removing a receptor; by contrast, my failure to find a receptor for *P. lophurae* merozoites was probably a result of the much greater susceptibility of duckling red cells to *P. lophurae* merozoites so that enzyme stripping failed to remove the receptor. Perhaps if McGhee and I had used a less susceptible cell for stripping, a receptor on the red blood cell might have been identified a decade earlier.

Miller and coworkers took advantage of the availability of human red cells that differed genetically in their surface molecules (something lacking for duckling erythrocytes) to show that *P. knowlesi* merozoites could invade human red cells that have the Duffy factor antigen (called Duffy positive) on their surface but that human red cells without Duffy antigen were refractory (331). Then, using enzyme stripping (329), they found that when Duffy antigen was removed from human red cells by chymotrypsin (but not trypsin or neuraminidase) treatment the erythrocytes became Duffy negative and resisted invasion by *P. knowlesi* merozoites. These laboratory studies were extrapolated by Miller to explain the absence of *P. vivax* in West Africa (where most of the population is Duffy negative). Indirect evidence in support of Miller's proposal came from the following observations. (i) West Africans and black Africans who are 95% and 70% Duffy negative, respectively, are resistant to *P. vivax*. (ii) Duffy-negative volunteers did not become infected when bitten by mosquitoes carrying *P. vivax*. (iii) Of 13 black Americans who had *P. vivax* malaria in Vietnam, all were Duffy positive (330). Final and direct proof of Miller's postulate came when a short-term in vitro culture system for *P. vivax* showed that these parasites could not invade Duffy-negative red cells (33).

Louis H. Miller (born in 1935) received a B.S. degree from Haverford College (Pennsylvania) in 1956 and an M.D. from Washington University (St. Louis) in 1960. His interest in tropical medicine was piqued by reading the autobiography *Out of My Life and Thoughts* by Nobel laureate Albert Schweitzer. After an internship and residency in internal medicine at the Mt. Sinai Medical Center in New York (1960 to 1964), he spent a year at Cedars-Sinai Medical Center in Los Angeles as a renal metabolism fellow, receiving training that would stand him in good stead for later studies of renal pathology in human and experimental malarias. Returning to New York City, he worked with Harold Brown (Head of Parasitology and Tropical Medicine at Columbia University School of Public Health, who had spent most of his life working in tropical medicine), receiving an M.S. in parasitology in 1965. After being drafted into the U.S. Army as a Captain in the Medical Corps, he was sent to WRAIR's SEATO Laboratory in Bangkok, Thailand, where he intended to study tropical sprue; however, after encountering only a single case in 2 years, he felt it wise to change his career. At that time, chloroquine-resistant malaria was a significant problem for the troops fighting in Vietnam, and so Miller, the clinician, turned his attention from tropical sprue to malaria. Knowing little about malaria, he spent a month in the library in Bangkok learning all he could. He performed comparative studies of pathology and renal physiology in monkeys and rodents, as well as deep vascular schizogony and sequestration, with Robert Desowitz (who had been a student of the great malariologist Colonel Henry E. Shortt at the London School of Hygiene and Tropical Medicine). After completion of his military service, Miller joined the faculty at the Columbia University School of Public Health as an Assistant Professor, where he began to move further away from the clinical aspects of the disease and to focus on more basic questions that he felt would have an impact on malaria control. To follow up his work on sequestration, Miller contacted Craig Canfield, a friend from Bangkok and then Director of Experimental Therapeutics at WRAIR, to obtain two *Aotus* monkeys to be infected with *P. falciparum*. The result was the discovery of the presence of knobs on the surface of red cells in organs where sequestration occurs. In 1971 Miller, by then an Associate Professor, left Columbia University to establish a malaria laboratory at the NIH, where he eventually became Chief of the Laboratory of Parasitic Diseases.

Shortly after joining the NIH, Miller attended a conference organized by Elvio Sadun at WRAIR, where he met Sydney Cohen (from Guy's Hospital in London), who was carrying out short-term cultures of *P. knowlesi* to study invasion. This led to Miller's seminal studies of host cell specificity and the discovery that *P. knowlesi* was unable to invade Duffy-negative red cells. A visit to the library that night made him realize that he had found the missing factor for the resistance of West Africans to *P. vivax*. Later, he was able to show that *P. vivax* did not invade Duffy-negative red cells. For the next 20 years the biochemistry and molecular mechanisms of invasion became the major focus of his research. Strategies to block mosquito transmission and development of blood stage vaccines are the current focus of his research, and he is now the Chief of the Malaria Vaccine Development Branch (MVDB) at NIH. The MVDB is similar to a small biotech company that focuses on taking promising vaccine candidates from concept through scale-up. The MVDB is involved with antigen purification, quality control, immunological studies, and early phases of clinical trials at the Malaria Research and Training Center in Bamako, Mali. Currently, the leading vaccine products are the blood stage vaccines AMA-1 and FMP-1; the TBVs Pfs25, Pfs48, and Pvs25; and a pre-blood vaccine.

Human Duffy antigen is a protein with two sugar molecules; the sugars are removed when the cells are treated with neuraminidase. However, under natural conditions Duffy negativity is not due to an absence of the sugars but results from a gene mutation (cytosine to thymine) on the Duffy gene that does away with the binding site for a transcription factor. As a result, transcription of Duffy mRNA in red blood cells but not other cell types is abolished (508). Although the Duffy determinant is crucial for the entry of *P. vivax* merozoites (509), recent observations indicate that this pathway is not absolute; *P. vivax* infections were found in two Duffy-negative individuals from South America (78) and in four Duffy-negative individuals from East Africa (412). This suggests that *P. vivax* merozoites are able (albeit infrequently) to infect Duffy-negative red cells, and in this species, as in *P. falciparum*, alternate pathways for erythrocyte invasion do exist.

Contemporaneous with the findings of receptor heterogeneity, Miller also attempted to identify the "glue" of the merozoite for the sialic acid-containing glycophorin on the red cell surface (and which is removed by exposure to neuraminidase and chymotrypsin) by using classical biochemical methods (208). When Camus and Hadley (69), working at WRAIR, extracted merozoites and allowed the extract to react with immobilized glycophorin A, only one protein, with a molecular weight of 175,000, bound. This protein (named EBA-175, for "erythrocyte-binding antigen") was the first merozoite glue identified. When schizont-infected red cells were extracted and precipitated with serum from a Nigerian donor, only EBA-175 bound to normal human red cells but not to neuraminidase-treated ones (which have sialic acid residues of glycophorin removed). Adding further support for the notion that EBA-175 is the glue of *P. falciparum* merozoites was the demonstration that when antibodies were made against purified EBA-175, they prevented the invasion of red blood cells by *P. falciparum* merozoites by 90% (365).

EBA-175 is a micronemal protein encoded by a single-copy gene located on chromosome 7. Knockout of the EBA-175 gene in a strain whose invasion is sialic acid dependent was associated with a switch to a sialic acid-independent pathway of invasion (152). However, when the gene for EBA-175 was knocked out in another line that had been selected on neuraminidase-treated red cells, the knockouts invaded the neuraminidase-treated cells, as did the nonknockout. Such findings suggest that *P. falciparum* without EBA-175 can invade red cells (183). In field isolates from The Gambia and Brazil, the sialic acid-dependent pathway is predominant, whereas in India and Kenya neuraminidase- and trypsin-sensitive, chymotrypsin-resistant receptors (characteristic of EBA-175-mediated invasion) were rarely used (119).

Recent studies have shown that there are two Duffy-binding proteins (DBP), PvDBP and PkDBP, found in the merozoites of *P. vivax* and *P. knowlesi*, respectively (85). PvDBP specifically binds to the human Duffy blood group antigen, and when this occurs the merozoite can dock and invade. PkDBP, on the other hand, binds to both the human and rhesus Duffy blood group antigens. Since it is possible to produce large amounts of the recombinant receptor-binding domain of PvDBP (565) and since rabbit antibodies raised against a recombinant PkDBP inhibited the invasion of human and rhesus red cells by *P. knowlesi* merozoites, it has been suggested that PvDBP may serve as a possible vaccine candidate. Indeed, this potential for a vaccine has been supported by the finding of antibodies to PvDBP in children in Papua New Guinea in whom the levels of these antibodies remain stable for at least a year; the antibodies also prevent invasion of reticulocytes, and although they do not completely abrogate infection, the infections were less severe and infected children were less likely to develop clinical disease (275).

   P. vivax, unlike *P. falciparum*, preferentially invades young red blood cells, reticulocytes. Two *P. vivax* reticulocyte-binding proteins, PvRBP-1 and PvRBP-2, have been identified, and the genes for these have been cloned and sequenced. The RBPs, localized to the apical tip of the merozoite, have been postulated to form a complex that mediates adhesion and recognition of the reticulocyte independent of the Duffy antigen. Since there is no evidence for the Duffy-like glue being exposed at the surface prior to initiation and formation of a tight junction, it has been suggested that merozoite binding may be necessary to trigger (signal) the timely release of sequestered proteins so that a tight junction forms and entry can proceed (179). In this way, by acting prior to RBP binding and junction formation, i.e., during apical-end orientation, the *P. vivax* merozoite is able to target reticulocytes, which usually represent less than 1% of the total red blood cell population. The receptor molecule on the reticulocyte to which the PvRBP binds remains unknown. Although we do not know whether a vaccine based on RBP blocking invasion will reduce the parasite burden, such a possibility seems unlikely.

   Over the past 30 years a staggering array of molecules involved in invasion have been identified (111). In addition, some species such as *P. falciparum* and *P. knowlesi* (unlike *P. vivax*) use multiple invasion pathways. The use of multiple pathways may provide those species with a "survival advantage when faced with host immune responses or receptor heterogeneity in host populations" (85). Because of the heterogeneity of merozoite glues, exploiting specific invasion pathways as a vaccine target may be an unrealistic goal; however, other mechanisms involved in merozoite invasion (e.g., processing by protein-digesting enzymes) might be suitable targets for drug and other therapeutic interventions.

A Live Blood Stage Vaccine?

The pioneering work of Edward Jenner and Louis Pasteur demonstrated that dead or attenuated pathogens are able to act as effective vaccines (see p. 77–80 and 80–87). Indeed, whole-organism vaccines, both live-attenuated and killed, represent 75% of presently licensed formulations for bacterial and viral infections including polio, rabies, cholera, influenza, whooping cough, typhoid, plague, and diphtheria; this successful use attests to their safety and efficacy. For malaria, however, whole-parasite vaccines have been viewed as impractical due to the complexity of the parasite's life cycle as well as to logistical and safety concerns relating to manufacture. Ever since the classic failure of Heidelberger (see p. 108–109) to protect against infection by using killed *P. vivax*, there has been a tendency to shy away from the use of whole blood stage parasites. As a consequence, even with the possibility of producing large numbers of *P. falciparum*-infected red cells in the laboratory from Trager-Jensen cultures, the last 30 years has seen most of the effort devoted to developing subunit vaccines (recombinant, synthetic, or gene-based constructs) that would protect through immunization. However, since such strategies have not yet proved very successful, there has been a resurgence of interest in immunization using radiation-attenuated or genetically attenuated sporozoites (see p. 247 and 280) and blood stage parasites followed by antimalarials. In the latter case, a strong cellular immunity was induced that led to a delay in blood infection. Following this lead, Michael Good proposed and tested a strategy for vaccination using blood stages (188, 373).

In the study, naive human volunteers (376) were infected four times with ultralow doses of 30 *P. falciparum*-infected red cells at 5-week intervals and treated with atovaquone-proguanil from day 8 after the first three challenges; after the fourth challenge they were given chloroquine on day 14. (With this regimen there were four rounds of parasite multiplication and no clinical symptoms were observed.) After challenge there was a strong immune response, characterized by a proliferation of CD4 and CD T lymphocytes and IFN-γ production in the absence of antibody. Three of the four volunteers were fully protected. It was suggested (but not proven) that the immunity was cross-reactive.

Major issues still to be addressed (390) are the hazards involved in the administration to many thousands of healthy individuals of a blood product for prophylaxis that could contain other blood-borne pathogens;

the fact that the parasites multiply in the body and could result in significant harm, including anemia and fever; and the fact that the vaccine may
select incoming parasites that are antigenically variant. Unresolved is
whether the immunity was sterile, whether the immunized individuals
remained protected against a subsequent infection with a higher dose of
parasites, and if there was protection against sporozoite challenge.

History has shown that the most easily developed vaccines have
been those that induce natural immunity after a single infection. For
malaria the situation is different: clinical immunity develops only after
several years of endemic exposure. This is generally assumed to be the result of parasite antigenic variation, as well as the ability of the parasite to
interfere with the development of the immune response. Vaccine strategies that are likely to be successful are those that combine many antigens
to induce a maximal response to a protective determinant that might not
normally be recognized after infection of immunologically naive individuals. It has been suggested that an alternative to multivalent vaccines
would be to use attenuated parasites or ultra-low doses of whole parasites; however, significant obstacles to these "back to the future" approaches remain. In the end, it may be that the greatest value in a whole-
organism vaccine approach is that it could allow the identification of
target antigens that are critical to protection, something that has not yet
been achieved.

11

Of Mice and Men

Louis Pasteur shouted across the laboratory to his assistants, "But don't you see what this means? Everything is found! Now I have found out how to make a beast a little sick—just a little sick so that it will get better from a disease . . . All we have to do is let our virulent microbes grow old . . . When the microbes age, they get tame . . ." Later, in a calmer voice, he informed the members of the French Academy of Medicine: "In this case I have demonstrated a thing that Jenner never could do in small-pox—and that is, that the microbe that kills is the same one that guards that animal from death." He bragged of his triumph and "held out purple hopes . . . that he would presently invent ingenious vaccines that would wipe out all disease from mumps to malaria" (124). Pasteur, now close to 60 years of age and past his prime, had made the chance discovery of turning microbes against themselves: first taming them and then using them as wonderful protective weapons against their own kind. He made vaccines against chicken cholera, anthrax, and rabies. He did not live to witness the discovery of the malaria parasite or its transmission by mosquitoes, and so he could not attempt to produce a malaria vaccine. However, he did establish the general principles that would inspire a century of malaria vaccine research.

Attenuation by Irradiation

The Sergent brothers, Edmond and Etienne, working at the Pasteur Institute of Algeria, carried out the first trial of a malaria vaccine. Following the practices established by their legendary master, Louis Pasteur, they attenuated sporozoites of *P. relictum* to immunize canaries (429). In 21 cases, after sporozoites were held in glass flasks for 12 to 24 h they no longer caused a blood infection; when 24 canaries immunized with these attenuated sporozoites were challenged by the bite of infected mosquitoes, 7 had a light or no infection, 1 infection was severe, and in 16 there was "evidence of immunity." Their report—a "proof of principle"—of a malaria vaccine was promptly forgotten.

By the early 1940s, with reports of the existence of protective antibodies in the blood of monkeys and birds (see p. 95–97 and 106) and the observation that the enhanced activity of serum used to treat *P. knowlesi* infections was directly correlated with the degree of "stimulation" of the immune system, there was a renewed interest in vaccination for malaria. Paul Russell and Major H. W. Mulligan at the Pasteur Institute of Southern India carried out the renewed attempts to produce a protective vaccine using sporozoites.

Paul F. Russell (1894 to 1983) received his bachelor's degree from Boston University in 1916 and his medical degree from Cornell University in 1921. Shortly thereafter, he joined the International Health Division of the Rockefeller Foundation and was posted to the Straits Settlements (1925 to 1928) and the Philippines (1929 to 1934). In 1935, he was assigned to India to carry out malaria control experiments. He became convinced that "the distribution of quinine . . . has never in any place in any country been effective in controlling malaria" and that "malaria control . . . must consist mainly in a direct attack on the adult malaria-carrying mosquito in its daytime resting places, or on the larva of this mosquito in its breeding places, or both at the same time" (212).

Russell never considered his work in India to be routine; rather, in every phase it was a program of research. As a firm believer that avian malaria was a reliable indicator for human malaria, Russell in collaboration with Mulligan developed a simple sporozoite-clumping (agglutination) test with chickens and *P. gallinaceum* (408). They found that the agglutinin titer of serum was elevated when the chickens were acutely or chronically infected. Then they went a step further: when the salivary glands of mosquitoes heavily infected with sporozoites were placed in a shallow dish and exposed for 30 min to the direct rays of a mercury arc Sun lamp, the sporozoites were inactivated. Chickens vaccinated either intramuscularly or intravenously with these attenuated sporozoites showed a considerable rise in the titer of sporozoite agglutinins. Further, of 14 chickens that received inactivated sporozoites and were then challenged with viable sporozoites (by mosquito bite), 7% did not become infected, 64% recovered spontaneously, and 28% developed severe infections and died. Those suffering from the mild infection had a very high agglutination titer. The authors wrote: "It seems fair to conclude, therefore, that repeated injections into fowls of inactivated sporozoites of *P. gallinaceum*, producing an agglutination titer of at least 1/32,000 render such fowls partially immune to the pathogenic effects of mosquito-borne infection with the homologous *Plasmodium*" (344). In another study they found that 92% of chickens with an agglutinin titer of 1/32,768 or higher showed spontaneous recovery and 8% died after challenge by the bite of infected mosquitoes whereas the mortality was 51% in fowl with normal or lower agglutination titers. Significantly, the immune birds did not resist a challenge with intravenous injections of the same strain of blood stage parasites. The authors concluded: "This suggests that trophozoites and sporozoites are not immunologically identical." In a subsequent report they (411) noted: "the combined mortality for 19 fowls was 21.1%, which was less than half that of normal fowls similarly infected. This is a significant measure of immunization, although in no case was infection prevented."

Russell returned to the United States in 1942. During World War II, he served as a colonel in the Army Medical Corps assigned to General MacArthur's headquarters in Australia. In his 1955 classic work *Man's Mastery of Malaria* (410), Russell provided a retrospective on the history of malaria and attempts at eradication; however, not a single sentence in the entire book refers to a malaria vaccine. Russell remained an administrator and technical advisor on malaria with the Rockefeller Foundation (409) until he retired in 1959. After his work in India, he never again was involved in vaccines.

As with the Sergent brothers, the vaccination work performed by Russell and coworkers was soon forgotten. In 1966, after decades of neglect of sporozoite vaccination, Wellcome's W. H. G. (Harry) Richards followed up the work of the Sergents and Russell. Richards was born in Camberwell in 1928 and raised in London. His early education was greatly affected by the 1939 to 1945 war. He and his sister remained with their parents in London throughout the war; they did not evacuate. For the first 2 years of the war there were no schools open as the majority of the teaching staff had accompanied the evacuated children. By 1941 some schools were operating, but with so few staff and with the ever-present air raids it was difficult to know which schools were open, when, and for how long. As a youngster, Harry found the war rather exciting, not realizing just how much of a hindrance this was to be in his educational opportunities. In the autumn of 1945 he left school with only an Oxford School Certificate and joined the Wellcome Laboratories of Tropical Medicine in London as a junior laboratory technician. Within a year of joining Wellcome, he was drafted into the Royal Air Force for National Service. He was sent for medical training and then was sent to the best RAF hospital to work with the senior burns and plastic surgeon. This experience allowed him to mature and grow up as an individual.

On release from the Air Force, he rejoined Wellcome to work in the Department of Protozoology with Len Goodwin, who was already a recognized leader in the field of tropical medicine. Goodwin became both mentor and friend to Richards and helped shape his career. Together they shared many successes in bringing drugs for the treatment of tropical diseases to market, and this allowed Richards to rub shoulders with future Nobel Prize winners and an international group of scientists. The hierarchy within Wellcome, however, dictated that technicians should not achieve independent status to conduct research, as this was the role of university graduates. Goodwin recognized this limitation and encouraged Richards to work for a bachelor's degree. This presented two major difficulties. The first was that he needed to pass a number of basic examinations before enrolling for a degree course, and the second was that by this time he was married and had a young family; he would therefore have to work for the degree part-time along with his job to provide a salary to support the family. Although Goodwin gave him support and encouragement, the university did everything it could to dissuade Richards. He received a B.Sc. (1963) degree after working at college half a day per week for 3 years. Goodwin and Richards continued to work together in the Department of Protozoology at Wellcome, with Richards running the malaria research program. In 1966, when the Wellcome Foundation decided to move the laboratories from London to their research facilities at Beckenham outside London, Goodwin left the Foundation. Although he asked Richards to join him at the London Zoo (where he was Director), he also recognized that Harry needed to stand a little apart, as he was often still regarded as not being independent of Goodwin.

Research on tropical medicine at Beckenham was carried out in the Departments of Protozoology and Helminthology. The move to Beckenham provided Richards with the opportunity to have purpose-built laboratories and a separate superbly equipped four-room insectarium. Although the emphasis of the Wellcome malaria research program was on the discovery of effective new chemotherapeutic agents, Richards now had the tools and equipment to explore vector-borne aspects of malaria control. By the early 1960s, resistance to antimalaria drugs had been reported. Mosquito vectors in some areas had also developed resistance to insecticides. In the face of this renewed threat of malaria, Wellcome not only encouraged further research into new therapeutic drugs but also allowed Richards to initiate a program within the Department of Protozoology to explore the field of malaria immunology. This research project was used as a basis for Richards' Ph.D. thesis (*Antigenic Studies of the Class Sporozoa with Particular Reference to Species of* Plasmodium, University of London, 1969).

At the start of the Wellcome program, Richards was aware that any antigen to be used would most probably be too weak to elicit a lasting immune response and that therefore an adjuvant would be required to increase the efficacy of the antigen. Not many adjuvants were available at the time, and the most commonly used, FCA, had the major disadvantage of causing necrotic lesions. In consequence, Richards used saponin as an adjuvant because it had been employed in a number of bacterial vaccines. The early studies were on *P. gallinaceum* in chickens, with the addition of an adjuvant to the blood stage antigen (394). The administration of antigen alone demonstrated an ability to protect chickens from challenge with viable blood parasites; the immunity of the chickens increased with the the number of vaccinating doses given. In no case was the protection complete, and the challenge never failed to produce a blood infection. In all the birds that had been immunized with the blood stage antigens, the development of a blood infection was delayed to a greater or lesser degree; this prolonged delay was more pronounced with successive immunizing doses. The disappearance of the parasites from the bloodstream of the survivors was a gradual process, taking 2 to 9 days. No great difference was observed between the routes of inoculation when antigen was used on its own and when it was used with adjuvant. The antigen prepared by freeze-thawing gave very poor results which were not improved even with the addition of an adjuvant. The results with antigens prepared from 0.1% formalin-treated infected red cells and by UV-irradiated infected red cells were comparable. A maximum protection of 40% was reached in chickens that were able to survive after challenge with homologous blood stage parasites, and this was increased to 60% when an adjuvant was used.

The sporozoite antigens, either with or without adjuvant, were all found to be completely ineffective in protecting birds against challenge with blood parasites. Of the four preparations of sporozoites examined, the freeze-thawed vaccine was the least effective when challenged with viable sporozoites. The other three, after three immunizing doses, gave 50% protection compared with 10% found in the untreated controls. The addition of saponin to the *P. gallinaceum* sporozoite antigen provoked an immune response, which was quite striking when the animals were challenged with 500 sporozoites of the same isolate. The most notable protection was found with the antigen treated with 0.1% formalin. With the antigen alone there was, after three doses, 60% survival of the birds following the homologous challenge. After three doses of antigen plus formalin adjuvant, 90% of the birds were able to survive challenge with the homologous sporozoites. When the immunizing doses were given intravenously, the results were superior to those obtained after the subcutaneous immunization. Irradiated sporozoite antigens and those prepared using dried sporozoites gave similar results, but the formalin antigen preparations were always superior.

The protection afforded by the sporozoite antigens, although never complete, did (especially with the formalin plus saponin) provide a positive result. It was obvious that when a blood infection developed the birds survived longer; however, a fulminating infection always followed. Unlike the results of the work by Russell and Mulligan, there was no correlation between protection and circulating antibodies being able to agglutinate homologous sporozoites.

Richards extended his immunization studies to the mouse malaria parasite *P. berghei* (discovered in 1948 in central Africa), using techniques similar to those he had used with *P. gallinaceum* (393). White mice were vaccinated either subcutaneously or intravenously at 2-week intervals with three doses of *P. berghei* red cells plus saponin. Fourteen days after the last immunizing dose, the mice were challenged with viable infected red cells and sporozoites, using different and similar strains. The red cell challenge consisted of 25,000 infected red blood cells, and the sporozoite challenge used 500 sporozoites, which were always given intravenously. Two control groups were included in each test; one group received saponin and saline to test for any nonspecific action of the adjuvant, whereas the second group was the control for the susceptibility of the mice to develop a blood infection after the challenge and also for whether they succumbed to the infection. After a single immunizing dose, there was no obvious effect with either the red cell or the sporozoite antigen when the mice were challenged with the same stage. The freeze-thawed preparation of the infected red cells responded in exactly the same manner as the untreated controls. The UV-irradiated infected red cells also gave very poor results on challenge. Treatment of this antigen prepared with 0.1% formalin gave the most promising results, with little or no effect after 30 min of treatment; after treatment for 2 h or more, there were no detectable parasites in the blood in the mice after challenge. Fourteen days after the immunizing doses were given, the mice were again challenged with either viable blood stage parasites or sporozoites, and there was no protection in either of the groups. This obvious lack of effect after a single immunizing dose of antigen forced adoption of a three-dose vaccination schedule. The results showed a promising degree of protection, especially when the 0.1% formalin vaccine was used together with saponin, with both the subcutaneous and intravenous routes being equally effective against challenge with the same strain; however, when the mice were challenged with a different mouse malaria parasite, *P. chabaudi*, there was no noticeable protection.

Infections produced by sporozoites of *P. berghei* were among the most virulent in the whole of the series of experiments, with blood infections averaging 42% by the end of week 2 and mice dying soon after. Saponin on its own delayed the appearance of blood parasites by about 7 days. The sporozoite antigens plus adjuvants exerted a noticeable effect on challenge by viable sporozoites. The freeze-thawed antigen and the irradiated antigen behaved similarly. The intravenously vaccinated animals showed a delay in the appearance of a blood infection by more than 1 week compared with those given the adjuvant alone. In the subcutaneously vaccinated mice, infected erythrocytes were not detected until week 3 after challenge, a still longer delay. Both groups showed a prolonged period before the appearance of parasites in the blood and allowed some animals to overcome the challenge. The formalin-treated antigen plus adjuvant gave the greatest protection. Vaccination by the intravenous route delayed the parasitemia until week 3, 1 week more than for the animals treated with adjuvant plus saline; in the antigen-plus-adjuvant group, the blood infection declined to less than 1% by the end of the 6-week trial period.

Richards' work on avian and mouse malarias had shown that vaccination was able to affect the course of an induced infection to such an extent that the infected animals could recover from what in the control animals was a fulminating disease. The use of adjuvant, although not new, was able to enhance the power of the malarial antigen. Regrettably, after Richards' pioneering work on vaccines, Wellcome, and in particular the Department of Parasitology (formed by a merger of the Departments of Protozoology and Helminthology, and headed by Ralph Neal), decided not to pursue such work, and Richards moved into other areas. After 44 years within Wellcome he had reached the position of a senior executive in charge of all scientific aspects of the company's entomology, veterinary, and medical activities throughout 50% of the world. Richards retired in 1989 and now spends his time in rural Devon gardening and caring for ill-treated animals.

Across the Atlantic Ocean, in 1967, vaccination experiments were being conducted in the Department of Parasitology at the New York University (NYU) School of Medicine, using the rodent malaria parasite *P. berghei* to determine whether X-irradiation of sporozoites might provide a better source of antigen than the UV irradiation of sporozoites used by Russell et al. (411) and Richards (393). At the time, the NYU group was aware that X-irradiation of *Trichinella* larvae attenuated by exposure to a suitable dose of γ irradiation interfered with normal development to adulthood, induced a degree of resistance to a challenge infection with normal larvae, and stimulated the production of antibodies (413). Other studies, carried out at WRAIR, used the X-irradiated larval stages (cercaria) of the blood fluke *Schistosoma* (382). This vaccine not only prevented the worms from reaching the lungs and liver of the immunized host but also resulted in a strong resistance to challenge. In most cases, the mean number of worms recovered from animals immunized with 500 irradiated cercaria was 10% of the number recovered from nonimmunized controls. The optimum dose for attenuation was 24,000 to 48,000 rads, and the percent reduction in the number of worms recovered was related to the number of immunizations up to the fifth immunization, with the duration of acquired immunity persisting for 119 days in mice, 343 days in rhesus monkeys, and 1 year in sheep (243). Success in the use of X-irradiation for reduction in the worm burden with blood flukes led quite naturally to vaccination trials with the invasive stages of malaria parasites by the NYU group.

At the City College of New York (CCNY), Jerome Vanderberg (born in 1935), after taking courses in ecology, entomology, and field zoology with the most eminent lepidopterist in the United States, Alexander Klots, decided that he wanted to become an entomologist. A course in parasitology in 1954 convinced him that he wanted to work on things that had medical significance. Klots, like many of his generation, had spent World War II in the U.S. Army, in his case doing mosquito surveillance and control in the South Pacific. Therefore, Vanderberg worked in his laboratory at the American Museum of Natural History on the mosquitoes that he had collected during the war there. After receiving his B.S. degree in 1955 from CCNY, Vanderberg did graduate work at Penn State and then at

Cornell. At the 1960 national meeting of the Entomological Society of America in Miami, where he presented his Ph.D. thesis, "The Role of Gonadotropic Hormone in Protein Synthesis in *Rhodnius prolixus*," he wandered into a session on malaria and listened to a talk by Ian McGregor on the successful passive transfer of immunity by administration of immunoglobulin to children at risk. This work made it clear that vaccination against malaria might ultimately be carried out and convinced Vanderberg that this was an area in which he wanted to work. Human malaria, however, does not lend itself easily to basic research in immunology, so a convenient laboratory model for in vivo studies of mammalian malaria was desirable. The need was satisfied with the description of a rodent malaria parasite, *P. berghei*, by a Belgian physician (Ignace Vincke) and entomologist (Marcel Lips). As often occurs with important discoveries, these workers had not been searching for what they ultimately found. Vincke spent World War II doing malaria surveys in the former Belgian Congo (now the Democratic Republic of the Congo). In 1942 he observed sporozoites in the salivary glands of the mosquito *Anopheles dureni*, collected near a major mining center, Elisabethville (now Lubumbashi). Tests on the blood meal contents of the mosquitoes' midguts indicated that the mosquitoes had fed on rodents or insectivores. When Vincke and Lips examined the blood of a local tree rat, *Thamnomys*, they discovered a new species of malaria parasite. They postulated that the mosquito salivary gland infection with sporozoites observed years earlier and the newly described blood infection were caused by the same species. Vincke named the parasite *P. berghei* in honor of his close friend, Louis van den Berghe, of the Prince Leopold Institute of Tropical Medicine in Antwerp. However, it was not until 1950 that Vincke was able to show that sporozoites collected from these mosquitoes produced a typical *P. berghei* infection when injected into laboratory mice. Due to the graciousness of the Belgian workers, *P. berghei* was soon widely distributed throughout the world. Among the recipients of this rodent malaria parasite was Harry Most, who had worked on the development of chloroquine while in the U.S. Army during World War II and upon return to civilian life was Chairman of the Department of Preventive Medicine at the NYU Medical Center. During the 1960s Most served as Chairman of the Armed Forces Epidemiological Board and Director of its Commission on Malaria, a civilian advisory panel. With the support of the Commission and funded by the U.S. Army Medical and Research Command, a project on the biology of *P. berghei* was initiated at NYU.

After receiving his Ph.D. in 1961, Vanderberg worked as a postdoctoral fellow at Johns Hopkins University (1961 to 1963), studying mosquito physiology and transmission of *P. gallinaceum* in the laboratory of Lloyd Rozeboom. During a trip to New York City in May 1963, he learned that Harry Most was interested in hiring a medical entomologist trained in insect physiology. He contacted Most, was interviewed, and, before Most advertised the position, was hired. Starting in September 1963, he immersed himself in what was known about the biology of *P. berghei* in Africa. During the next couple of years, his research was focused on working out the parameters for sporozoite transmission of *P. berghei* to laboratory rodents and the characterization of these infections in laboratory rodents. In 1965, the immunologist Ruth Nussenzweig joined the NYU malaria research group (consisting of Most, Vanderberg, and Meier Yoeli).

Immunization studies with irradiated rodent malaria sporozoites at NYU showed that the optimal dose of X-irradiation needed to inactivate the sporozoites dissected out of mosquito salivary glands was 8,000 to 10,000 rads; of 46 mice, each immunized with 75,000 irradiated sporozoites by intravenous injection, only 1 developed an infection. The remaining mice were challenged 12 to 19 days later with 1,000 viable sporozoites (351). The percentage of animals infected varied from 14 to 73% (average, 37%), whereas in the controls the percentage of mice infected averaged ~90% and the protection was estimated to be 27 to 86%. There was no protection against an inoculum consisting of infected red blood cells, confirming the stage specificity observed earlier by Russell. In a subsequent study of 103 mice injected with 75,000 sporozoites irradiated with 8,000 rads, only 3 developed blood infections; when 10,000 to 15,000 rads was used none of 29 mice became infected. Unlike in the work done by Richards, when the blood of the X-irradiated immunized animals was inoculated into other animals, it failed to produce an infection and the immunized animals did not develop infections after removal of the spleen; this was taken to indicate that sterile immunity had been induced (527). Protection remained strong for up to 2 months and then declined. By repeated intravenous challenge of immunized mice at monthly intervals, there was a boosting effect and protection could be maintained for up to 12 months (360). The NYU group explored different routes of immunization (intraperitoneal, intramuscular, oral, and intracutaneous) and different methods (formalin fixation, heat inactivation, and disruption) for preparing sporozoites; however, not only was the reproducibility poor, but also it was not possible to induce high levels of protection.

One of Vanderberg's most valuable findings was his observation that upon in vitro exposure to serum from immunized mice, an antibody-mediated precipitation occurred around sporozoites and the precipitate projected from one end (524). This reaction was called the circumsporozoite precipitation (CSP) reaction. The terminology for the CSP reaction was suggested by the similar-appearing circumoval precipitation reaction around schistosome eggs incubated in immune serum. The abbreviation CSP originally referred to the circumsporozoite precipitation reaction; however, over time it has come to refer to the circumsporozoite protein. Because of the striking way in which serum from immune animals deformed sporozoites, this was postulated to be the basis for a humoral component of protective immunity against sporozoites. Indeed, incubation of sporozoites of *P. berghei* for 45 min with immune serum neutralized their infectivity and did not require complement (92). The CSP reaction was produced in mice and rats immunized by intravenous inoculation with irradiated sporozoites of *P. berghei* or by the bite of infected irradiated mosquitoes, as well as in animals injected intravenously with viable *P. berghei* sporozoites (92). After immunization with irradiated sporozoites there was a reduced circulation time for intravenously injected infective sporozoites (352). Antibodies to CSP appeared to be critical to protection; however, because protection was never complete in passively immunized animals, it suggested that additional mechanisms might be necessary to bring about complete protection against sporozoite-induced malaria infections.

The discovery of the CSP reaction could have led to an enduring and productive collaboration between the members of the group at NYU. Instead, it created a rift, the basis of which had nothing to do with the finding but which involved personalities and priority of authorship. Earlier, Vanderberg and Ruth Nussenzweig had an agreement that she would be the senior author on research publications on sporozoite immunity whereas he would head the author list on papers on sporozoite biology. Vanderberg, upon completion of his studies on the CSP reaction, prepared a draft manuscript. This was shown to Ruth Nussenzweig, who, although she had contributed nothing to the work directly, insisted on lead authorship, claiming that this was her research area. Vanderberg, on the other hand, said that the studies were conducted by him alone and concerned sporozoite gliding—a biological phenomenon. Most, as Department Chair, was asked to settle the conflict, and when he failed to support Nussenzweig's position, a stressful and hostile environment was created. Henceforth, they were unable to work together in harmony.

With the initial success of immunization of mice with X-irradiated sporozoites, Nussenzweig recognized that it would be important to define the correlation between the in vitro-detectable anti-sporozoite antibodies and protective immunity and to characterize the antigens involved. Emboldened by the results of the mouse vaccinations with irradiated sporozoites, she assembled a team of immunologists to carry out investigations to isolate and characterize the antigens responsible for protection and to extend the rodent malaria immunization studies to monkeys. Vaccination attempts were initiated using *P. cynomolgi* and *P. knowlesi* since in rhesus monkeys the former produces a mild and benign infection, similar to that caused by *P. vivax* in humans, whereas the latter is highly virulent and is more akin to *P. falciparum* in its effect on children. The results were disappointing: an initial study of two monkeys failed to protect against challenge after inoculation of irradiated *P. cynomolgi* sporozoites divided into five immunizing doses over a period of 146 days (92). In a follow-up experiment with 12 monkeys, protection was observed only after intravenous inoculation of large doses of irradiated *P. cynomolgi* over 9.5 and 13.5 months. However, only two of the animals were totally protected against challenge with 10,000 to 20,000 infective sporozoites (92). With rhesus monkeys immunized multiple times with 300 million to 400 million X-irradiated *P. knowlesi* parasites, two developed sterile immunity but the third did not. Unlike the situation with merozoite vaccination, there was no amplification of the antisporozoite protective response with *P. knowlesi* sporozoites emulsified with FCA and administered intramuscularly, nor was there protection against sporozoite challenge. Consequently, Nussenzweig abandoned, at least for the time being, further vaccination work on primates and directed her attention to characterizing the immune mechanisms seen with irradiated sporozoites in mice.

Vanderberg, in contrast to Nussenzweig, thought that it was timely to carry out studies with humans. A collaborative arrangement was set up with David Clyde (359) at the University of Maryland School of Medicine. Clyde (1925 to 2002) was born in India while his father was stationed there as an officer in the Indian Medical Service. At age 7 he was sent to England for schooling and then to the United States, where he lived with relatives and completed his high school and college education at the University of Kansas in 1946. He graduated in medicine from McGill University in Canada in 1948. Following an internship, he joined the British Colonial Service stationed in Tanganyika (now Tanzania), where he held positions from Medical Officer to Specialist Malariologist to Senior Consultant in Epidemiology. For his malaria work during his time in Tanzania, he gained a Ph.D. from the University of London. It was in Tanzania that Clyde "became convinced that a malaria vaccine against the scourge would be a key to controlling the disease." In 1966, he left Africa and joined the University of Maryland, where he was the director of a program studying drugs against sporozoite-induced malaria in human volunteers at the Maryland House of Correction in Jessup.

The irradiated-sporozoite immunizations of human volunteers and subsequent challenge with unirradiated sporozoites proved to be a long and sometimes frustrating effort because the first series of immunizations used X-ray doses that had previously been used with rodent malaria sporozoites. In those studies sporozoites had been directly obtained from dissected-out mosquito salivary glands; however, these could not ethically be injected intravenously into humans. An alternative approach was suggested by studies on mosquito-borne viruses. Serum surveys done after epidemics of infections with these viruses consistently showed that only a very small percentage of serum-positive individuals had actually experienced signs or symptoms of disease. Thus, from an epidemiologic standpoint, mosquitoes might be more important as vehicles of immunization than as vectors of disease. The use of *Plasmodium*-infected mosquitoes to induce infection was not entirely novel, since its practice dated back at least to 1926 with the treatment of 2,500 patients with late-stage syphilis in England at the Horton Mental Hospital in Epsom (see chapter 12, p. 287). This treatment, practiced in England, the United States, and other countries, persisted well into the 1950s, but with the widespread use of penicillin to treat syphilis, its usage declined. The human trials first carried out by Clyde (88) and later by Rieckmann (398) would be a reapplication of this method for the purposes of testing the protective efficacy of malaria vaccines.

Accordingly, an initial trial was conducted by Vanderberg with rodent malaria, using infected, irradiated mosquitoes as substitute "hypodermic syringes" to deliver sporozoites. The argument (525) made was as follows: "The technique that we presently use for immunization involves the intravenous injection of infected mosquito salivary glands which have been dissected out, ground up, and irradiated. However, this preparation contains considerably more extraneous mosquito debris than sporozoites, and the injection of such material into humans would possibly pose medical risks of embolisms and sensitization. Until sporozoite preparations can be purified it would seem prudent to avoid this. A more reasonable approach for the present would be to x-irradiate infected mosquitoes and then let them feed on volunteers, thus allowing the mosquitoes to inject the sporozoites in a relatively uncontaminated condition. Such a technique would have limited practicality, but it has the advantage of being performable now. If protective immunity could be demonstrated under such circumstances, it might encourage further work on attempts to establish purification procedures for sporozoite homogenates. The injection of irradiated sporozoites by mosquitoes should thus be viewed as an attempt to test the feasibility of vaccination in humans, which if successful could lead to trials using more practical techniques." The results showed that mice so immunized were completely protected from sporozoite challenges that caused blood infections and death in 100% of nonimmunized control mice. With the success of this approach, it seemed appropriate to move from mice to men.

The first series of immunizations of volunteers with irradiated-sporozoite-infected mosquitoes resulted in some breakthrough blood infections, so several months was lost while retooling. A second series was begun with several new volunteers and with higher doses of radiation against the infected mosquitoes (88). An average of 222 mosquitoes that had been irradiated (17,500 rads) fed on each man, and after 6 to 7 weeks each man was again fed on by an average of 157 irradiated mosquitoes. This time there were no breakthrough blood infections during the immunizations. The vaccinated individuals, along with unvaccinated control volunteers, were then challenged by bites of infected mosquitoes sufficient to have induced a blood infection in every single volunteer who had ever taken part in prior trials conducted by David Clyde and his associates. Upon challenge by the bite of mosquitoes infected with normal infectious sporozoites, one of three vaccinated volunteers was fully protected whereas all of the non-vaccinated volunteers developed a blood infection, as expected. After the challenge, sporozoite-infected mosquitoes and serum from the challenged volunteers were brought to New York. When Vanderberg tested these sera for reactivity with live sporozoites, he found strong CSP reactions in the serum of one of the vaccinated individuals. Vanderberg immediately telephoned Clyde and predicted that this individual would be protected. All were delighted when this prediction turned out to be true. It is perhaps relevant that the individual had taken part in the first vaccination trial, although he was not one of the vaccinees who had experienced a breakthrough blood infection. Thus, he had received an especially extended schedule of vaccinations. They concluded that a sufficient dose of radiation-attenuated sporozoites was necessary to attain a sterile immunity upon challenge.

In 1974, Clyde began experiments on himself to determine whether it was possible to immunize against *P. vivax* as well as *P. falciparum* and whether there was cross-protection (89). Clyde allowed nonirradiated mosquitoes infected with *P. vivax* or *P. falciparum* to bite him, and when he developed both infections he knew he was not immune. He described the first attack (7): "You shake like anything. You are very cold. You have a high temperature and a splitting headache. Then you start vomiting, and that is the most awful part of it. You have about 4 hours of absolute misery and then it gradually lets off for about another twelve hours. Then it starts again." Clyde went on to see whether different strains had immunological differences, and so he allowed irradiated mosquitoes from different geographic regions to bite him. Each time, he received scores of bites. "It was a damn nuisance and very unpleasant to have 6 cages of 350 mosquitoes hanging on to you but that's part of it." The welts from the bites itched, and he applied cortisone cream to relieve the irritation and to prevent himself from scratching. By the end of the experiment he had received over 2,700 bites. To test the efficacy of the "vaccine," he accepted a challenge of being bitten by unirradiated mosquitoes. He was protected. Unfortunately the protection was not long-lived—3 months for *P. falciparum* and 3 to 6 months for *P. vivax* (7).

Although for a decade Ruth Nussenzweig had been a driving force in the rodent malaria research program at NYU (350, 353), she was not an author of any of the landmark publications involving human trials. This was not so much a matter of her not being directly involved but, rather, was due to an earlier incident. In 1974, to mollify Victor and Ruth Nussenzweig, who were considering leaving NYU, and to respond to their claims to the Dean that Harry Most was becoming senile, the Department of Preventive Medicine was abolished and Most was transferred to the new Division of Tropical Medicine (as its sole member) within the Department of Medicine. The remaining members were reconstituted as the Division of Parasitology with Ruth as Head. Most retaliated to this coup by insisting that Ruth's name not appear on the publication that reported on human vaccinations; in this way, Most prevented her from receiving credit for the first successful trial in which humans were vaccinated with irradiated sporozoites.

Vaccination studies similar to those reported by Clyde, Vanderberg, and Most were conducted between 1971 and 1975 in a collaboration between the Naval Medical Research Institute and Rush-Presbyterian-St. Luke's Medical Center at the Stateville Correctional Center, Joliet, IL, under the direction of Karl Rieckmann (398). Five volunteers were bitten over a period of 4 to 17 weeks by fewer than 200 *P. falciparum*-infected mosquitoes irradiated with 12,000 rads, and during the immunization two volunteers developed blood infections (which were quickly cured by chloroquine). This indicated that the radiation dose was too low to inactivate all of the sporozoites. Of the four men challenged, none was protected. Another three volunteers were selected. One was exposed six times within a 2-week interval; mosquito dissection showed that he had been bitten by 440 irradiated mosquitoes. Two weeks after the last immunization, he was bitten by 13 infected nonirradiated mosquitoes; he did not become infected. A second volunteer was exposed eight times to a total of 984 irradiated mosquitoes, and although the intervals were not exactly 2 weeks, he too was protected against challenge 2 to 8 weeks after the last immunization; however, he showed no protection when challenged at 17 and 25 weeks after the last immunization. A third volunteer was exposed to 987 irradiated mosquitoes with immunization at irregular intervals during a 38-week period, and when challenged 8 weeks later with a strain different from that used in the immunization he was protected; however, when challenged with the same strain 18 weeks after the last immunization, there was no protection.

Thus, vaccination with irradiated sporozoites was "an encouraging step towards the goal of immunizing man against malaria" (Vanderberg, personal communication). The limited success of these vaccination studies served to establish what had been hoped for, namely, a clear "proof of concept" demonstrating that production of sterile immunity to malaria in humans might be biologically feasible and was deserving of further efforts. However, not everyone appreciated the value of this first modest step. Robert Desowitz (127) in his 1991 book, *The Malaria Capers*, in a chapter entitled "The Sporozoite Follies," described the result: "Two of the three volunteers came down with malaria. No protection whatsoever." To the contrary, the NYU and University of Maryland groups had been gratified that even one of the volunteers was completely protected. It was felt that the previously stated goal, namely, "The injection of irradiated sporozoites by mosquitoes should thus be viewed as an attempt to test the feasibility of vaccination in humans, which if successful could lead to trials using more practical techniques" (90), had been fulfilled, whereas Desowitz expected a fully developed, functioning vaccine on its first try in humans. Indeed, subsequent studies with this approach would show that when sufficient numbers of irradiated sporozoites were used for immunization, there was protection in 33 (94%) of 35 challenges, with protection lasting for at least 10 months and with effectiveness against multiple strains of *P. falciparum* (224).

Desowitz went on to criticize a field trial conducted by R. S. (Bill) Bray on two "vaccinated" children in The Gambia (53), stating that "The irradiated-sporozoite vaccine did not work in the endemic setting for which it had been intended" and later that "it was a complete failure" (127). Aside from the fact that Bray was not a physician and the study was clearly unethical in that the children were injected with an uncharacterized suspension of debris from mosquito salivary glands and did not give informed consent, the trial failed because he gave a single immunizing dose followed by a second dose too soon afterward for it to have had an authentic booster effect. Therefore, it was not, as Desowitz asserted, that the vaccine did not work but simply that the immunization dose was too low to have possibly been effective under any circumstance.

At the time of these human vaccine trials, critics alleged that conducting research with volunteers in prisons was unethical. This was addressed by Clyde et al. (90), who noted that the study was conducted in strict accordance with regulations established by medical ethics committees of the University of Maryland and the supporting agency and conformed to the guiding principles of the Declaration of Helsinki. The operating principle was from the Nuremberg Code of 1947, enacted after the war crimes trials of Nazi doctors, which declared, "The voluntary consent of the human subject is absolutely essential." In 1974, the American Civil Liberties Union filed suit to prevent volunteer studies in all prisons, including the Prison Volunteer Research Unit at the University of Maryland, where the vaccination work had been conducted. The American Civil Liberties Union lost the case in federal court because it was unable to prove any instance in which the volunteers had been coerced. The presiding judge praised the studies and agreed that the highest ethical standards had been maintained. Nevertheless, the University of Maryland program closed in 1974 because of ongoing opposing legal actions on behalf of prisoner volunteers. David Clyde said at the time that problems arose when he moved from *P. falciparum* to *P. vivax* vaccination studies. Because the black prisoners were Duffy negative, and thus resistant to vivax malaria, they were necessarily excluded from these studies; however, these prisoners alleged that this exclusion was discriminatory toward them. At the same time, according to Clyde, the white prisoners complained that they were being used as "guinea pigs" in a project from which the black prisoners had been excused. Therefore, because of these issues, vaccination studies were discontinued within prisons and later continued with non-prisoner volunteers (163).

One of the problems encountered in these early human studies with irradiated infected mosquitoes was that it was difficult to carry them out within a defined interval or with a constant number of mosquitoes since immunization depended on the availability of gametocyte donors; in addition, there was the problem of estimating the number of sporozoites released during feeding. In vitro culture and production of *P. falciparum* gametocytes on demand, so to speak, solved one of these problems. To better delineate the requirements for the irradiation dose and the number of sporozoites introduced, studies were carried out from 1989 to 1999 (224). One hour before immunization, the female mosquitoes were exposed to 15,000 rads of γ radiation from a ^{60}Co or ^{137}Cs source (lower irradiation doses did not attenuate sporozoites). The sporozoite dosage was calculated in the following way: 50 mosquitoes were dissected to estimate the percentage of mosquitoes with sporozoites, and the number of mosquitoes with a score of >2 (usually 50 to 75%) was multiplied by the total number of mosquitoes taking a blood meal to calculate the number of immunizing bites. Using this method of calculation, four volunteers who received >1,000 immunizing bites were protected against seven challenges or rechallenges from 5 to 10 mosquitoes infected with a different strain of *P. falciparum* from that used in immunization; one individual was not protected despite having the same immunizing dose when challenged by 90 mosquitoes infected with a different strain. Protection was evident at least 9 weeks after primary challenge and for at least 23 to 42 weeks against a rechallenge. Overall, 33 of 35 of those individuals challenged within 42 weeks after receiving >1,000 immunizing bites were protected whereas only 5 of 15 of men challenged after receiving >378 and <1,000 immunizing bites showed protection.

A Sporozoite Antigen Identified

Over 90% of adults living in The Gambia, an area of high malaria endemicity, were found to have detectable levels of antisporozoite antibodies (92, 345), suggesting that these antibodies may be related to the acquisition of natural resistance by continued exposure to the bites of malaria-infected

Anopheles mosquitoes. Furthermore, observation of rodent, human, and primate malarias showed that protective immunity mediated by immunization with irradiated sporozoites was associated with antibody induction. A monoclonal antibody raised against the surface of *P. berghei* sporozoites neutralized their infectivity (92). Passive transfer with the monoclonal antibody protected mice against sporozoite challenge (566). Sporozoites could also be neutralized by Fab fragments, suggesting that agglutination was not required to inhibit parasite infectivity and that it was the surface epitope recognized by this monoclonal antibody that was involved in sporozoite penetration of liver cells (378).

The sporozoite antigen of *P. berghei* recognized by the monoclonal antibody was stage and species specific and was named the circumsporozoite protein (CSP) (92). The involvement of CSP in inducing immunity was also suggested by the observation that irradiated oocyst sporozoites, having less CSP, afforded minimal protection whereas those that had matured in the salivary gland contained larger amounts of CSP. Monoclonal antibodies against sporozoites of *P. knowlesi* and *P. cynomolgi* were also prepared; these were used to precipitate similar CSPs. As with *P. berghei*, sporozoite neutralization was associated with these monoclonal antibodies.

The CSPs from a variety of *Plasmodium* spp. are encoded by one gene (of the 5,300 malaria parasite genes). Ruth Nussenzweig, now joined by her husband Victor, was convinced that the CSP would be an attractive target for the development of a protective vaccine. Victor Nussenzweig (born in 1928) entered Medical School at the University of São Paulo in 1946 just after the end of World War II. Like many of his friends, he was impressed with the important role of the Soviet Union in the defeat of Nazi Germany and was attracted to socialist ideals and Marxist ideology. During his first 3 years of Medical School, he was deeply involved in student politics. His future wife, Ruth, who was born in Austria and whose family emigrated to Brazil, was in the same class, and when the two started dating she convinced Victor that science was much more interesting than politics. After receiving M.D. degrees, Ruth and Victor went to Paris (1958 to 1960) with their two young children. Ruth worked at the Collège de France studying the metabolism of thyroid hormones, and Victor was accepted in the laboratory of Pierre Grabar at the Pasteur Institute, working on antigen processing. He discovered that fragmentation of fibrinogen by plasmin generated peptides, called C and D, with entirely different antigenic determinants. This was a surprise to Elvin Kabat, a prominent immunologist and a student of Michael Heidelberger, who at the time was at the Pasteur Institute on a sabbatical leave and who insisted that proteins were like carbohydrates and had repetitive identical epitopes.

Back in Brazil, Ruth and Victor worked with the prominent Brazilian immunologist Otto Bier. In 1963, Victor received a Guggenheim Memorial Foundation Fellowship and came to the United States with Ruth and their three children, Michel, Andre, and Sonia. Ruth worked with Zoltan Ovary and Victor worked with Baruj Benacerraf at the NYU School of Medicine. It was an exciting time, a time when the structure of immunoglobulin was being solved. Although they tried to go back to Brazil in April of 1964, they were deterred because the military had just overthrown the elected government and started a witch-hunt at the University of São Paulo. Therefore, they returned to New York City in October. In 1965 Ruth was appointed Assistant Professor in the Division of Parasitology at NYU Medical School, where she began work on a sporozoite-based malaria vaccine. Victor, who had joined the Department of Pathology of NYU Medical School, had an interest in the complement system, discovering the complement receptors in leukocytes, studying the control of complement activation, and demonstrating that complement played a central role in the clearance and metabolism of immune complexes. Prior to Harry Most's transfer to the Division of Tropical Medicine (see p. 248), he made overtures to the Pentagon to request that Victor, who was about to be deported back to Brazil for his prior socialist activities, remain in the United States since the Vietnam War was ongoing and therefore it was in the national interest for him to work on malaria. The entreaties worked. Victor was not deported, and his career changed: he began work with Ruth on the Department of Defense-funded rodent malaria project. The quest of Ruth and Victor for the protective antigen would lead, in succeeding years, to the cloning of the gene for the CSP, studies of CSP function in the initial stages of invasion of liver cells, the inhibitory role of IFN-γ in the development of the liver stages, and the demonstration that a CSP-based human vaccine against *P. falciparum* was possible.

In 1981, a practical impediment to the use of CSP as a vaccine was that its only source was the mature sporozoite. This difficulty was overcome by the cloning of the CS gene and the ability to deduce the amino acid sequence of the protein from the DNA base sequence. The first CS gene cloned was from *P. knowlesi* (159). To clone the gene, several thousand *Anopheles* mosquitoes were raised and fed on a monkey infected with *P. knowlesi* (this was done in collaboration with Robert Gwadz at

NIH). At NYU the mosquitoes were hand dissected, the salivary glands were separated, and mRNA was extracted from sporozoites by a graduate student (Joan Ellis) in the laboratory of the molecular biologist Nigel Godson. Three cDNA clones were obtained, and the region that coded for the immunoreactive region was identified and sequenced. The cloning of the CS gene from *P. knowlesi* was quickly followed by the cloning of the CS genes from several human malaria parasites, i.e., *P. falciparum*, *P. vivax*, and *P. malariae* (115, 160, 284, 321) by a number of laboratories including those at the NIH and WRAIR.

On 12 February 1981, NYU filed a patent application on behalf of the Nussenzweigs and Godson for their cloning of the CS gene. After filing the patent, NYU notified the funding sources, including USAID, NIH, and WHO, and indicated that they were entering into negotiations with a genetic engineering company, Genentech, to produce CSP. When Genentech asked for exclusive licensing to market the vaccine, objections were raised by WHO, which indicated that their support required "public access" and that under U.S. patent law USAID held the patent for the work it supported at NYU. The conflict dissipated in 1983 when the NYU-based research had moved ahead, accomplishing the work that was to be done by Genentech. The bargaining over the market rights was discouraging to the Nussenzweigs, who were falsely accused, among other things, of having a financial stake in the patent. The legal wrangling, however, continued for years. In the end it was resolved and the achievement by the NYU group was heralded in a 13 August 1984 *New York Times* (48) headline, "Malaria Vaccine Is Near." One scientist quoted in the article boldly predicted: "This is the last major hurdle. There is no question now that we will have a vaccine. The rest is fine tuning." In 1989 NYU licensed the CSP patent nonexclusively to GSK, royalty free; this ultimately would lead to RTS,S.

The gene sequences of all CSPs code for 300 to 400 amino acids, with the central region consisting of tandem repeats of amino acids rich in the amino acids asparagine (N) and proline (P) separated by the amino acid alanine (A) and flanked by two regions of highly conserved amino acid sequences, designated region I and region II (region II has been suggested to be involved in sporozoite binding to and invasion of liver cells). The CSP of *P. falciparum* has six repeats and is often written as $(NANP)_6$. Screening a large number of sporozoites from different areas of the world showed that all isolates had the same repeats although variations occurred at other regions.

When human volunteers were immunized three times with a conjugate of tetanus toxoid and (NANP)$_3$ in alum, there was a good correlation of the antibody titers of antipeptide and antisporozoite antibodies (220). However, when three of the individuals with the highest titers were challenged by the bites of five heavily infected mosquitoes, two underwent a prolonged period (11 and 7 days) before parasites appeared in the blood and only one individual did not develop a blood infection (221).

In another study under the auspices of WRAIR, volunteers were immunized with a recombinant NANP protein (R32NS181) formulated with a more potent adjuvant (monophosphoryl lipid A [MPL], a cell wall skeleton of mycobacteria and squalene). Of the 11 volunteers, 6 had high CSP titers; 2 of these did not develop an infection when bitten by five *P. falciparum*-infected mosquitoes and 2 had a delay in the appearance of parasites in the blood (225). WRAIR also conducted a trial of a recombinant protein produced in *E. coli*, consisting of 32 NANP repeats and 32 irrelevant amino acids. When this protein was used to immunize 15 individuals, 12 developed antibodies to the NANP repeats but the titers were low (1:500 and 1:1,000). Six of those immunized were challenged with sporozoites; no parasites were found in the blood of the volunteer with the highest antibody titer, and the appearance of blood parasites was delayed in the other two (25).

Although the work with human volunteers suggested that a CSP-based subunit vaccine might be feasible, it was clear that "not everyone with high titers of antibodies to the repeats was protected; however, those who were protected had high antibody titers" (466). In addition, it was evident from the NANP-tetanus toxoid vaccine study that although the antibody response was dose dependent, there was a limitation: the levels of the carrier protein, tetanus toxoid, could not be increased, as they were toxic.

RTS,S

The 25-year quest to develop a safe and effective subunit malaria vaccine is instructive in delineating the hurdles faced in the production and testing of a vaccine, the critical importance of having good collaborations, and the empirical nature of vaccine development. In early 1984, after the scientists at WRAIR had cloned and sequenced the *P. falciparum* CS gene, it became possible to develop a subunit vaccine. WRAIR entered into collaboration with GSK to produce CSP by using the GSK recombinant *E. coli* expression system. Although efforts to produce a full-length CSP were unsuccessful, four constructs were expressed, purified, and tested for immunogenicity in animals, and one (R32Ltet32) was selected for clinical development. Combined with alum (as an adjuvant), the vaccine, FSV-1, was tested on volunteers in 1987 (25). Ripley Ballou, then a young U.S. Army physician, and five colleagues (including Stephen Hoffman from NMRI [see p. 265]) taped a mesh-covered cup containing five infected *Anopheles* mosquitoes to their arms. The mosquitoes were allowed to bite them, and afterward, to make certain that the mosquitoes in the cup were infective, their heads were lopped off in Ross-like fashion with a pair of tweezers and the salivary glands were examined under a microscope. Ballou and the other volunteers had been injected a year earlier with FSV-1, and now it was time to be challenged with infectious sporozoites to assess protection. Nine days after the infected mosquitoes had fed, the first unvaccinated control had parasites in his blood and was given chloroquine to clear the infection. The second control and three vaccinated volunteers also came down with blood infections, and on day 11, Hoffman, who was confident that the vaccine would work and who had traveled to San Diego to give a presentation on the vaccine, fell ill with malaria. On day 12, Ballou also succumbed. Only the sixth volunteer, Daniel Gordon, was still healthy and remained so. The efficacy of the vaccine was disappointing; however, for the first time an individual had been completely protected by a subunit vaccine. When the vaccine was tested by WRAIR for safety and immunogenicity in western Kenya, the majority of vaccinated subjects had antibodies to CSP.

Over the next several years, a series of recombination constructs of CSP ($R32NS1_{81}V20$, $R32NS1_{81}$) were produced by GSK that incorporated the $NS1_{81}$ protein from the influenza virus to stimulate helper T cells; when these were tested with volunteers at WRAIR, the immunogenicity of $R32NS1_{81}V20$ was low and further clinical development was not pursued. In parallel, WRAIR tested $R32NS1_{81}$ in combination with the only acceptable adjuvant for use in humans, alum (the vaccine was called FSV-2), and although it was more immunogenic, this too failed to protect any of the volunteers. Over the next few years FSV-2 in combination with other adjuvants including MPL, an emulsion of MPL, mycobacterial cell wall skeleton, and squalene (Detox; Ribi Immunochem), and cholera toxin were also tested at WRAIR, but again the results were disappointing (24).

In 1990, Gray Heppner volunteered for a vaccine trial at WRAIR. Heppner, who was raised in Lynchburg, VA, did his undergraduate work at the University of Virginia and, after completing his M.D. at the University of Virginia Medical School, did an internship and residency at the University of Minnesota. In the late 1980s at Minnesota he had assisted a malaria researcher in growing *P. falciparum* and in the process became intrigued with malaria. Although Heppner joined the Army Reserves while in Minnesota, he did not sign on for active duty until 1990. That brought him to WRAIR as an infectious-disease officer. At WRAIR the CSP lacking a central repeat region, called RLF, had been expressed in *E. coli* and, when encapsulated in liposomes, was found to be immunogenic in mice; furthermore, the anti-RLF serum reacted with the surface of intact sporozoites and was able to inhibit their invasion of liver cells in vitro (550). The safety and immunogenicity of the RLF vaccine were tested in experiments with 17 malaria-naive volunteers, Heppner being one of them. Although RLF formulated with alum or MPL was well tolerated and immunogenic upon sporozoite challenge, all the immunized volunteers developed malaria (217).

In 1987 the GSK malaria vaccine program transferred its laboratories from Philadelphia, PA, to its vaccine division in Rixensart, Belgium. Joe Cohen, who had taken over the project at the same time that Ballou and colleagues at WRAIR had been using themselves as human guinea pigs, had another plan for using CSP as an antigen. Cohen, drawing on experience gained from GSK's successful development of a recombinant hepatitis B vaccine, Energix-B, decided to couple CSP with the surface antigen protein from hepatitis B virus grown in yeast (*Saccharomyces cerevesiae*), where at high concentrations the protein spontaneously formed viruslike particles; when this antigen is used as an immunogen, antibody formation is enhanced. Cohen hoped that if the NANP repeat from CSP was fused to the hepatitis virus surface antigen (consisting of 226 amino acids), similar particles now festooned with the active epitope of CSP would be made and would be able to elicit antibodies targeted to the sporozoite surface. To address the possibility that antibodies alone would not suffice, Cohen added a fragment from the tail end of CSP to stimulate T-cell responses. This construct would provide, as Cohen said, a "double whammy," with 19 NANP CSP repeats (R), along with T-cell epitopes (T) fused to the hepatitis B surface antigen (S), coexpressed and self-assembled with unfused S antigen. It was named RTS,S (24). In 1992, the first clinical trial of RTS,S for safety and efficacy was carried out with volunteers at WRAIR. Malaria-naive volunteers received RTS,S with either alum or alum plus MPL. Both formulations were well tolerated and immunogenic; however, after challenge with sporozoites, none of six in the alum group and two of eight in the alum-MPL group were protected. Significantly, one protected subject had an increased CTL activity against CSP (191). These results were considered to be encouraging enough to warrant further improvement in the vaccine to enhance both humoral and cell-mediated immunity (218). Taking clues from the study, Heppner cooked up a formulation with adjuvants that would produce the right sorts of cell-mediated immune responses. He suggested formulating RTS,S with an oil-in-water emulsion plus the immunostimulants MPL and QS21 (a proprietary GSK saponin derivative from the Chilean soap bark tree, *Quillaja saponaria*) (AS02), and by 1996—12 years after the first trials—RTS,S/AS02 was tested in human volunteers and protected six of seven vaccinees. When the volunteers were challenged 6 months later, one of five volunteers was still protected. Two or three vaccinations were necessary to produce sterile or partial immunity (i.e., delayed appearance of parasites in the blood) in most vaccinees. In liquid form RTS,S/AS02 had a limited shelf life, and so RTS,S was freeze-dried and then reconstituted with AS02. When 40 volunteers received the vaccine on a 0-, 1-, and 3-month schedule or on a 0-, 7-, and 28-day schedule of vaccinations, protection against sporozoite challenge was seen in 45% and 38% of the vaccinees, respectively (484).

In the summer of 1998 trials were held in The Gambia, with 250 men receiving RTS,S/AS02 or a rabies vaccine on a 0-, 1-, and 3-month vaccination schedule. The RTS,S vaccine showed a 34% reduction in the first appearance of parasites in the blood over a 16-week period. During the surveillance period, 81 of 131 men immunized with RTS,S had parasites in the blood whereas in the control group 80 of 119 tested positive. The following summer the men were given a booster shot, and this showed that the vaccine was acting in two ways: protecting against infection and weakening the symptoms in those who became infected (24).

Although GSK was encouraged, the company felt it needed funding assistance to move RTS,S into trials with infants (24). Ballou wrote a proposal to the Gates Foundation, which provided $50 million through the Malaria Vaccine Initiative (MVI) through PATH (Program for Appropriate Technology in Health). Ballou was asked to lead the program, but instead he took a position with a Washington, DC, biotechnology firm, MedImmune, to work on other vaccines. In 1999, Heppner succeeded Ballou as Chief of the WRAIR Malaria Program.

With the MVI on board, GSK and WRAIR collaborated with Pedro Alonso of the Barcelona Center for International Health in Spain, who had developed a research site in Mozambique. Alonso's site would be the biggest RTS,S trial, enrolling 2,022 children between the ages of 1 and 4 years. By 2003 Ballou had rejoined the effort, having left MedImmune to join GSK at Rixensart. The trials with children in Mozambique showed that the vaccine conferred a 35% efficacy against the appearance of parasites in the blood and a 49% efficacy against severe malaria that was maintained for 18 months after the last vaccination. Another formulation specific for children (RTS,S/AS02D) has undergone tests in 214 infants in Mozambique in preparation for licensing by 2011. Infants were given three doses of RTS,S/AS02D or the hepatitis B vaccine Energix-B at the ages of 10, 14, and 18 weeks (15). Early on 17 children in each group had adverse reactions, and later 31 had serious adverse reactions, none of which seemed to be related to the vaccination. Vaccine efficacy for new infections was 65% over a 3-month follow-up after completion of the immunizations. The prevalence of infection in the RTS,S group was lower than in the controls (5% versus 8%). The reason why 35% of the children did not respond was not clear, nor was the reason why the vaccine protected only 34% of adults and had an effect for a shorter duration. These reactions may stem from problems associated with induction of immunological memory.

It has been suggested that both humoral and cell-mediated immunity contribute to protection by the RTS,S/AS02 vaccine, with their relative importance depending on the preexisting immune status of the individual and other, currently undetermined, host factors. The reason why antibody alone does not eliminate incoming sporozoites is not understood; however, it may be that continuous shedding of CSP-antibody complexes allows the highly motile sporozoites to escape. RTS,S vaccination may reduce but not completely prevent emergence of the merozoites from the liver, so that vaccinated children experience an attenuated low-dose blood stage infection that allows a more effective immune response to blood stages (560).

One of Heppner's rivals, Louis Miller of the NIH, questions the idea of fighting off a malaria infection altogether. His group is devoted to creating vaccines designed to reduce the number of parasites in the blood, thus moderating the severity of the disease. He says: "RTS,S hasn't shown in clinical trials that it can stop most or all parasites. It is important to remember that malaria is not like measles and mumps; it isn't something people, young or old, necessarily get once in their lives and it never comes back. This is more like flu; many people are going to get malaria several times in a lifetime . . . So how long is the RTS,S vaccine or any other vaccine going to protect people? How often will they need to be re-vaccinated in order to stop all the parasites?" (289). With Miller's vaccines (MSP-1 and AMA-1) the aim is to keep children alive long enough for their bodies to build immunity to severe disease. Heppner counters (289) that RTS,S will work like a Patriot missile, with the vaccine shooting down the incoming parasites, and then it will survey the infected liver cells so that the parasite within can be attacked and killed. "RTS,S will be the first attacker. The other vaccines will get the remaining parasites at the next stage."

In the Mozambique infant study (15), the primary end point was safety and a secondary end point was efficacy against infection. The efficacy in delaying the time to the first infection was calculated to be 65.9%, with efficacy defined as 1 minus the hazard ratio, adjusted for distance to a health facility. A total of 20 of 93 infants in the RTS,S group and 46 of 92 infants in the control group had at least one infection, to give a difference of 52.6%. To assess vaccine efficacy, these investigators used time-to-event analysis; i.e., in the 1- to 4-year-old children in Mozambique the efficacy was calculated as 45% for delaying first infection over the first 6 months. However, as NMRI's Epstein (161) noted, many malariologists "are more accustomed to thinking of vaccine efficacy as being based on a proportional analysis i.e., the proportion of vaccinated participants versus controls with the outcome during a defined follow up." Using proportional analysis of the same data, the vaccine efficacy at 6 months was calculated at 11%. It is not entirely clear whether time-to-event or proportional analysis will better predict the long-term effects of the vaccine.

A recent RTS,S trial with toddlers and infants (5 to 17 months of age) in Africa was hailed by Christian Loucq of the PATH MVI, writing in the *New York Times* in December 2008, as "we are closer than ever before to developing a vaccine" because the efficacy achieved was 60% against all episodes of *P. falciparum* malaria. During the 6-month period after immunization with three doses, the incidence of the first and only malaria episode in the control group was 16% (66 of 407 subjects) whereas the incidence in those vaccinated with the newer and more immunogenic adjuvant, AS01E, was 8% (32 out of 402) (40a). Where there was more than one malaria episode, the differences were 3% and 1%, respectively. Although a reduced risk of infection and an increased antibody titer to CSP were correlated, no reduction in the incidence of malaria was associated with the increase in titer. In another study with infants 8, 12, and 16 weeks old, RTS,S was given along with vaccines against polio, diphtheria, tetanus, pertussis, and *Haemophilus influenzae* B without lessening the effectiveness of the vaccines (1). Clearly, these results are a "hopeful beginning" (103a); however, the real test of the RTS,S vaccine will come when it is used in large areas of Africa with moderate to intense transmission (unlike that in this trial).

Protection using the subunit RTS,S vaccine is far from perfect, and perhaps with better formulations of adjuvants and with viral vectors (450, 482) there will be enhanced efficacy (270). However, there remains a nagging question: is immunity actually dependent solely on a CSP construct? Indeed, recently the critical role of CSP in induction of sterilizing immunity has been called into question by researchers studying rodent malarias. In one instance, in a study of *P. berghei* infection, sterile immunity was obtained despite there being no immune response specific to CSP (201); in another study a mouse strain completely immunotolerant to CSP was fully protected after three immunizations of irradiated sporozoites and challenge with infectious sporozoites (280).

Are We There Yet?

In the past decade research on malaria vaccines has moved along two parallel paths: those involving recombinant proteins formulated with novel adjuvants and those involving DNA plasmids, recombinant viral vectors, and DNA-prime/virus-boost combinations. Studies of the preblood stages of malaria parasites begun by Clay Huff at NMRI in 1947 and continuing until his retirement in 1969 were followed by those of Richard Beaudoin (1931 to 1990) until his untimely death and continued by Stephen Hoffman from 1985 to 2001. In the early 1990s the main focus of the vaccine program at NMRI was on DNA vaccines with special emphasis on preblood stages and immunity to these stages. Indeed, when the program was started, DNA vaccines seemed to overcome the impediments of more traditional recombinant protein subunit vaccines. DNA-based vaccines induce both CD8 T-cell responses (to produce IL-12 and IFN-γ) and T-helper cell responses (to release Il-4, IL-5, IL-6, and IL-10), are simple to design and modify, can be combined easily, are stable, do not require refrigeration, and can be produced on a large scale. In 1993 a plasmid DNA encoding a portion of the CSP from *P. yoelii* was used to immunize mice (426); as predicted from experience with other systems, the levels of CD8 T cells and IFN-γ increased. A year later, when mice immunized with a plasmid encoding another *P. yoelii* antigen were shown to be protected against sporozoite challenge, it encouraged the development of DNA-based vaccines for use in humans. Indeed, the combination of DNA plasmids was pursued by NMRI in partnership with Vical Inc. in a program called MuStDo (MultiStage DNA Vaccine Operation) (133). The initial clinical trial tested the efficacy of a five-plasmid mixture of preblood antigens, and as with the mouse malaria vaccines, antigen-specific CD8 T cells and IFN-γ responses were found in the volunteers. However, the CTL and IFN-γ responses in the volunteers were suboptimal for protection, and despite earlier work showing that the vaccine induced antibody responses in mice, rabbits, and monkeys, there was no such response in humans (133). Further, although the vaccine did provide evidence for priming in that the immune responses were boosted in the vaccinees after exposure to sporozoite challenge, there was no protection (543).

Such disappointments with plasmid DNA vaccines (519) have led to another approach: using virus vectors. Viruses, when modified, have the advantage of being able to serve as vaccine carriers that are able to deliver DNA more efficiently than when it is in plasmid form; however, they may have a drawback in that they may divert the immune response to irrelevant virus antigens and may neutralize preexisting antibody responses to viral components. The NMRI vaccinologists elected to take advantage of both plasmid DNA and viruses. In a complex regimen, they used a priming dose of plasmid DNA followed by boosting with a recombinant virus. A cytokine-enhanced multiantigen DNA plasmid encoding two sporozoite antigens (CSP and a thrombospondin-related antigenic protein, TRAP) plus two blood stage antigens (AMA-1 and MSP-1$_{42}$) for priming, and a canarypox virus as a booster, was used with *P. knowlesi*; 2 (18%) of 11 monkeys vaccinated showed sterile protection and 7 (77%) of 9 vaccinated rhesus monkeys spontaneously recovered (400). In a subsequent and more detailed study involving priming with three doses of DNA plasmids containing a mix of the same four *P. knowlesi* antigens and then boosting with the poxvirus, it was concluded that (i) the timing of vaccination and formulation were critical to protective efficacy, (ii) protection could be achieved in some animals against both preblood and blood stage infections, and (iii) in a multistage vaccine the individual antigens may interfere with one another. A perplexing finding was that although the vaccine was protective, the IFN-γ responses correlated with an earlier appearance of parasites in the blood (547). The fowlpox virus (FP9) and a modified vaccinia Ankara virus, MVA, expressing peptides from TRAP and AMA-1 fused to a multiepitope string of T- and B-cell epitopes including copies of NANP, named FP9-MVA-TRAP, developed by the University of Oxford and shown to reduce by 90% the number of parasites emerging from the liver when tested in The Gambia, showed only 10.3% efficacy relative to controls. When PEV3A (Pevion Biotech in collaboration with the Swiss Tropical Institute), an influenza virus carrying the NANP B-cell epitope and an AMA-1 peptide, was tested on its own or in combination with FP9-MVA-TRAP after sporozoite challenge, there was neither sterile protection nor clear evidence for a reduction in the number of liver stage parasites with the combination; however, for PEV3A on its own there was evidence of reduced multiplication of asexual stages in the blood and abnormal-appearing parasites (called "crisis forms," first described in monkey malaria by the Taliaferros in 1934 and later found in mouse and human malarias). None of the volunteers was completely protected (505).

One of the more promising vaccine approaches has been to use adenovirus vectors, which induce both humoral and cell-mediated responses simultaneously and which can protect after a single immunization. For clinical evaluation, NMRI in collaboration with GenVac Inc. is testing a combination of two adenovirus vectors, one carrying CSP and another carrying AMA-1, in one- and two-dose regimens and subsequently in prime-boost combinations in humans to determine their feasibility as a multistage vaccine (293). Other groups, particularly one led by Adrian Hill at the Wellcome Trust Centre for Human Genetics, University of Oxford, are in the process of testing other virus vectors able to induce strong antigen-specific antibody responses (143, 391, 392, 474).

A priori the obvious way to develop a protective vaccine for malaria is to employ multiple antigens against multiple stages; however, to date, no multiantigen, multistage vaccine tested in clinical trials has been able to induce strong immune responses against each component simultaneously in more than a few volunteers, nor has it been possible to simultaneously induce both a humoral response and a cell-mediated immune response in humans. There is, however, no shortage of potential vaccine candidates—composed of either a single antigen or several antigens—but of the 23 vaccine candidates under clinical testing and listed in the 2006 WHO portfolio, most are based on just three antigens (CSP, MSP, and AMA), and these were cloned two decades ago. Because there is suggestive evidence that the repeat sequences in CSP and TRAP may act as a smoke screen, allowing the parasite to evade the immune response, it remains imperative that efforts be directed at identifying other target antigens from among the 5,300 *P. falciparum* putative proteins.

So are we there yet? On an optimistic note, Epstein et al. (162) wrote: "The recent advancement of several subunit vaccine candidates into test-of-concept studies as well as the now vigorous efforts to develop whole organism vaccines, the goal of a licensed vaccine appears closer than ever . . . In spite of less than optimal and even disappointing trial outcomes, important lessons have been learned regarding vaccine construction and stability, potentially relevant immunological assays, clinical trial design and site specific epidemiology. The package of lessons should eventually enable vaccinologists to develop, license and deploy a malaria vaccine." The time frame for accomplishing this is undetermined; however, in the opinion of some independent-thinking malariologists (who are also un-compromising vaccinologists), it is at least several decades away.

Back to the Future

Since the 1980s subunit malaria vaccines have received the most attention, and it is sobering to note there is only a single recombinant protein vaccine on the market for any disease, and there are no vaccines based on synthetic peptides, recombinant viruses, recombinant bacteria, or DNA plasmids (see Table 1, p. 89). In stark contrast to subunit vaccine formulations for malaria, protective studies with irradiated sporozoites of rodent and human malaria parasites seem to have fared better. In 1989, faced with the disappointing trials with CSP-based vaccines, γ-radiation-attenuated sporozoite immunization studies with volunteers were begun at NMRI and WRAIR (163, 224). After 10 years of clinical experience with these immunizations, Luke and Hoffman (301) concluded: "immunization with radiation-attenuated *P. falciparum* sporozoites provides sterile protective immunity in >94% of immunized individuals for at least 10.5 months against multiple isolates of *P. falciparum* from throughout the world." They went on to write: "Given the . . . need for an effective . . . malaria vaccine . . . we believe that an attenuated sporozoite vaccine should be produced and tested for safety and protective efficacy as soon as possible." And, echoing the words Paul Silverman (see p. 124) used decades earlier to convince USAID that a vaccine against blood stages was an achievable goal, i.e., that the hurdles associated with a whole par-asite vaccine were "simply technical problems" that could be overcome by a major, well-funded research program, Hoffman wrote that "although technically and scientifically challenging such an approach has an enor-mous advantage over other approaches." Shortly thereafter (in 2002),

Hoffman founded and became the Chief Executive Officer of the only company in the world dedicated solely to developing a radiation-attenuated sporozoite vaccine for malaria. The name of the company, Sanaria, meaning "healthy air," is a clever counterpoint to the Italian word mal'aria, meaning "bad air." In 2005, Hoffman filed a patent (no. 200050220822) for "aseptic, live, attenuated sporozoites as an immunologic inoculum" and began the difficult task of putting theory into practice.

Stephen L. Hoffman (born in 1948), currently the Chief Executive and Scientific Officer of Sanaria, is a lean, green-eyed, exuberant optimist, heralded in the 11 December 2007 *New York Times* as "the soul of a new vaccine." Impatient and intolerant of negativity, he is following the centuries-old precepts of Louis Pasteur to turn a crippled malaria parasite into a wonderful protective weapon against itself. Hoffman grew up on the Jersey Shore, where he went to Asbury Park High School, received a B.A. degree from the University of Pennsylvania (in 1970), and enrolled as a medical student at the Cornell University Medical College. At the time, Cornell required a 60-contact-hour course in tropical medicine; it was this course that led to a fellowship allowing Hoffman to spend the summer studying lactase deficiency in infants in Bogotá, Colombia. Following a year in Colombia, during which he was hospitalized for 10 days in southern Ecuador with typhoid fever and had amebic dysentery and giardiasis multiple times, he was dedicated to pursuing a career in tropical medicine. After completion of his M.D. degree in 1975, Hoffman did a residency at the University of California San Diego and 6 months at the London School of Hygiene and Tropical Medicine, where he received a Diploma in Tropical Medicine and Hygiene in 1978. He returned to San Diego, where he founded the Tropical Medicine and Travelers Clinic, which is still in operation; 2 years later he joined the U.S. Navy specifically to conduct research on tropical diseases. He arrived at the Naval Medical Research Unit 2 (NAMRU-2) in Jakarta, Indonesia, in 1980, where he studied typhoid fever, filariasis, cholera, and malaria. His most striking success was the demonstration that treatment with high-dose dexamethasone reduced the mortality of severe typhoid fever by 82%, enabling his team to reduce the typhoid death rate from 15% to 1% in the Infectious Disease Hospital in Jakarta; a similar approach had no impact on patients from Papua with cerebral malaria. Upon rotation back to the States in 1985, Hoffman was based at NMRI; from 1987 to 2001 he was the Director of its Malaria Program, where he built a team of over 100 individuals in the United States and overseas working on all aspects of malaria but especially vaccine development. In 1995, a time when Hoffman was actively engaged in DNA-based vaccines for malaria, he read Craig Venter's papers on the genomes of *Haemophilus influenzae* and *Mycoplasma* and mused, "wouldn't it be interesting to do the same with *Plasmodium falciparum*?" He paired with Venter to organize a consortium consisting of The Institute for Genomic Research, NMRI, the Sanger Centre, and Stanford University with funding from NIH, the Department of Defense, The Wellcome Trust, and the Burroughs Wellcome Fund to sequence the genome of *P. falciparum*, and in 2002 the feat was accomplished. By 2001, Hoffman had to come to the conclusion that it would take many, many more years to develop a highly effective malaria vaccine based on recombinant DNA technology. He retired from the Navy and in 2001 joined Celera Genomics as Senior Vice President of Biologics with the goal of utilizing genomics and proteomics to produce new biopharmaceuticals, especially immunotherapeutics against cancer. While at Celera he organized a Keystone Symposium on malaria vaccine development, but the meeting left him frustrated by the realization that biotechnology was unlikely to produce a highly effective malaria vaccine for at least another 25 years. However, in putting together data from 10 years of work in immunizing people with radiation-attenuated sporozoites, he came to the conclusion that there already was a way to make a malaria vaccine. The approach had been pioneered decades earlier by the NYU group (Nussenzweig and Vanderberg) using rodents and later tested with human volunteers in Maryland and Illinois (by Clyde and Rieckmann). Thinking that this approach would provide a highly protective malaria vaccine, he left Celera in 2002 and founded Sanaria in his kitchen with a single goal in mind: to develop a malaria vaccine for infants, young children, and women by using a disarmed version of the whole parasite. Malariologists had always considered radiation-attenuated sporozoites as the "proof of principle" that a malaria vaccine could be developed, but none thought that a vaccine composed of radiation-attenuated sporozoites that met regulatory and cost-of-goods requirements could be manufactured. The accomplishments made by Sanaria during the succeeding years have shown that an irradiated sporozoite vaccine can be manufactured. Clinical lots have been produced, and the PfSPZ vaccine has now been shown to be safe in preclinical animal studies. Hoffman will once again be conducting clinical trials and anticipates that Sanaria will be able to provide the 100 million to 200 million doses of vaccine needed to protect infants in Africa.

Hoffman is no newcomer to malaria vaccines. Indeed, in the spring of 1987 as a Commander in the Navy, he was part of a team of military physicians involved in the test of a CSP-based subunit vaccine. At that time he was so confident in the vaccine (FSV-1) that he allowed himself to be bitten by 3,000 infected mosquitoes and then 10 days later went off to a medical conference in San Diego, CA, to deliver what he thought would be a triumphant message. The morning after he landed, however, he was already shaking and feverish, and shortly thereafter he was suffering with a full-blown attack of falciparum malaria. One of his fellow test subjects, Ripley Ballou, a Major in the U.S. Army from WRAIR, who now heads malaria vaccine research at GSK, shared his fate (see p. 256). Ballou has spent the past 20 years working on the development on a subunit vaccine and, unlike Hoffman, still thinks the original vaccine (RTS,S combined with AS02A and tentatively named Mosquirix) is the most promising vaccine candidate. He is critical of Hoffman's attenuated-sporozoite approach, stating that it is impractical: large numbers of irradiated sporozoites must be repeatedly injected to provide solid protection; radiation attenuation is not precise and a proportion of the damaged parasites are ineffective; there are safety risks associated with the sterility and manufacture and use of such a vaccine; and there are huge challenges to the production of such a vaccine on a commercial scale, including the need for thousands of liters of infected blood, the problems associated with raising and infecting 200 million pathogen-free *Anopheles* mosquitoes, and the tedious and time-consuming manual extraction of sporozoites from the salivary glands of the mosquitoes (23). Further, Ballou argues that the dose and schedule will need to be worked out, there is no guarantee that vaccinated infants living in an area of endemic infection will not be boosted, and there is a significant risk that a vaccine that completely prevents infection would shift the occurrence of severe disease to older age groups. Another critic of Hoffman's approach is Pierre Druilhe of the Pasteur Institute, who says, "Even calling it a vaccine is a compliment. It has no chance of offering protection. It is like Captain Ahab trying to kill Moby Dick with a knife" (325). Aside from the risk of underirradiated mosquitoes causing a blood infection or the possibility that the irradiated mosquitoes may not reach the liver, Druilhe says that a frozen vaccine will never be practical in tropical areas of endemic infection that are without proper facilities for refrigeration.

Not all agree with Ballou and Druilhe. Hoffman's proposal to develop a whole sporozoite vaccine has obtained $15 million in backing from the U.S. Army, a San Francisco nonprofit pharmaceutical company, and the Institute for One World Health, as well as another $29 million from the Bill and Melinda Gates Foundation to allow the building of an assembly line to mass-produce the vaccine. This involves raising mosquitoes aseptically, supercharging them with far more parasites than nature does by membrane feeding the mosquitoes on blood containing in vitro-grown gametocytes, allowing 2 weeks for the sporozoites to mature in the mosquito salivary glands, irradiating the infected mosquitoes, and finally dissecting out the sporozoites from the salivary glands. It has been claimed by Hoffman that four trained dissectors working in two biosafety hoods can aseptically isolate sporozoites from at least 75 mosquitoes per dissector per day; he thinks this would be enough for 1,200 three-dose immunization regimens. With two shifts per day and 310 workdays a year, a small factory with 50 full-time dissectors per shift could produce 110 million doses of vaccine per year. These irradiated sporozoites would then be placed in suspended animation in liquid nitrogen. The first clinical trial has been scheduled for 2009, with cooperation by the U.S. Navy and the University of Maryland. The PATH MVI and the Seattle Biomedical Research Institute are teaming up to build a new facility for testing of the safety and efficacy of potential vaccines on paid volunteers. Christian Loucq, the PATH MVI Director, optimistically states (43): "The target is that by 2025 there will be a vaccine with at least 80 to 85% efficacy. It is possible we will combine the Sanaria vaccine with protein-based vaccines . . . and we are going to be in a position to test it here in Seattle and later on in Africa." Even if the Sanaria vaccine is successful, questions remain. Can Sanaria actually produce enough irradiated sporozoites to deliver 200 million doses a year? What will the dosing regimen be? Will the vaccine create dangerous side effects, or will it induce a malaria infection? Will there be boosting when the immunized are exposed to natural infections by the bite of infected *Anopheles* mosquitoes?

Looking to the Liver

The hunt for a vaccine against the liver stages of malaria began more than 75 years ago when Clay Huff (1900 to 1982), then a young Assistant Professor in W. H. Taliaferro's Department of Microbiology at the University of Chicago (see p. 92–93), made an unexpected discovery: malaria parasites could develop in the tissues outside of red blood cells. Huff had been trained as a medical entomologist, receiving his Ph.D. in 1927 at the newly organized School of Hygiene at Johns Hopkins University for studies of the susceptibility and immunity of mosquitoes to bird malaria. However, in 1930, upon finding malaria parasites developing outside of red blood cells, he redirected his research to these stages. He called these stages exo-erythrocytic (EE) forms, and in describing a new species of avian malaria (*P. elongatum*) he wrote (248): "The segmenting forms are abundant in hematopoietic tissues" and in these stages "pigment may be absent." One of his closest friends at Chicago, the eminent histologist William Bloom, became excited on seeing Huff's slides showing parasites in the bone marrow, and for the next 40 years Huff devoted his energies to learning more about these forms. Particularly, he wondered whether these cryptic stages could be immunogenic. In 1935, Huff and Bloom described multiplication of *P. elongatum* EE forms in the canary and, in 1944, using the popular chicken malaria parasite *P. gallinaceum* (see p. 91–92), it was possible for Huff and his graduate student Frederick Coulston to monitor the entire development of the parasite from sporozoite to EE form to the stages growing and reproducing in the red blood cell (249).

With Huff's discovery of EE forms, he unwittingly initiated a hostile environment in the Department of Microbiology, and there was a souring of his relationship with its Chairman, W. H. Taliaferro. This was due in large part because Taliaferro had expected to be given coauthorship, and thus to bask in the glory of the finding of EE forms; however, Huff denied him such an accolade. There was a final break between the two when, in 1947, Huff left the University of Chicago and moved to NMRI as the only parasitologist; he spent the remainder of his career at NMRI until he retired in 1969. From the outset, Huff recognized that an impediment to studies of EE forms was the lack of sufficient material. Over the next 10 years, he labored to harness the newly developed methods of tissue culture for growing EE forms. Success came with *P. fallax* growing in turkey brain tissues, and beginning in 1960 the Huff laboratory used the EE forms growing in tissue cultures to describe for the first time the morphology, development, and behavior of a living intracellular parasite (244). The availability of large numbers of EE forms of *P. fallax* in cell cultures free from red blood cells prompted him to prophesy that this technique may be expected to make possible the analysis of the antigenic potencies of the two forms of asexual stages of the parasite (247).

In 1964, Huff invited me to visit Bethesda and to join him in studies of the EE stage antigens at NMRI, but having just established my own research laboratory at the University of California I declined his offer to move. Another factor was that having served 2 years in the U.S. Army, I was reluctant to become, so to speak, a "sailor" and to endure the routine weekly field days, the inspections, and the formalities attendant on military rank. Huff, in addition to being an exceptionally gifted scientist, was a kind and gracious man; he had special warmth not only for his beloved EE forms but also for younger scientists as well as colleagues, and for years afterward I felt bad for declining his job offer. Subsequently, Huff offered the position at NMRI to Richard L. Beaudoin, who, after spending 2 years with the Navy as a line officer, returned to graduate school at Notre Dame and then earned a Ph.D. in parasitology at the University of Iowa. Beginning in 1965, under Huff's tutelage Beaudoin began to study the morphological effects of antimalarials on the tissue-grown EE forms.

In 1974 Beaudoin, now Head of the NMRI malaria laboratory, came to realize that although the avian model for cultivating EE forms had been used for 30 years (thanks to the efforts of Huff), little progress had been made with mammalian parasites. He therefore initiated a program of research to grow EE forms in liver cells, first using avian malaria parasites. When Michael Hollingdale joined the group, success was achieved. Although in vitro cultures of mammalian malaria parasites initiated from sporozoites were reported as early as 1976, most of these were only partially successful. The discovery that the human embryonic (WI38) cell line was susceptible to *P. berghei* led to the first successful culture of mammalian EE stages (233). Following this, in vitro cultivation of various rodent species of *Plasmodium* in liver cells was achieved; however, the necessity for repeated isolation of these cells, which do not replicate or retain their differentiated functions, restricted the value of such culture systems. This was soon overcome by the use of "immortalized" liver cells such as the human hepatoma line HepG2-A16, which supports the development of *P. berghei*, with about 8% of the sporozoites becoming EE forms (234).

The EE stages of malaria living outside of red blood cells lack hemozoin, and unlike the pathology-producing and rapidly dividing pigmented parasites in the blood, they are clinically silent. As a consequence, their significance for vaccine development went unappreciated until 1971, when Leslie Stauber (1902 to 1973), a Professor in the Department of Zoology at Rutgers University who had in 1937 received a Ph.D. (*The Factors Affecting the Synchronous Periodicity of the Reproductive Cycle in an Avian Malaria Parasite*) from the University of Chicago under the direction of W. H. Taliaferro and had been friendly with the young Assistant Professor Clay Huff, spent a sabbatical leave in Huff's NMRI laboratory. Stauber had a special interest in malaria immunity stemming from his experiences during World War II, when he was at the Squibb Institute for Medical

Research, where he had developed the first simple serologic test for an antimalarial antibody. Stauber initiated vaccination trials using the EE stages of *P. fallax*; however, his death before the work could be finished prompted two of his students, Harry Graham and Thomas Holbrook, to carry it to completion (227). Graham et al. were able to passively transfer protection by intravenous administration of antisera specific for EE stages of *P. fallax*; the protective activity was found in the IgG fraction. Later, Holbrook et al. found that 100% of turkeys vaccinated with formalin-killed EE stages of *P. fallax* survived a challenge infection with EE merozoites from culture when blood stages were drug suppressed. When EE-immunized birds were challenged with EE merozoites without drug treatment, blood infections were the same as those in controls. Challenge of EE-immunized turkeys with infected red cells resulted in blood infections similar to those in controls.

In 1976, Beaudoin attended a WHO meeting in Geneva, Switzerland, and was asked whether cultivated EE parasites might be considered a source of antigens for malaria. He cited the work of Graham et al. and Holbrook et al. and replied: "there is a recent demonstration that EE merozoites are immunogenic" and from these findings "it is reasonable to suspect that these stages may stimulate protective immunity" (38). He went on to speculate that perhaps anti-EE immunity will be found to be related to the protection stimulated by irradiated sporozoites. Indeed, in the report of his very first study using irradiated sporozoites, he suggested: "the results of the present study do not imply that liver stages were unimportant in initiating the immune response" (40). In effect, he questioned the prevailing view, held since 1967 from experiments using X-irradiated sporozoites in rodents and humans for protection, that the dominant sporozoite antigen, CSP, was key to the induction of sterilizing immunity, and as such it alone could serve as the basis of a protective vaccine. The gauntlet had been thrown down, and the research groups at NYU and NMRI as well as those at the Institut Pasteur in Paris soon became adversaries.

Using the mouse malaria model, *P. yoelii*, it was shown that after mosquito inoculation the irradiated sporozoites are deposited in the skin (176, 526) and then move into the regional lymph nodes, where, after dendritic cell presentation, T cells were primed to recognize the parasite (362). The activated T cells then moved on to the liver, where sporozoite-infected liver cells were eliminated upon subsequent challenge. Only antigens expressed by both the sporozoite and the liver stage parasites, and not those expressed solely by the liver stages, it is claimed by the Zavala group, are involved in protection. Further, they argue that CSP, expressed in both the liver and sporozoite stages, is the immunodominant target of the T-cell response and that the antisporozoite CD8 T-cell response originates not in the liver but in the lymphoid tissues at the site in the skin where the mosquito has bitten. Afterward, when the CD8 T cells are activated, they travel to the liver, where the sporozoite antigen is recognized by the liver-resident CD8 T cells (209). This occurs rather quickly, and within a few hours IFN-γ is produced within the liver, killing the parasites (79).

Important as CD8 T cells are to preblood stage immunity, the immune response is strictly dependent on CD4 T cells. This occurs several days after CD8 T-cell activation. The IL-4-secreting CD4 T cells are crucial in mediating the interaction and play a triple role in protective immunity against liver stages: helping B cells secrete antibody, assisting in the induction of CD8 T-cell responses, and inhibiting the development of the liver stage parasites (136, 232).

Maximal activation of CD8 T cells occurs within 8 h after immunization with irradiated sporozoites and then declines to imperceptible levels by 96 h. In spite of this rather brief presentation of sporozoite antigen, it is sufficient to trigger T-cell proliferation lasting several days (210). Of some interest was the observation that the magnitude of the T-cell response was not enhanced by repeated bites by infected mosquitoes to increase the amount of antigen presented. Zavala contends these mouse studies with *P. yoelii* mimic the manner by which immunity to sporozoites is acquired under natural conditions in areas of endemic infection and is similar to the sterile immunity to *P. falciparum* induced in human volunteers. Unexplained, however, is why sterile immunity does not occur in humans under natural endemic conditions, subjected to the bites of infected mosquitoes.

Fidel Zavala (born in 1952) grew up in southern Chile and received his M.D. in 1977 from the University of Chile in Santiago. From the very beginning of his medical studies, he was more interested in research than in clinical medicine. As a result, in medical school he frequently skipped clinical duties (especially surgery and obstetrics-gynecology) to work in an immunogenetics laboratory. After graduation, he was a pediatrician for a few months and then became an Instructor in the Department of Cell Biology and Genetics at the University of Chile School of Medicine. In 1980, he was offered a position with Ruth and Victor Nussenzweig at the NYU School of Medicine as a postdoctoral fellow in immunology. Although he had absolutely no knowledge about malaria (in Chile there is no malaria), he thought, "just another antigen," and he expected to spend just a couple of years as a fellow learning a bit of immunology—something he is still trying to do (and in the process has forgotten his medical training). Instead of 2 years, Zavala remained at NYU for 23 years; he became a Professor there, and only in 2003 did he move to the Johns Hopkins School of Public Health as a Professor of Molecular Microbiology and Immunology.

Zavala has contributed significantly to a better understanding of the basic mechanisms of protective immunity against malaria. His current work with rodent malarias involves the characterization of the cellular and molecular mechanisms underlying the induction of protective T-cell-mediated immunity by using strains of transgenic mice bearing T-cell receptors specific for parasite antigens recognized by $CD8^+$ or $CD4^+$ T cells (518). In addition, his research focuses on the characterization of the molecular and genetic events involved in the development and maintenance of anti-parasite T cells, and the manner by which $CD4^+$ and $CD8^+$ T cells interact during the development of immunity to malaria. His laboratory is currently investigating the mechanisms involved in tissue trafficking of activated $CD8^+$ T cells and identifying the nature of the molecular processes leading to the development of memory. He is convinced that studies of immunity with rodent malaria parasites are important since they have the potential to identify important elements and to raise critical questions that can then be applied to the human disease. Indeed, he opines, almost everything involved in the RTS,S vaccine (see p. 255–261) now in clinical trials is based on the large amount of work done with rodents and carried out by a legion of people including himself. He laments, "that is something not readily apparent when one reads papers on clinical trials with the sporozoite vaccine is that the seminal findings in rodents have been ignored and oftentimes it appears that basic knowledge sprang de novo. Alas, life is unfair . . ."

Zavala acknowledges that the amount of immunologic knowledge concerning malaria gained over the past two decades has been truly amazing. However, more impressive is that in looking over its history, it can be seen how naive most investigators were in believing that, after gene cloning, development of a malaria vaccine would be a piece of cake. At the time few realized how little was actually known about immune mechanisms, as most of what was known came from immunological studies of ovalbumin, keyhole limpet hemocyanin, and bovine serum albumin—hardly "horrible infections" that affected nobody. Immunologists in the 1970s and 1980s did not work with real microbes because the models were not clean enough for use in what was considered modern immunology. Things have changed a bit, and presently microbial immunology is something else: a different immunology that is also worthy of study.

Zavala is confident that a vaccine is possible and in fact quite likely. He feels that the fundamentals are sound and very strong: humans can be protected by immunization with irradiated sporozoites, and individuals who develop immune responses to blood stages have an attenuated infection or disease. Wisely, he proffers no time line.

Not all malaria researchers who have studied pre-blood-stage immunity place their entire faith in CSP as the key pre-blood-stage antigen. Indeed, CSP as a dominant antigen interferes with the development of T-cell responses to antigens of different specificity, and thus the result is an immunological response which is too narrow in its specificity and much too regulated since it does not seem to increase with time and repeated exposure. Clearly, antigens other than the CSP must exist, but it has been difficult to find them because of the immunological dominance of CSP (Zavala, personal communication).

The suspicion that liver stage antigens could act in the induction of immunity came to be raised by more and more observations first in mice and then with human malarias (223, 502). With mouse malarias, it was found that the X-irradiated sporozoites had to be alive for some time in the liver: too low a dose of radiation led to a blood infection, whereas too high a dose resulted in an absence of protection. At the irradiation dose inducing protection, sporozoites were found to transform into young liver stage parasites that were blocked in development and unable to divide but still able to secrete antigen. Indeed, in the rodent malarias, young liver stage forms could persist for up to a year after immunization with X-irradiated sporozoites, thus allowing prolonged antigenic stimulation. Similarly, if live nonirradiated sporozoites were used to inoculate mice and then the mice were treated with a drug that blocks liver stage parasites (α-difluoromethylornithine) or one that prevents blood stages (chloroquine), the animals were protected against a challenge with sporozoites but not with blood stage parasites (41). However, despite such evidence, for decades investigators were discouraged from studying liver stage antigens in mammalian malarias. The presumption was that because the liver form was clinically silent, it could have no immunological importance.

There was another impediment to their study: EE forms are present in minuscule numbers within the very large liver. By way of example, if a human were to be bitten by an infected mosquito and received a maximum of 100 sporozoites, this could result in 100 EE forms being present in a liver containing 100 billion cells. Although the liver of a mouse is much smaller than that of a human and contains only 2 billion cells, the chance of finding EE forms still is very slim. Further, because CSP was (and to some malaria researchers still is) considered to be the sole major antigen, the focus was almost entirely on it; other antigenic proteins that might have been expressed by the sporozoite or liver stage were neglected. The disappointing results with CSP-based vaccines, coupled with the early observations on irradiated sporozoites and drug treatments (see above), prompted some investigators to search for liver stage antigens (81, 144).

Following clues provided by the studies of Verhave (529), Beaudoin et al. (39) made a seminal finding: if during the course of immunization they were treated with chloroquine to suppress the blood infection, and if, prior to sporozoite challenge, they were given a curative dose of primaquine to eliminate the liver stage parasites, they resisted a challenge with sporozoites but not to blood stages. The authors concluded that unaltered sporozoites were immunogenic and stimulated protection similar to that achieved by immunization with irradiated sporozoites. Vaccination studies by others (again using a mouse malaria parasite, *P. yoelii*) with live sporozoites followed by chloroquine treatment showed that antibody was not implicated in protection. However, depletion of CD8 and CD4 T cells did result in a partial reversal of protection against a challenge infection with sporozoites (and this was mirrored by elimination of liver stage parasites). In this they agreed with Zavala et al. that the CD8 T-cell-derived IFN-γ, as well as nitric oxide and other factors, affected liver stage parasites. However, they went on: "The immunity induced by this procedure (i.e. live sporozoites with chloroquine treatment) is mainly directed at the liver stage parasites, which is also the case for irradiated sporozoites" and "the observations imply that a very short exposure to the immune system to liver stage parasites could result in efficient induction of protection . . . and only T cells are responsible" (41).

It was suggested by some investigators that the antigens expressed in the maturing and mature liver stage parasites were potent inducers of protective immunity. They also had the audacity to claim that the protection afforded by the sporozoite vaccine might in fact depend on liver stage antigens! There are, however, skeptics who doubt the importance of liver stage antigens in preblood immunity because liver cells do not prime T cells and because the liver is an organ that suppresses rather than induces T-cell responses. The non-CSP antigens that may be effective in the induction of immunity are those expressed both in sporozoites (so that they can be induced after immunization) and in liver stages (so that they can make the liver the target of activated T cells). Reinforcing the proposition that liver stage antigens as well as CSP are involved in preblood stage immunity is a recent report concerning Kenyan children with high levels of IgG with antibodies to LSA-1 (liver stage antigen) as well as CSP and TRAP. All these levels correlated with a lowered risk of developing clinical malaria; however, the antibody to LSA-1 was the most strongly associated with protection (263).

By the late 1980s, Beaudoin's group had identified multiple non-CSP antigens in sporozoites that might be associated with liver stages (137, 138, 232). However, after attempts to produce monoclonal antibodies for liver stages in mouse malarias failed, another approach was tried: screening of human sera to find those that reacted only with liver stages. This high-risk, innovative approach was pioneered by Pierre Druilhe of the Pasteur Institute. Druilhe, born (in 1946) and educated in Paris and with M.D. and postgraduate degrees in hematology and medical parasitology (in 1972) and immunology (in 1975), respectively, is currently the Head of the Biomedical Parasitology Unit of the Pasteur Institute in Paris. The analysis of the interaction between humans and *Plasmodium*, especially the importance of liver stage antigens, and the development of laboratory methods able to assess clinical protection for use in identifying potential protective antigens as vaccine candidates have been central themes of his research. Druilhe revels in the knowledge that as an immunologist/parasitologist he has marched to the beat of a different drummer. Rather than employing the more traditional methods of moving from bench experiments to using the mouse immune system to screen for putative protective antigens, he used human IgG passive-transfer experiments to identify blood stage candidates (145, 146).

In vivo transfer of antibody from immune adult Africans (in a state of premunition) reduced the severity of blood infections, and in laboratory studies the antibodies acted indirectly to kill the parasites in an antibody-dependent cellular inhibition (ADCI) involving monocytes. The ADCI mechanism involved two of the seven IgG subtypes (IgG1 and IgG3), and these were used to select protective blood stage antigens, one of which was MSP-3. Antibodies raised in mice as well as from humans immunized with MSP-3 mediated killing of *P. falciparum* in vitro (ADCI assay) as well as in vivo (SCID mouse model). In humans living in a region in Senegal where malaria is endemic, a strong correlation was observed between protection against malaria attack and the levels of IgG1 and IgG3; the level of IgG3 against MSP-3 was predictive of clinical outcome. However, human vaccine trials have yet to be conducted.

Druilhe was able to successfully culture *P. yoelii*, *P. vivax*, and *P. falciparum* liver stages in vitro, and using these he showed that anti-sporozoite antibodies affected not only invasion but also development within the liver cell. Since preblood stage immunity depends on the successful transformation of irradiated sporozoites into liver stages, Druilhe, ever the iconoclast, has contended that what has been long considered "anti-sporozoite immunity" is in fact "liver stage-dependent immunity." Druilhe has 18 patents—3 for blood stage antigens and 15 for liver stage antigens—and is the author of 275 publications.

The sera Druilhe used to identify liver stage antigens came from individuals who had been living in Africa for 15 to 26 years and who had taken daily doses of chloroquine to suppress the appearance of blood parasites (145). A serum (designated PM) from one of these individuals, a priest known as Père Mauvais, who had not suffered from a blood infection but who had repeatedly been bitten by infected mosquitoes, was used to screen a *P. falciparum* genomic DNA expression library (see p. 165). The priest's serum identified many new antigens, several of which reacted only with liver stages. Two of the antigens, LSA-1 and LSA-3, have received special consideration as vaccine candidates because individuals living in areas where malaria is endemic show a high prevalence of immune responses (antibody, CTL, and helper T cells) to them.

LSA-1 is a very large protein consisting of a large central repeat (approximately 89 repeats of 17 amino acids that make up 77% of the protein) flanked by two nonrepeats at the ends and is exclusively found in the liver stage forms (222). The gene for LSA-1 was cloned in the late 1980s (202); however, it was only recently that scientists at WRAIR were able to develop an LSA-1-based recombinant protein (designated FMP011) (222). When FMP011 was used to immunize human volunteers along with the adjuvants AS02 and AS01, it was shown to be safe; however, both preparations failed to show protection against challenge with sporozoites and there was no delay in the appearance of parasites in the blood.

LSA-3 is an abundant large protein made up of nonrepeated sequences flanking glutamic acid-rich repeated regions; it is expressed in both sporozoite and liver stages. It is thought to be involved in the reproduction of EE forms and their release from liver cells. LSA-3 was able to confer strain-independent protection against *P. falciparum* sporozoite challenge in immunized chimpanzees and *Aotus l. griseimembra* monkeys (116, 117, 370). Of nine chimps immunized with LSA-3, either as a lipopeptide or as a polypeptide in saline or with the adjuvant Montanide or SBAS2

(SmithKlineBeecham), six showed full sterile protection on first challenge. In an *Aotus* trial, LSA-3 was adsorbed to polystyrene beads and the animals were vaccinated three times and challenged as late as 5 months after the last immunization by intravenous inoculation of 100,000 sporozoites; sterile protection was achieved in three of five monkeys. Protection appeared to be associated with high levels of IFN-γ production. LSA-3 is now undergoing clinical trials in human volunteers for safety.

Over the past three decades, rodent malarias have been the most extensively studied of all malarias. Research using these models has been based on the principle that the findings obtained using the readily available and genetically defined laboratory rats and mice could be extrapolated to monkey models and then to human disease. To be sure, much has been learned from the use of rodent models; however, one malariologist has cautioned that they have received more than their fair share of attention and counsels that a hazard in using such models, for better or worse, may be that they become the object of study in themselves. Furthermore, care must be taken when extrapolating the findings from mice to humans since there are now numerous examples where the observations of mice have little or no relevance to the disease in humans. Indeed, it is seldom admitted by those racing to discover "the sporozoite vaccine" that the number of doses of irradiated sporozoites needed for protection of humans is particularly large and that sterile immunity is not as consistent in humans as in rodents. The induction of protection by irradiated sporozoites is complex; it is highly dependent on the host-parasite combination, the dose of radiation, the number of sporozoites inoculated, and the number of immunizations. Simply put: since mice are not men, a successful vaccine for mice is unlikely to protect humans. And so the hunt for a protective vaccine for humans based on sporozoite and liver stage antigens goes on.

Is an Antidisease Vaccine Feasible?

In 1866 Camillo Golgi suggested that a toxin caused the malarial paroxysm. In 1899—after Behring and Kitasato had shown the therapeutic benefits of immune serum against diphtheria and tetanus—Koch deduced the existence of a malaria toxin and proposed the development of an antitoxic or clinical immunity. Half a century later, Brian Maegraith of the Liverpool School of Tropical Medicine proposed that fever and other pathologic processes in malaria were the result of a plasmodial toxin and suggested that the clinical symptoms of malaria were similar to those of bacterial sepsis, which is caused by a toxin. Developing an antitoxicity or antidisease vaccine would be a distinctly different approach from developing vaccines that are designed to kill malaria parasites. An antitoxicity vaccine would reduce malaria mortality and morbidity directly by immunizing against the parasite products that are the chief cause of host pathology. In 1993, Schofield and Hackett (424) identified the malaria toxin as GPI. GPI activates macrophages and induces the expression of acute-phase cytokines such as TNF, IL-1, and IL-6 and a chemokine cascade leading to local, organ-specific, and systemic inflammatory responses. In a number of studies done in areas where malaria is endemic, human antibodies were shown to recognize GPIs purified from *P. falciparum*; this raised the possibility that naturally occurring antibodies might neutralize GPIs and influence the outcome of the disease. GPI toxicity requires both lipid and carbohydrate components; as a proof of principle for an antitoxicity vaccine, a nontoxic *P. falciparum* glycan [NH_2CH_2-PO_4($Man\alpha1$-2)6$Man\alpha1$-2$Man\alpha1$-6$Man\alpha1$-4$GlcNH_2\alpha1$-6myoinositol-1,2-cyclic-PO_4] was synthesized, conjugated to a protein carrier to increase its immunogenicity, and then used to immunize mice (423, 425). The mice were protected against severe malaria pathology, with reduced death rates; i.e., 75% survived to day 12 compared to 0.0 to 8.7% survival in controls. Parasite levels in blood were the same in both groups, indicating that the vaccine did not act on parasite multiplication. The antibodies did not cross-react with uninfected red cells, and the serum neutralized TNF production by mouse macrophages.

Proponents of an antitoxicity vaccine have been encouraged by these findings and have suggested that such a vaccine might alleviate severe malarial anemia and that reducing the blood infection would diminish the risk of early childhood mortality, allowing the development of normal acquired immunity. However, others are concerned (51) about an antidisease vaccine, suggesting that individuals may feel less sick and present to a hospital later with extremely severe malaria. Furthermore, inhibition of GPI-mediated early proinflammatory responses may favor a rapid multiplication of blood parasites. Some of these concerns may be offset by combining the antitoxicity vaccine with an antiparasite vaccine; however, this would require that the antiparasitic effect be close to 100%—a difficult target to achieve. In addition, the antiparasite immune response elicited by such a combined vaccine would need to be of longer duration than that of the anti-GPI immune response—also difficult to achieve. Theoretically, vaccination against GPI may also interfere with any protective effect of NO. Finally, given the similarity of GPIs across many species, it would be important for an anti-GPI vaccine to avoid inducing an autoimmune response (52). Despite these perceived potential problems, Ancora Pharmaceuticals is developing an antidisease vaccine using the GPI glycan. It will be interesting to see whether such a vaccine will live up to the expectations of its proponents.

Attenuation by Gene Knockout

Irradiated sporozoites penetrate the liver cells, where they are inhibited from becoming fully mature. Studies carried out with rodent malarias show that these radiation-attenuated sporozoites persist in the liver cells for up to 6 months. When these stages are eradicated by drug trestment, protection is abrogated, suggesting that continued synthesis of pre-erythrocytic (sporozoite and EE) antigens is necessary. Indeed, it has been found that the correct irradiation dose is critical to protection since under-irradiation results in breakthrough blood infections and overirradiation results in loss of protection. This being so, attempts have been made to mimic the protection afforded by irradiated sporozoites by inactivating genes that permit the sporozoite to invade liver cells but not allow further development (326).

Although *P. berghei* was the first rodent malaria isolated and is still one of the most commonly used by investigators, other species were subsequently discovered in the Central African Republic: *P. vinckei* in 1952, *P. chabaudi* in 1965, and *P. yoelii* in 1966. *P. yoelii* is highly infectious to BALB/c mice, and fewer than 10 sporozoites can produce an infection. This models the high infectivity of *P. falciparum* sporozoites for humans better than does *P. berghei,* which requires at least 10 times as many sporozoites to infect mice. Hence, some investigators contend that *P. yoelii* is the more relevant model for malaria vaccine studies. The big question has been: Which of the parasite's 5,000 genes to pick for inactivation? It had been known for some time that sporozoites that reside in the salivary glands gain high infectivity for the mammalian host; however, the molecular basis for this phenomenon remained unknown until Stefan Kappe studied it.

Stefan Kappe (born in 1965) always had a strong interest in biology, majoring in it during high school and spending many happy days collecting beetles in the woods around the German town, Bad Homburg, where he grew up. Subsequently, he received a diploma thesis (Masters equivalent) in Parasitology (1991) at the Rheinische Friedrich-Wilhelms University in Bonn, Germany, working under Rolf Entzeroth on host cell invasion by two relatives of malaria parasites (*Eimeria* and *Sarcocystis*). In 1992 he applied to the University of Notre Dame, South Bend, IN, for doctoral work and became John Adams' first graduate student. He studied the molecular events in red blood cell invasion using rodent malaria models. Upon completion of his Ph.D. in 1998, he applied for an advertised postdoctoral position in the laboratory of Victor Nussenzweig at the NYU School of Medicine. The Nussenzweig laboratory was very focused on CSP, but in collaboration with Robert Ménard the group had recently used reverse genetics to analyze the function of TRAP (thrombospondin-related antigenic protein). Kappe joined the ongoing project on TRAP and studied its role in sporozoite gliding motility and invasion. The group at NYU was exceptional and was a breeding ground for new ideas and great projects. However, as Kappe recalls, nearly everything revolved around CSP and TRAP. He remembers coming into the laboratory one morning and glancing into Victor's office. Victor sat at his desk with his shoulders slumped and his head nearly on the desk, not moving. Worried because Victor had not moved for a while, Kappe walked over to check on him. It turned out that Victor was not suffering from an illness but instead was completely entranced by the recently published crystal structure of a thrombospondin domain, also found in CSP and TRAP.

At that time it was obvious that there had to be many interesting molecules other than CSP and TRAP that were important in sporozoite biology; however, no one knew how to identify these molecules. So Kappe took up the challenge and made the first high-quality cDNA library of *P. yoelii* salivary gland sporozoites. He remembers telling Steve Hoffman about having succeeded, and hearing his response: "impossible!" In collaboration with Kai Matuschewski (now at the University of Heidelberg) he used cDNA subtraction screens to identify genes that were specifically expressed in highly infective salivary gland sporozoites—genes now called UIS (for "up-regulated in infectious sporozoites"). Gene chips were also used recently to elucidate the complete set of UIS genes (326a). These gene expression screens also identified numerous other proteins potentially involved in liver infection. When the *P. yoelii* P52 and P36 genes (both coding for a secreted protein of the sporozoite) were deleted (knocked out), the malaria parasite was arrested in its early development in the liver. When as many as 100,000 of these knockout sporozoites were injected intravenously and the immunized BALB/c mice did not develop a blood stage infection, this suggested that the knockout sporozoites might be used as a vaccine. Three immunizations with 10,000 of these knockouts conferred protection against challenge for at least 30 days (283). In another mouse strain (C57BL/6), after three vaccinations with 1,000 *P. berghei* P52 single-knockout attenuated sporozoites, there was resistance to challenge by 10,000 sporozoites (142). Inoculation of these knockouts into muscle or under the skin resulted in complete protection of the animals against challenge; however, the dosage had to be increased.

In other studies, again with rodent malarias (262, 500), parasite development in the liver was blocked by another set of genes, UIS3 or UIS4, uniquely expressed in both sporozoite and liver stage parasites, when these genes were knocked out. Sterile protection was induced by a single intravenous immunizing dose of 50,000 *P. yoelii* UIS4-deficient sporozoites, whereas two doses were needed with the UIS3 knockout. Mice immunized three times with 10,000 UIS4 knockouts were completely protected for at least 6 months. Mice injected three times with 10,000 *P. berghei* UIS3 knockouts were protected for at least a month against a challenge of 10,000 sporozoites.

In 2002 the Seattle Biomedical Research Institute offered Kappe a position with their fledging malaria research program. Feeling that it was important for him to leave the NYU "nest" and start a truly independent program, he moved to Seattle in 2003. His rodent malaria work enabled the identification of a handful of genes that, when deleted, cause a complete block in liver infection. The fact that mutant parasites infect the liver but cannot develop into a pathologic blood infection allowed for testing of their potential as live-attenuated vaccines. Kappe therefore (as well as Kai Matuschewski and A. K. Mueller) filed U.S. patents 7,261,884 and 7,122,179 and an international patent application PCT/US2004/043023 entitled "Live Genetically Attenuated Malaria Vaccine."

Kappe and coworkers have identified the equivalent genes in *P. falciparum* and created the first gene deletion strains of P52-P36 double-gene knockouts in collaboration with Alan Cowman's group at WEHI. These strains are currently under assessment for growth in liver cell culture and undergo an arrest in development. It is anticipated that human safety and efficacy tests with one of the genetically weakened *P. falciparum* vaccine strains will be undertaken this year (2009) at WRAIR.

The Kappe laboratory is currently engaged in the systematic identification of liver stage-expressed proteins and so far has identified ~800 proteins. In collaboration with other laboratories, the Kappe group is evaluating many new antigens as targets for cellular responses by using immune cells taken from animals that have been exposed to the live-attenuated sporozoites.

Genetically attenuated sporozoites, it has been claimed, differ from the irradiated ones in their consistent production, genetic stability, and higher potency. Protection is strictly dependent on CD8 (killer) T cells. Encouraging as these trials in mice are, and assuming that protection in humans is achieved, it is still not clear whether it will be possible to produce sufficient numbers of genetically attenuated sporozoites in a viable state for vaccinating 200 million people. Even if these problems are solved and the vaccine is successful in humans, if there were no boosting of the response in individuals living in an area of endemic infection, such a vaccine would benefit only short-term visitors. Regardless of their eventual practicality, studies of how genetically altered parasites induce immunity could provide information on the nature of the protective antigens and in this way serve as a guide for future vaccine developments.

12

Informed by the Immune System

Curiosity stokes the fires of imagination. However, discoveries about the natural world are more often than not the result of serendipity; regrettably, they are simply unpredictable. That is the way it was with my first encounter with malaria immunity. It was a dismal and dreary day in December. The cold wind coming off the lake chilled one to the bone. I trudged through ankle-deep snow to my destination: the animal room in the basement of a building on the Northwestern University campus. At the time, I was a beginning graduate student/research assistant and caretaker of several dozen white Leghorn chickens that served to maintain avian malaria infections. The task was simple enough: obtain a drop of blood from a chick, make a thin blood film on a glass microscope slide, fix it in methanol, place it for 1 h in Giemsa stain, rinse it with water, dry it with a hair dryer, and then place the slide on the stage of a microscope. Once the slide was in place, I added a drop of oil on top of the blood film, swung the oil immersion objective lens into place, and lowered it just enough that it touched the oil droplet. I switched on the microscope lamp and opened the diaphragm of the substage condenser to provide better illumination. Peering through the eyepiece lenses, I carefully rotated the fine-adjustment knob to focus on the stained slide. With the microscope's magnification of $\times 1,000$, infected and uninfected red blood cells could easily be seen. By counting their relative numbers, it was easy to determine the severity of infection. If the parasite numbers were large enough, I would withdraw some blood into a syringe and inject a fresh batch of chicks with a small amount. Infections passed by syringe required no mosquitoes, and the time of appearance of parasites in the blood as well as the infection's severity was determined by the number of parasites inoculated. Some infections were lethal to the chicks, whereas others were mild and the birds recovered.

That day the warmth of the dingy, windowless animal room with dozens of feathered cacklers was welcoming. I removed my fleece-lined jacket, put on a lab coat, reached into a cage, retrieved a chick, and went through the familiar passage routine. Finished with the job, I was ready to remove my lab coat, and if the weather outside had been better I would have left that smelly and noisy animal room, but not this day. Looking at the cage containing several recovered animals, and purely out of curiosity, I wondered whether these birds would resist another infection. I made blood films from samples from two birds; microscopically, no parasites were visible in their blood. I injected the recovered chicks with a syringeful of infected blood left over from the passage. Immediately after inoculation, I made a blood film and examined it for parasites but saw none. Since no parasites were visible, the infected material introduced into the recovered bird had been removed from the circulation. The following day I returned to the animal room. This time I took blood from the recovered chicks, passed it into 4-day-old chicks, and each day thereafter I made blood films from samples taken from the recipients. After a week, I saw small numbers of parasites in their red blood cells. Apparently, the recovered birds, though showing no sign of infection, continued to harbor a few parasites undetectable by microscopy. They were immune. Thus began my pursuit, as it has for a thousand other investigators over the past century, namely, to identify the parasite molecule(s) that elicits immunity, to characterize the mechanisms of immunity, and to develop a noninfectious agent, a vaccine, that could provoke immunity. Although animal models have contributed greatly to our understanding of the immune response to malaria parasites, in the main the picture of human immunity has been provided by the treatment of those with late-stage syphilis, characterized by dementia and insanity, with malaria parasites and by studies of natural infections in populations in areas where the disease is endemic.

Clues from Therapy by Malaria

Syphilis—"the great pox" as the English called it—was a disease that from 1493 onwards swept first over Europe and then across the rest of the world. By the end of the 19th century it was estimated that 10% of the population of Europe was infected, and by the early 20th century mental institutions noted that one-third of all patients could trace their neurologic symptoms to syphilis (442). As early as 1530, Girolamo Fracastoro (1483 to 1553) recognized that this disabling, disfiguring, and dangerous disease was not only disgusting but also contagious. Although Fracastoro was able to describe the earliest stages of syphilis, he could not identify the infectious agent. Indeed, for 400 years no one knew what actually caused the symptoms. Then in 1905 two German microbe hunters, Fritz Schaudinn (1871 to 1906) and Erich Hoffmann (1869 to 1959), identified a slender, corkscrew-shaped bacterium, a spirochete ("spiro" meaning coiled; "chete" meaning hair), in the syphilitic chancres—painless, pea-sized ulcers that appear at the site of spirochete inoculation. Because of its shape, Schaudinn and Hoffmann called the spirochete *Treponema* ("trep" meaning corkscrew and "nema" meaning thread in Greek) and because it stained so poorly they named its kind *pallidum* (from "pallid" meaning pale in Latin). When Hideyo Noguchi isolated the bacterium from the brains of patients with insanity and paresis (incomplete paralysis) due to tertiary syphilis, it became clear that all of the stages of the disease were linked to one kind of spirochete. *T. pallidum* has no spores, cannot be cultured in petri dishes with ordinary culture media, can be grown only in experimental animals such as rabbits and guinea pigs, and divides slowly. Humans are the only natural host for *T. pallidum*; the spirochete is quite fragile, requiring a moist environment, and is sensitive to heat.

If untreated, the chancre usually disappears, leaving a small inconspicuous scar. The secondary or disseminated stage usually develops after the chancre. Spirochetes are present in all the tissues but especially the blood. There is a general tissue reaction—headache, sore throat, mild fever, and, in 90% of patients, a skin rash. In 40% of the cases the central nervous system is involved. Then the patient enters the early latent stage, in which he or she appears to be symptom free. The infection, however, continues to progress, and after about 2 years the late latent or tertiary stage develops. The disease then becomes chronic. In 40% of untreated patients, 2 to 25 years after the initial infection the spinal cord and brain become involved, leading to general paralysis of the insane or more simply to paresis, characterized by disturbances in memory and judgment, megalomania, delusions, headaches, epileptic seizures, impotence, and pains in the joints. Most untreated patients die within 5 years after showing the first signs of paresis.

Treatments for "the great pox" varied with the times. In 1496 Giorgio Sommariva of Verona, Italy, tried mercury, and soon thereafter mercury was applied topically to the suppurating sores or was taken in the form of a drink. The treatment came to be called salivation because the near-poisoning with mercury salts tended to produce copious amounts of saliva. In 1909 Paul Ehrlich (1854 to 1915) developed an arsenical derivative, salvarsan, which was effective in reducing the severity of the disease.

Salvarsan—Ehrlich's "magic bullet"—was not without its problems. It was toxic, required years of treatment, and was difficult to administer. Only 25% of patients ever completed the therapy.

The Viennese psychiatrist Julius Wagner-Jauregg (1857 to 1940), finding that the insanity and dementia of those with late-stage syphilis did not respond to Ehrlich's magic bullet, shunned treatment with arsenicals and became convinced that raising the body temperature would benefit paretics (253). Initially he used Koch's tuberculin to induce fever; however, this was found to be ineffective and dangerous. He also injected patients with typhoid vaccine, streptococci, and staphylococci, but none of these worked. In the summer of 1917 while treating a soldier with tertian (*P. vivax*) malaria, he believed he had the perfect treatment—malariatherapy (537). He suspected that the malaria-induced fever would roast the spirochetes without killing the patient. He inoculated nine paretic patients with blood from the soldier and in 1918 reported a remarkable success rate of 67%. In 1922, after having inoculated 200 patients, he reported that 50 patients had experienced complete remission (56). His methods were soon adopted throughout Europe and elsewhere. Indeed, in the 1920s malariatherapy became "state of the art" for the treatment of the insane with late-stage syphilis, and in 1927 Wagner-Jauregg received the Nobel Prize for this fever therapy (537). The first detailed report of laboratory-based malarial studies associated with treatment of neurosyphilitic patients in England (at the Horton Laboratory at Epsom) was that by Sydney Price James (a member of the League of Nations Malaria Committee) (452) and P. G. Shute in 1926 (453). Malariatherapy was first used in the United States at St. Elizabeth's Hospital in Washington, DC, in 1922. Success with malariatherapy led to the establishment of special centers around the world. In the United States, the Syphilis Division at the Johns Hopkins Hospital, the Station for Malaria Research in Tallahassee, FL (established in 1931), and the U.S. Public Health Service Laboratories in Columbia, SC (established in 1931) and Milledgeville, GA (established in 1946), carried out treatments.

In the early years, 2 to 5 ml of blood was taken from the vein of an infected donor and was injected below the skin into the paretic individual; later, intravenous inoculations of ca. 10 ml were used. When the donor and recipient were not side by side, blood was drawn into a tube containing a substance to prevent clotting. This could be held for 48 h in an ice bath before being used. Initially, since the parasites could neither be grown in vitro nor be preserved indefinitely, the therapy required fresh parasitized blood.

After approximately 12 bouts of fevers and chills, patients received quinine to terminate the infection; however, just before the quinine was administered, blood was removed from the paretic to inoculate another paretic. At the time physicians were not worried about HIV/AIDS, but they did fear transmission of syphilis and hepatitis; despite these drawbacks, the direct-blood-injection route was the predominant form of malariatherapy in the United States and Britain. Although it was recognized that using infected mosquitoes more closely mimicked natural infections, this was much more difficult than blood transfer. Eventually, permanent laboratory colonies of mosquitoes were established and then transmission could be effected via the bite of infected *Anopheles*.

Between 1922 and 1936, 2,158 patients with advanced syphilis were given malariatherapy at St. Elizabeth's Hospital, and by the end of 1929, 3,155 patients at Horton had been given malariatherapy to relieve their symptoms (253). Malariatherapy was not without some risk to the patient from the malaria infection itself: anemia, anorexia, nausea, malaise, headache, jaundice, falling blood pressure, and renal disease were not uncommon, and when *P. falciparum* was used for treatment there was the occasional death (258). On the positive side and in general, the therapy resulted in very few deaths and was less expensive than salvarsan, and clinical improvement was achieved more rapidly than with drug treatment. However, by the late 1920s concerns were being raised over the ethics of malariatherapy and there were issues of lack of informed consent; later, familial permission was sought before proceeding with therapy. Although penicillin was introduced for the treatment of syphilis in 1943 and effectively replaced salvarsan, this fever therapy by malaria continued to be used in many hospitals throughout the world until 1963.

It has been claimed that one-third of paretics treated with malaria went into full remission of variable duration, another third or more experienced some remission, and the remainder improved marginally or not at all or died. Retrospectively, these cure rates have been questioned on various grounds including spontaneous remission, the fact that the studies were neither blind nor double-blind, and the fact that variability in response was not considered. Further, the postulated mechanism, i.e., "cooking" the spirochetes, would appear to not hold up since their thermal death point (258) is about 41 to 42°C or higher for 6 h, and these temperatures were too high for the patients to bear.

In 1931 the Rockefeller Foundation funded the malariatherapy work of Mark F. Boyd and Warren Stratman-Thomas at a Florida mental hospital with the direct aim of studying malaria, not syphilis. When they found that patients reinoculated with the same strain were free of disease, they proved the existence of short-term immunity to a particular strain of *P. vivax*. If, however, the patients who had received one strain were challenged with a different strain, they responded as vigorously as if they had never had malaria. Thus, the authors accurately concluded that immunity to vivax malaria was strain specific. Boyd and Stratman-Thomas also discovered something of importance that James in England had not. They reported having a "devil of a time" infecting black patients with vivax malaria (253). At first they thought the problem was due to a bad batch of mosquitoes, but they later found that only three of eight black patients had any symptoms at all in spite of receiving a large dose of sporozoites; as a consequence they expected little clinical benefit from vivax malariatherapy. (White patients treated similarly came down with infections.) Boyd and Stratman-Thomas had discovered that African Americans had an innate resistance to *P. vivax* malaria. Their report on innate immunity was neglected, and it would take until 1975 to uncover the underlying mechanism of resistance, i.e., Duffy negativity (see p. 229). The relative immunity to vivax malaria by African Americans prompted Boyd to give them the more dangerous falciparum malaria. Of a group of 72, 60 developed malaria, 4 died, and 49 had to be treated with quinine after several fever episodes. Thus, *P. falciparum*-induced fever therapy was as effective with paretics as was *P. vivax*-induced fever therapy. Boyd and Stratman-Thomas failed to consider the ethical lapses inherent in their work. Indeed, they saw themselves as providing a valuable therapeutic option to the patients, and thus they justified their use of experimental subjects without informed consent and the fact that they had used a selective target, i.e., blacks (253).

A retrospective examination of fever and blood smears acquired between 1940 and 1963 at the NIH centers at the South Carolina State Hospital in Columbia and the Georgia State Hospital at Milledgeville, where malariatherapy was practiced (104–106) has revealed the following:

1. In both blood- and sporozoite-induced infections, *P. falciparum* and *P. malariae* appeared to have common antigens such that clinical protection was produced when infection with *P. falciparum* was followed by infection with *P. malariae*.

2. A previous infection with *P. vivax* or *P. ovale* provided no protection against *P. falciparum*.

3. Patients previously infected with *P. falciparum* were susceptible to reinfection with the same and other strains. Following reinfection, clinical immunity (measured by frequency of fever as well as lower parasite counts) was enhanced (106).

4. In the same individual infected with *P. falciparum*, there was a consistent decrease in parasite counts, suggesting that increasing immunity reduced but did not eliminate subsequent parasite populations (106). Therefore, the level of conserved immunity becomes more effective against succeeding parasite populations.

5. After multiple challenges the individual develops some immunity to all strains of *P. falciparum*; however, it is unknown whether sterile immunity to all potential strains can be developed (106).

6. Patients who have reached equilibrium with one *P. falciparum* strain, i.e., low parasite counts and no fever, may be subject to renewed episodes of fever and elevated parasite counts when a new strain is introduced (104).

7. During relapse, parasite counts fail to reach high levels, suggesting that greater immunity to *P. falciparum* develops during chronic infection than from newly introduced strains or variants (106).

8. Immunity to reinfection was in evidence with the same strain of *P. vivax* but less so with a different strain. Therefore, vaccines developed against the blood stages of a particular *P. vivax* strain may be more effective against the same strain than against different strains (108).

9. Previous infection with *P. ovale* did not prevent reinfection; however, there was a reduced level of fever and parasite counts.

Malariatherapy made contributions to malaria research as well as to the treatment of those suffering from neurosyphilis (84, 453); the advances in malaria research included the discovery of the preblood cycle of malaria; the source of relapse and the reason for the failure of quinine and certain other drugs to effect cure in the vivax malarias; a questioning of Schaudinn's observation of direct penetration of red cells by sporozoites; the establishment of the first permanent laboratory colonies of mosquitoes used in transmission; isolation and identification of *P. ovale*; isolation, use, and study of various strains of malaria parasites, notably the Madagascar strain of *P. vivax*, which for 30 years was used to infect over 10,000 patients and 68,000 mosquitoes at the Horton Laboratory and the Chesson and St. Elizabeth strains used in the United States; and the realization that biological, clinical, and therapeutic differences in malaria depend on whether the infection is blood or sporozoite induced. Perhaps most important for vaccine development malariatherapy showed that immunity to malaria develops with time and is mostly strain and species specific. Despite its exceptional value, however, therapy by malaria still left unknown the nature of the antigens that produced protection against a single strain of a single kind of malaria or those able to transcend the species barrier.

Clues from Naturally Acquired Immunity

The tuberculin therapy against tuberculosis announced with great fanfare by Robert Koch at the 1890 Berlin Medical Congress not only proved to be useless but at times was lethal (54). Koch, the once hard-working and eager young microbe hunter who had virtually single-handedly established the field of microbiology, had ceased to do any research on tuberculosis and was becoming increasingly despondent. At his lowest point, when the crusty and opinionated Koch was sitting for an official portrait, he noticed a painting of an attractive young girl on the wall of the artist's studio. He was smitten, and since he was no longer living with his wife he pursued the 17-year-old art student, Hedwig. They soon became intimate, and in 1893, immediately after his divorce, Koch married Hedwig. She was 21 and he was 50. Such a May-December romance did not make the couple popular in the conservative Prussian society of Berlin, and as a consequence they became social outcasts. Despite the failure of tuberculin therapy, Koch was still in demand for his microbe-hunting services, and in 1896 he was dispatched by the government for 4 years to study tropical diseases in the German-controlled colonies of East and Southwest Africa, Cameroon, and Togo as well as northeastern New Guinea.

In Africa he investigated a viral disease of cattle called rinderpest, in India it was bubonic plague and cholera, and in Italy and New Guinea he set his sights on malaria. Hedwig was his constant companion. For the rest of his life Koch devoted himself to tropical medicine. In 1899, he studied malaria in a valley in Java. Using stained blood films to determine the frequency and density of blood parasites, he observed that while clinical illness and parasites usually occurred in the blood of young children, adults did not appear to suffer illness and frequently were parasite free.

Ronald Ross regarded this as the first real proof that the immunity shown by adults in an area of highly endemic infection was an acquired and not an inherited characteristic. It showed, he said, "the blood of those who survive gradually produces something which after a number of years has the power of reducing and perhaps extinguishing the parasite invasion." Ross pandered to Koch's ego by claiming that Koch was the first malaria immunologist.

By the late 1920s, a significant body of information regarding natural acquired immunity in humans, both from field studies and from malaria therapy, had already been accumulated (132). It was described as being species and stage specific, effective in adults after uninterrupted and life-long exposure to infected mosquito bites, dependent on the degree of exposure, and lost upon cessation of exposure. In evidence was a decreased risk and extent of morbidity, a lower density of blood stage parasites, and premunition (i.e., protection against new infections by maintaining a low-grade and generally asymptomatic blood infection). In light of this evidence, some immunologists have suggested that when researchers try to develop a safe and effective vaccine, they should ensure that it mimics natural acquired immunity so that infants may enjoy the feeling of well-being demonstrated by their older siblings and parents despite their harboring blood parasites that would be lethal to a malaria-naive visitor. A prerequisite for the design of such a vaccine would be identification of the appropriate antigens and their formulation in a presentation to elicit an antidisease response. As yet, this has not been achieved.

Two hypotheses have been proposed to explain the slow onset of acquired immunity (i.e., it may take 10 to 15 years of roughly five infections per year) (132). The first and the most popular is that natural acquired immunity results from the cumulative effect of exposure to infections that have a diverse repertoire of strain-specific differences. By this mechanism—an accumulation of antigenic memory lasting in some cases 8 to 15 years—the parasite is able to evade host immunity by varying the antigens exposed on the surface of the red cell responsible for antigenic variation (see p. 218). In humans, the variant antigen receiving the most attention is PfEMP1, encoded by one of ~60 variant (or *var*) genes (see p. 218). In studies carried out with Kenyan children, a variant of PfEMP1 expressed during a clinical episode was less likely to be recognized by preexisting antibodies in the same child than by antibodies in other children. In malaria-naive subjects experiencing a single infection with *P. falciparum*, there were antibodies reactive with six different lines, each of which expressed different PfEMP1 variants (132). Further, in Ghana the levels of antibodies to PfEMP1 were correlated with protection from clinical malaria.

The evidence from field studies that antibodies provide a modicum of variant-specific protective immunity has led to the proposal that PfEMP1 (and other variant antigens) should be a component of a malaria-vaccine. To develop such a vaccine, it will be necessary to identify and combine multiple variant epitopes of PfEMP1 or to define a conserved epitope not normally recognized after infection, but one that is an antibody target. Such an epitope would, under ordinary circumstances, be cryptic. Cryptic epitopes are ordinarily poorly immunogenic; however, after infection they can be recognized by the immune system either naturally or artificially. Indeed, monoclonal antibodies to PfEMP1 (and to band 3 protein, see p. 316) that can agglutinate multiple strains have been produced, and these may be targeting cryptic epitopes; once identified, such epitopes could provide a basis for developing a vaccine that could induce a nonnatural immunity (562).

An alternative hypothesis for acquired immunity is that with increasing age and exposure, individuals naturally exposed to malaria develop cross-reactive antibodies able to recognize an increasingly broad range of *P. falciparum* strains. Evidence for this so-called strain-transcending immunity comes from studies with patients undergoing malariatherapy: when individuals who had 18 months earlier been infected with a Colombian strain of *P. falciparum* were infected with a Thai strain, the parasites were quickly cleared from the blood with a marked rise in the level of antibody to the Colombian strain, and a much lower titer of antibodies to the Thai strain (132). And, in the classic study of antigenic variation in *P. knowlesi*, Brown and Brown (see p. 114–117) observed a strain-transcending immune response. Despite this evidence, the importance of strain-transcending immunity in natural acquired immunity, which is directed predominantly against the asexual blood stages and not against sporozoite or liver stages, is unknown, as is the nature of the antigens. It is suspected that the primary target is the blockade of invasion by merozoites (see p. 123). However, it is also entirely possible that it may involve antibody-dependent cellular killing or binding of antibody to host molecules on the surface of the parasitized red blood cell, leading to greater clearance of these cells (see p. 316–318). If natural acquired immunity is 100% effective against severe disease and death in heavily exposed adults, as some immunologists contend, then understanding the mechanisms that trigger it could be a step on the path toward a protective vaccine. However, we are not there yet.

There is no single measure of a protective immune response for a complex pathogen such as the malaria parasite. Indeed, because of the difficulties in analyzing the mechanisms of malaria immunity in humans, immunologists and vaccinologists have resorted to using animal models, especially mice. Valuable as these malaria models are, interpretation and extrapolation must be made with caution: rodent parasites are not natural mouse pathogens, and much of the knowledge gained has come from infections initiated by unnatural routes. If the ultimate aim is to produce an effective vaccine for humans, the mouse models should recapitulate the human infection as far as possible and should involve the full parasite life cycle initiated by a mosquito bite (286). This aside, after a blood-induced infection in mice, immunity can be induced by drug cure. After one or two infections the mice are immune, although they harbor small numbers of parasites. Antibodies taken from these mice can be passively transferred to naive mice, but even if the antibody-secreting B cells are knocked out or antibody synthesis is suppressed, the mice survive challenge but are unable to completely clear the infection. This suggests that immunity to a blood stage malaria infection involves additional immune mechanisms.

Protection by antibody and synthesis by B cells have been and continue to be a cornerstone of malaria immunity. In the1970s, however, there began to be evidence for the involvement of T lymphocytes. When CD4 (helper) T cells from an immune mouse were transferred to a naive mouse (thereby protecting it) and there was no evidence of antibody, it was concluded that these T cells, independent of antibody, could control parasite growth and reproduction (200). These studies in mice were also supported by work with chickens infected with P. gallinaceum (see p. 98). Subsequent research revealed that the respective contributions of cell- and antibody-mediated immunity differed between Plasmodium spp. and different rodent hosts: in P. chabaudi blood stage immunity is primarily cell mediated whereas in P. yoelii it is primarily antibody mediated. A clue to the possible role of T cells in malaria immunity in humans comes from the observation that concurrent infection with HIV appears to have little effect on the course of a P. falciparum infection. Although this suggests either that CD4 T cells do not play a major role in naturally acquired immunity or that only a small number of T cells are required (189), there is other evidence to indicate that cell-mediated immunity does play a role and that it can be either detrimental or beneficial to the infected individual. It is a matter of cytokine balance and control.

Children with mild P. falciparum malaria have higher levels of plasma IL-12—a key cytokine that initiates T-helper cell responses by triggering

IFN-γ release from NK (natural killer) and CD4 T cells—than do children who suffer from severe infections. Direct contact between infected red cells and NK cells (in addition to IL-12 and IL-18) is required for optimal production of IFN-γ. IFN-γ is thought to be important in controlling blood stage malaria, and children with severe malaria also show lower levels of this cytokine and are more susceptible to reinfection. In naive subjects infected with *P. falciparum*, there is a coordinated increase in the levels of IFN-γ, IL-12, and IL-8 in the serum at the time the parasites first appear in the blood (481). TNF-α released from macrophages and induced by IFN-γ is more abundant in children with *P. falciparum* malaria and correlates with lower blood sugar levels and higher mortality; TNF-α is also found in the brains of patients dying from cerebral malaria. The formulation of malaria vaccine with adjuvant must take into account the fact that the protective and pathologic effects of TNF-α depend on the amount, timing, and location of expression (560) and that the adjuvant should enhance the former effects and minimize the latter. A possible strategy for enhancing the immunogenicity of recombinant malaria antigens might be the inclusion of cytokines and microbial products that activate cell-mediated immunity (481).

Mary M. Stevenson (born in 1951) grew up in Bethesda, MD, the daughter of a pharmacist and a chiropractor, with a natural curiosity about biology and medicine. She majored in biology in college, with an interest in pursuing a career as a physical therapist, but she had the opportunity during her senior year to undertake an honors research project on the immunology of bone marrow transplantation in a laboratory in the National Cancer Institute (NCI) at the National Institutes of Health (NIH) in Bethesda. She became fascinated with immunology and decided to pursue graduate studies with the goal of conducting research in cellular immunology. As a Ph.D. student in the Department of Biology at The Catholic University of America in Washington, DC, she had the opportunity to do off-campus research in the laboratory of Monte S. Meltzer at the NCI, where the focus of her doctoral research was to characterize macrophage dysfunction in tumor-bearing mice and to investigate the use of immunostimulating agents such as *M. bovis* BCG and *Corynebacterium parvum* (*Propionobacterium acnes*) to correct the defects in macrophage function. Stevenson had her introduction to malaria research as a postdoctoral fellow (1979 to 1981) in the laboratory of Emil Skamene at the Montreal General Hospital Research Institute (now the Research Institute of McGill University Health Centre) in the Department of Medicine at

McGill University. She had been recruited by Skamene to study the role of macrophages in genetic control of resistance to *Listeria monocytogenes* in inbred mice because of her strong background in cellular immunology, especially macrophage biology. Fortuitously, there was also an opportunity to work with John J. Lyanga (1939 to 2005), a Resident in the Department of Clinical Immunology and Allergy at the Montreal General Hospital, who had elected to spend a year in the laboratory doing bench work for a project on malaria. Lyanga, originally from Cameroon, was all too familiar with the suffering and devastation inflicted by malaria, particularly malaria due to *P. falciparum* in sub-Saharan Africa, but he had little or no laboratory research experience. Stevenson was very intrigued with the idea of doing research on such a devastating disease as malaria. At this point in her career, her only knowledge about malaria was that it involved mosquitoes and red blood cells and was endemic in developing countries, particularly in Africa and Asia. Stevenson and Lyanga were a perfect team!

They established a rodent model of blood stage *Plasmodium* infection that could be used to identify resistant and susceptible inbred mouse strains, something Eugui and Allison had published on with the AS strain of *P. chabaudi* (165). Working together, Lyanga and Stevenson successfully reproduced and extended the observations of Eugui and Allison by identifying 5 of 11 strains of inbred mice as resistant and 6 as susceptible by monitoring the course of blood parasitemia and survival. By crossing a resistant inbred mouse strain as one parent and a susceptible strain as the other parent, F_1 hybrid mice were produced, and these mice were even more resistant to the infection than was the resistant parent. This finding indicated that resistance to *P. chabaudi* AS infection is inherited as a dominant trait. Much like the experiments performed with peas by Gregor Mendel (1822 to 1884), the "father of modern genetics," they next performed a segregation analysis of backcross and F_2 progeny derived from a resistant and a susceptible parent. They found that resistance to *P. chabaudi* AS was genetically controlled by an autosomal, dominant, non-histocompatibility-linked gene(s) (480). Furthermore, female mice were observed to be more resistant than male mice to *P. chabaudi* AS, similar to the superior resistance of women, except when pregnant, to *P. falciparum* malaria. It was also shown that resistant mice rapidly produced larger numbers of red blood cells in response to the anemia induced by malaria infection than did susceptible mice.

At the time of the initial publication of their genetic work, Stevenson was offered a position as an Assistant Professor in the Department of

Medicine at McGill University. Having no formal training in genetics, she defaulted to her training as a cellular immunologist and decided to pursue work in the field of malaria immunology. Initially she was disheartened upon reading the malaria immunology literature of the early 1980s, where the role of humoral immunity and antibodies was the dominant theme. Hope was renewed, however, when she came across several publications by Ian A. Clark and his colleagues, showing that mice are protected against infection with various species of *Plasmodium* and *Babesia* by prior infection with BCG or *P. acnes* (86, 87). Her previous research experience as a Ph.D. student came into play since these agents were the same ones that restored macrophage function in tumor-bearing mice. Clark et al. eloquently argued that "neither an increased specific immune response nor phagocytosis, but a non-specific soluble effector substance" best explained their findings and highlighted the seminal work of George Mackaness on nonspecific macrophage activation (305). They further proposed that a nonspecific mediator(s) produced by macrophages, possibly similar to mediators elicited in BCG-infected mice and known to suppress the growth not only of microorganisms but also of a mouse sarcoma, might be involved. Stevenson was struck by the similarity of the mechanisms of cellular immune responses known to effectively destroy tumors and to kill malaria parasites being proposed by Clark. Even at this early point, efforts to produce an effective malaria vaccine were disappointing. Stevenson hoped that by targeting cellular immunity rather than antibody-mediated immunity, an effective malaria vaccine may be devised. Little did she anticipate that 25 years later, even with increasing knowledge of the immune response to *Plasmodium* and with so many technological advancements, would the development of an effective and safe malaria vaccine remain so elusive.

In 1981, William P. Weidanz, now a Professor in the Department of Medical Microbiology in the School of Medicine and Public Health at the University of Wisconsin–Madison, published a seminal paper with Jim Grun, showing that while nude mice which lack T cells were unable to control acute infection with *P. chabaudi adami*, infection was successfully resolved in B-cell-deficient mice (200). Based on their observations, Grun and Weidanz proposed that a T-cell-dependent, antibody-independent immune mechanism, possibly involving activated macrophages, was required to control an acute malaria infection with some *Plasmodium* species and suggested that elucidation of the postulated cellular mechanism could provide a novel target for developing a malaria vaccine. This assertion contravened the prevailing thinking that an efficient malaria vaccine should be targeted toward induction of protective antimalarial antibodies! Shortly after the publication of their paper, Stevenson met Weidanz, who generously shared his vision as well as his vast knowledge and expertise in malaria immunology with the budding malaria immunologist.

Although in the intervening years Stevenson has moved into other research areas, including investigating the role of the inflammatory response and T-cell responses in the destruction of lung tissue in patients with cystic fibrosis, her interest in malaria research has not waned and the focus of her laboratory remains the identification of the underlying cellular mechanisms and soluble mediators (cytokines) involved in the development of immunity to blood stage malaria parasites (481). It is now generally appreciated that multiple immune mechanisms involving innate (dendritic cells and NK cells) and adaptive cellular immune responses as well as antibody-mediated immune responses contribute to the development of immunity to malaria (255). She is also investigating the contribution of suppressed erythropoiesis to malarial anemia and the role of host- and parasite-derived factors in malarial anemia (80). She is also exploring the effectiveness of a whole-parasite vaccine delivered with novel immunostimulatory adjuvants against challenge with *P. chabaudi* AS (490) and the effect of coinfection with a nematode (*Heligmosomoides polygyrus*) on immunity to blood stage malaria and on malaria vaccine efficacy (489). She has been actively collaborating with Philippe Gros (who was a Ph.D. student under Emil Skamene's supervision during her time as a postdoctoral fellow) and his graduate students in the Department of Biochemistry at McGill University to identify susceptibility genes to blood stage *P. chabaudi* AS infection (170) and, more recently, to cerebral malaria due to infection with *P. berghei* ANKA (335).

Many of the lessons learned early in her career regarding the host response to malaria continue to be major themes in her work. She is convinced that with the explosion of technology and with the interest in malaria research by an increasing number of scientists with expertise in other fields, an effective and safe malaria vaccine will rapidly become a reality; she believes that targeting multiple immune mechanisms including cellular immunity will assist in this endeavor, with a whole-parasite vaccine being a promising approach.

The role of T cells in liver stage immunity was first reported in 1977, when B-cell-deficient mice but not T-cell-deficient mice immunized with irradiated sporozoites were protected against sporozoite challenge in the absence of a CSP reaction or sporozoite neutralization (138). Subsequently it was shown that transfer of spleen cells but not serum from sporozoite-immunized mice protected naive recipients. In the mouse model, protective immunity induced by irradiation of sporozoites was abrogated by depletion of CD8 T cells but not of CD4 T cells. Although it is clear that protection is directed against the parasite developing within the liver, the precise mechanism by which the CD8 T cells are able to detect the minute number of infected liver cells among the billions of uninfected ones is not known (176); it may be that the malaria-specific T cells are aided by the trail of sporozoite antigens left behind. The manner by which development is inhibited is also not known. CD8 T cells are considered the primary effectors in the irradiated-sporozoite-induced protection, and the CD4 T cells that recognize the surface of the infected liver cell are also thought to play a role. IFN-γ and TNF-α have also been implicated in liver stage immunity, and in the case of IFN-γ the activity (at least in vitro) was dependent on nitric oxide (NO).

Despite there being no direct evidence for the involvement of CD8 T cells in protective immunity in human malaria, their role is suspected because CD8 T-cell responses to all liver stage antigens (CSP, LSA-1, and LSA-3) have been detected in human volunteers immunized with radiation-attenuated *P. falciparum* sporozoites or volunteers naturally exposed to the bites of infected mosquitoes.

It is not clear why sterile protective immunity against liver stages is not induced upon natural exposure to the malaria parasite nor why T-cell responsiveness is so low compared with the response to experimental immunization; i.e., vaccination with RTS,S is correlated with CD4 T-cell and antibody responses, not with CD8 T-cell responses (132). It may relate to the persistence of the liver stage parasites, since naturally acquired immunity decays rapidly once exposure to the parasite is terminated and CD4 and CD8 T-cell responses to liver stage antigens disappear within 2 years of exposure to sporozoites.

Denise Doolan was born in 1965 in the highlands of Papua New Guinea and lived there for the first decade of her life, accompanying her father, a Patrol Officer and District Commissioner, and venturing into the wilds of the Papua New Guinea jungle. Her childhood in Papua New Guinea had a significant impact on a career choice by introducing her to the wonders of other cultures, providing an appreciation for the impact of disease on public health, and giving her an interest in malaria control. It was likely abetted by her dislike of the weekly Sunday morning ritual of those little bitter-tasting antimalarial tablets! After graduation from the University of Queensland in 1985, she accepted a position with the Australian government Commonwealth Scientific and Industrial Research Organization in the field of virology, studying bovine ephemeral fever virus (a viral disease of cattle and water buffalo with a high morbidity and significant economic impact). She was able to characterize the viral proteins biochemically and immunologically; for that work she was awarded an M.Phil. (Science and Technology) from Griffith University, Brisbane. In 1989, she moved from veterinary medicine to human health and tropical disease research by becoming a research technician with Michael Good, Head of the Molecular Immunology Laboratory at the Queensland Institute of Medical Research. Subsequently she enrolled in a doctoral program under Good's mentorship and was awarded a doctorate in 1993 on the molecular and immunological characterization of the *P. falciparum* circumsporozoite protein. After completion of her Ph.D., she was awarded a National Research Council Associateship (U.S. National Academy of Science; 1993 to 1996) to undertake a fellowship with Stephen Hoffman at NMRI. She spent a total of 12 years affiliated with the U.S. Navy, beginning as a postdoctoral fellow, then becoming Director of Basic Research, later becoming Director of Basic and Preclinical Research & Development, and finally becoming Scientific Director.

At NMRI Doolan performed preclinical studies of mice and nonhuman primates, demonstrating that DNA vaccines against malaria are immunogenic and are protective in mice. She completed the cloning, sequencing, and characterization of a new malaria antigen expressed during the liver and blood stages and showed that it was protective (135); she subsequently demonstrated that a DNA vaccine based on this antigen was effective in one strain of inbred mice but ineffective in other strains; a vaccine based on a second antigen had a different pattern of protection (140). Immunization with a combination of the two vaccines induced additive or synergistic protective immunity against challenge. Subsequently, with a colleague, she characterized the T-cell responses to a *P. falciparum* DNA vaccine in humans (542), providing the first demonstration in any system that DNA vaccines could be immunogenic in humans. Although highly effective in rodents, however, DNA vaccines largely failed to live up to their promise in larger species, and subsequent efforts focused on other molecular biology-based vaccine technologies.

Another major aspect of her research has been the identification, from genomic sequence data, of novel *Plasmodium* antigens that could be good targets of an effective malaria vaccine. Using a peptide-epitope-based screening strategy called ImmunoSense, she found novel *P. falciparum* antigens recognized by T cells from volunteers immunized with radiation-attenuated *P. falciparum* sporozoites, as well as noting differential recognition by cells from volunteers who were protected or not protected against sporozoite challenge (141). Importantly, some of those new antigens were more immunogenic than the well-characterized *P. falciparum* antigens currently in clinical evaluation. More recently, she has expanded her research into the application of chip technology to identification of novel antigens (139).

A prevailing theme throughout Doolan's research has been the promise of rational vaccine design: the importance of understanding the mechanisms and antigenic targets of protective immunity and identification of the appropriate vaccine delivery system able to induce the required immune responses to specific targets. She is convinced that a pathogen as complex as *Plasmodium* will likely require a multistage, multi-immune response vaccine based on more than one parasite antigen and/or their epitopes. She has coauthored over 75 articles, contributed chapters to books in the field of malaria vaccine development, and edited a book on malaria methods and protocols; she holds a number of patents in the areas of antigen discovery, immune assays, and vaccination strategies.

The message to designers of a malaria vaccine from studies of naturally acquired immunity is clear: a vaccine that affords protection against the disease will require stimulation of both CD4 and CD8 T cells and B cells; however, the repertoire of antigens able to produce such an outcome remains elusive.

For the Love of a Vaccine

At the time of my rediscovery of the phenomenon of premunition (see p. 285), interest in the immunology of malaria was virtually dead. One contributing factor may have been the failed vaccination trial in 1945 by Michael Heidelberger with 200 human volunteers injected with formalin-fixed blood carrying *P. vivax* (see p. 109). In the early 1960s, however, two observations prompted a change in the degree of interest in malaria immunity. In The Gambia, McGregor and Cohen found that children exposed to malaria from birth showed clinical evidence of effective acquired immunity by the age of 3 years, and the acquired immunity was correlated with the development of significantly elevated levels of γ-globulin in serum.

These observations prompted a passive-transfer study in which the γ-globulin from immune adults was administered to young children experiencing acute clinical malaria (323). The results were spectacular: over 4 days the number of parasites in the blood diminished sharply, as did illness. This study, as well as other observations in areas where malaria is endemic, found that for the first few months of life children were resistant to severe disease, although frequently challenged and infected, and that this protection was generally associated with the presence of maternal antibodies acquired by the fetus as well as in breast milk. As the levels of immunoglobulin in blood decreased over the first year of life, these infants become more susceptible to severe disease and death. This convinced physicians and immunologists alike that acquired malaria immunity can be protective and antibody-mediated mechanisms can be important. At the time, and even today, it is widely felt that vaccination against malaria is possible if only the protection-inducing antigens can be identified.

Another important event in the 1960s provoked a renewed interest in controlling the mortality and morbidity from malaria via vaccination: the appearance of chloroquine resistance by *P. falciparum* in Southeast Asia and South America. Indeed, at the beginning of the Vietnam War (1959 to 1975) the number of U.S. soldiers evacuated as a result of malaria was as great as the number evacuated after being wounded by enemy forces. The need to protect and treat U.S. forces led the U.S. Congress to expand funding for research into malaria, and in 1963 the U.S. Army Research Program on Malaria was launched at WRAIR. Most of this program was devoted to a search for new antimalarials; by 1974, 26 new drugs or combinations had been developed and 11 clinical trials had been completed, with mefloquine being the flagship response to chloroquine-resistant *P. falciparum*. However, a substantial proportion of the funding was allocated to basic research, including immunology.

In 1963, Elvio Sadun (1918 to 1974), Chief of Medical Zoology at WRAIR and Special Assistant to the Director for Basic Research in Malaria, hosted an International Panel Workshop on the Cultivation of Plasmodia and the Immunology of Malaria to summarize available information, delineate some of the most pressing problems, and explore new approaches to current problems (414). Sadun was born in Livorno, Italy (546), and because of virulent anti-Semitism under Benito Mussolini was forced to emigrate to the United States in 1939. He received an M.S. degree at Harvard under L. R. Cleveland (who was also the major professor for William Trager), served in the U.S. Army in North Africa and Italy, and then completed a Ph.D. at the Johns Hopkins University, studying immunity in chickens to the roundworm *Ascardia galli*. In 1951, Sadun accepted a commission in the U.S. Public Health Service and was assigned by US-AID to Thailand, where he conducted surveys on worm infections. Returning to the United States, he was appointed Head of Helminthology at the CDC, where he evaluated new antigens and developed serologic tests for trichinosis and hydatid disease. Following a 2-year stint in Japan, he was appointed (in 1959) Chief of Medical Zoology at WRAIR, a position he held until 1973. Although a "worm immunologist" by inclination and training, Sadun recognized that one of the unfortunate consequences of the illusion that insecticides and drugs could eradicate malaria was a decline in biological research; he also noted that DDT went further toward the eradication of malariologists than the eradication of mosquitoes. Consequently, he organized a program at WRAIR to conduct research in malaria, to attract highly capable scientists from various disciplines to carry out malaria research, and to produce a continuing flow of highly trained people able to contribute to the management and control of this disease. One of those who joined WRAIR to study the immunology of malaria was Carter Diggs.

Carter Diggs (born in 1934) grew up in sight of the Piankatank River, which empties into the western side of the Chesapeake Bay, where he saw (and ate) a lot of marine biological specimens. Using a primitive microscope, he saw for the first time paramecia collected from jars of standing water. Truly awed, he became a microbe hunter. As an undergraduate at Randolph-Macon College in Ashland, VA, and still on the trail of microbes, he declared himself a biology major; since most of his colleagues majoring in biology were headed for medical school, he followed the path of least resistance and applied too!

After graduating from Randolph-Macon College in June 1956, he enrolled at the Medical College of Virginia (MCV) in the fall. He soon found that he did not want to practice clinical medicine but instead wanted to do some kind of research. His first experience in a research laboratory was with Robert Marston, with whom he worked on isolation of viruses from children with aseptic meningitis and found that most cases that summer in Richmond were due to coxsackievirus B-5. (Marston would go on to become Director of NIH.) During the summer of his junior year at MCV, he worked on several projects, but the one that sticks out in his memory was the (accidental) discovery of serum trypsin inhibitor: a very exciting moment until he found out that it had been discovered some decades earlier!

These experiences in the laboratory reinforced the idea that he wanted to be involved in research, and so he applied for an internship and residency in pathology at MCV.

The year following receipt of his M.D., he worked in the morgue doing autopsies and in the pathology laboratory reading slides from tissues. The experience was valuable, but he knew he did not want to devote his career to it. During his second year of pathology research, he received a letter, which began: "Greetings from the President of the United States." It directed him to show up at a certain date for induction into the U.S. Army, but included a note at the bottom indicating that since he held an M.D. degree, he did not have to show up on that date if he had applied for a commission before then. His boss, George Margolis, knew some people at WRAIR, and an appointment was set up. As a result, he obtained a position in the Division of Communicable Diseases and Immunology (DCD&I) directed by Colonel Abraham Benenson.

After basic training at Fort Sam Houston, Carter showed up at WRAIR early in 1962, reported to Benenson, and discovered that there was a Department of Immunochemistry in DCD&I headed by Elmer Becker. Becker's area of research was complement (see p. 67–68); however, it was rumored that he might have an interest in the function of the lymphocyte, a subject about which essentially nothing was known. Diggs had been looking at lymphocytes infiltrating sites of chronic diseases during his year in the pathology laboratory, and this provoked a keen interest in finding what they were for. When Benenson asked Diggs what he would like to do, Diggs immediately responded that he would like to work with Becker on lymphocyte function. Benenson said he was sorry but there were no slots available in that laboratory and suggested visiting all the laboratories in DCD&I and then come back to talk some more about an assignment. After visiting the Departments of Virology, Rickettsiology, Bacteriology, Medical Zoology, Applied Immunology, Molecular Biology, and Serology as well as Immunochemistry, Diggs told Benenson that he was still interested in working with Becker on the lymphocyte. Benenson said he was very sorry that there was still no opening there but there was one in Medical Zoology and he needed someone to start thinking about vaccination against malaria.

The next week Diggs reported to Sadun in Medical Zoology. Sadun was a major figure in schistosomiasis research in the 1960s. The Army's interest in schistosomiasis was not great, and WRAIR was just kicking off its malaria drug development program, which was taking funding from other areas of less military importance including schistosomiasis. Sadun was under pressure to pay attention to malaria; however, he was able to maintain the program on schistosomiasis, perhaps due in part to advocacy by the scientific community, within which he was very prominent. He was out of the drug development loop, but the malaria immunology/vaccine mandate provided an avenue for him to participate in research on this important disease. He instigated a meeting to kick off the malaria effort (Cultivation of Plasmodium and Immunology of Malaria) and supported Diggs in his fledging efforts at malaria immunology. At the time, Diggs was the only one in the department working on malaria! The first thing Diggs did was to get some *P. berghei* from NIAID where there was some work going on (albeit not in malaria immunology, he remembers). A considerable amount of Diggs' time during 1962 to 1964 was devoted to isolating parasites, immunizing rabbits with the crude parasite extract, and doing electrophoresis and immunoprecipitation to try to identify malarial proteins. He got lots of lines of precipitate (antigens), but did not progress much beyond that.

By this time, convinced that he needed more training in immunology, he applied to the Army for a Ph.D. program at the Johns Hopkins School of Medicine. In due course this application was approved and he moved to Baltimore in time for courses in the fall of 1964. His mentor at Hopkins was Abraham Osler, a very bright, knowledgeable, and meticulous immunologist specializing in allergy. Although not engaged in malaria research, Osler was a great humanitarian, and Diggs' desire to learn how to engage in malaria vaccine development tweaked his interest from that point of view. In addition, Osler's own training was under Michael Heidelberger, who had actually produced and tested malaria vaccines derived from lysed infected cells (see p. 108–109).

After leaving Johns Hopkins, Diggs was assigned directly to the SEATO laboratory (now the Armed Forces Institute of Medical Sciences) in Bangkok to head the Department of Parasitology. During his 2 years there (aside from administrative chores), he introduced some serologic techniques and supported the ongoing fieldwork that had been started by others. In 1970, he returned to WRAIR and the Department of Medical Zoology, where he had the pleasure of working with Bruce Wellde. Wellde came to WRAIR in uniform as a corpsman and quickly developed into an independent scientist. With Wellde, Diggs studied aspects of immunization against malaria with irradiated parasitized erythrocytes (548), a little-celebrated phenomenon, at about the same time as the NYU group was showing protection induced by irradiated sporozoites. Diggs and Wellde also demonstrated a delayed appearance of parasites in the blood of *Aotus* monkeys passively given IgG from Nigerian donors at the time of infection with blood stage *P. falciparum* (129). At the same time, pioneering clinical trials, supported by the U.S. Army and Navy, had shown that immunization of human volunteers by bites from irradiated mosquitoes protected the subjects from challenge by infected *Anopheles* (see p. 247).

In 1973, a number of events converged to change the administrative structure at WRAIR. As WRAIR moved more and more to applied research and development, the Department of Immunochemistry was judged to be too focused on basic research to be justified. Another Department, Serology, was phasing out as personnel retired. At about the same time, Sadun developed hepatocarcinoma, to which he eventually succumbed. The upshot of it all was that a new Department, Immunology, was formed out of the three existing departments, and Diggs was appointed Head. In spite of the name, much of the focus in the Department of Immunology was on parasitic diseases, especially malaria. Carl Alving of liposome fame was a part of the department, and it is of interest that Carl's father, Alf Alving, ran the malaria unit at Joliet prison, where much of the U.S. Army's World War II clinical antimalarial drug testing had taken place. Carl, who is still at WRAIR, is still working on liposomes as vaccine formulation elements; however, currently his work is with HIV, not malaria. Interestingly, GSK's lead adjuvant, AS01, is a liposome formulation.

In time, malaria vaccine development became the sole mission of the Department of Immunology at WRAIR. In 1975, in collaboration with a young physician, David Haynes, Diggs cultured *P. falciparum* for 10 days. The key to their success was gassing the cultures with CO_2. Haynes had done cell culture at Yale University and was familiar with the requirement for CO_2 as a component of the buffer system in the medium in a number of systems. He had approached the problem very systematically and appreciated the anaerobic nature of malaria parasites. As often happens in science, Haynes and Diggs were not the only ones who cultured *P. falciparum* that year. Trager and Jensen published their work before Haynes and Diggs. This was the scoop of a lifetime, and while the Trager-Jensen discovery was celebrated (see p. 137–139), the Haynes-Diggs culture system was forgotten.

After Diggs made a site visit to the laboratory of Miodrag Ristic to consider Department of Defense funding of ongoing USAID-supported malaria vaccine work at the University of Illinois, where "exoantigens" were being used as the basis of vaccines, it was discovered that the "exoantigens" were a "gimisch" and efforts to correlate protection with a component in the vaccine were not promising. The project was not recommended for funding; however, now Diggs entertained the idea of looking at the gimisch to see whether it contained any interesting antigens. At the time, a senior fellow from France, Daniel Camus, had come to work at WRAIR, and Diggs asked him to look at what was in the spent medium (i.e., the exoantigen) by using immune serum. Together with a young investigator, Terence Hadley (who had been in the Army, worked at WRAIR, and was making a transition to NIH to work with Louis Miller), the two had other ideas: What was in the spent medium mixture that would stick to red cells? The experiment involved exposing fresh red cells to the medium, washing the cells, and determining whether the immune serum agglutinated the cells. It did, and a new antigen was discovered: erythrocyte-binding antigen (EBA-175) (69).

In 1979, Franklin Top vacated the Directorship of the Division of Communicable Diseases and Immunology to become WRAIR Deputy Director and Diggs was asked to fill the slot while continuing as Chief of Immunology. One of his first jobs was to pick a successor to head the Immunology Department. The choice was obvious: a young officer, Wayne Hockmeyer, who had just finished a tour as Director of the WRAIR unit in Nairobi. Keen to develop an immunoassay for detection of *P. falciparum* sporozoites in mosquitoes, Diggs contacted Ruth Nussenzweig's laboratory, which had just published on the development of monoclonal antibodies against CSP; with those, he reasoned it would be a snap to develop such an assay. The only problem was that Nussenzweig had the monoclonal antibodies and would not part with them. When asked for the monoclonal antibodies or the cell line, she assured WRAIR that they could have them . . . but not just yet. She wanted to be sure her students had time to submit their papers. Or, there was too little supply right now. Or, there was some other problem. This went on for about a year. At that point, the group at WRAIR decided to make their own monoclonal antibody, not a trivial task in those days. With effort and an excellent team, WRAIR soon had CSP monoclonal antibodies and a sporozoite immunoassay. At about the same time, Fidel Zavala at the NYU School of Medicine, who was working in the Ruth Nussenzweig laboratory, developed a sporozoite radioactive immunoassay, much to the chagrin of the WRAIR team.

At this time, the NYU group identified the *P. knowlesi* CS gene. Diggs recognized that this might allow for a well-characterized antigen for vaccine development, and so the decision was made at WRAIR to go all out to develop a *P. falciparum* vaccine by recombinant DNA methods. However, WRAIR had no molecular biology capacity at that time, but it did have a monoclonal antibody! Then Diggs got a call from Louis Miller of NIAID, who said that his group had developed a *P. falciparum* genomic library and proposed that the cDNA library be probed with the WRAIR monoclonal antibody to look for the CS gene. The arrangement, to which Miller agreed, was to be a full collaboration, and WRAIR would have access to any materials that came out of the shared work. Haynes took the initiative to be the main WRAIR hands-on collaborator, working with John Dame and Tom McCutchan at NIAID. Within a week, plaques of bacteria were producing CSP.

Once CSP was available, the work on a vaccine took off, thanks to a very active collaboration between GSK and WRAIR in which Wayne Hockmeyer took the lead for WRAIR. The vaccine (called FSV-1) was developed in record time, being much less hampered by regulatory constraints than is the case now. Hockmeyer (who was a Ph.D., not an M.D.) arranged for infectious-disease fellows in training at Walter Reed Hospital to do their research elective in the department, thus building a clinical vaccine-testing capacity. The subjects were primarily the staff, and all but one subject who received the vaccine developed malaria on challenge (25). Although the trial was considered a failure, in retrospect, that one subject who did not become infected was an important lead. He had the highest antibody titer, was repeatedly challenged, and still did not become infected (although he finally did). The protected volunteer was not Diggs, even though he only had a chill the night before becoming slide positive and felt only a little tired before the chloroquine took effect whereas other subjects were extremely sick!

Not long after the initial FSV-1 trial, and now more than 5 years beyond the obligatory 20 years in the U.S. Army, Diggs had reached the pinnacle at WRAIR (with the title of Associate Director for Plans and Overseas Operations), and he began to search for his next career move. It was at this time that the USAID MVDP had run into serious trouble, and it seemed that this presented a greater opportunity to make a contribution to malaria vaccine development than staying at WRAIR. Phillip Russell, Commanding General of the U.S. Army Medical Research and Development Command and a former colleague at WRAIR, had the ability to assign Diggs to USAID while still on active duty.

The first year (1988) at USAID was stressful for Diggs. At that time, health activities were under the Bureau for Science and Technology, and it appeared that there was open warfare with the rest of the Agency! Some at USAID believed that the entire malaria vaccine development effort should be abandoned. In addition, there was a GAO audit of the program going on as well as an investigation by the Inspector General (see p. 146–157). When Diggs took the helm of the MVDP, most people were looking at the first clinical trial of an NYU peptide-based vaccine (221) as a failure, although there was significant delay in the onset of the appearance of parasites in the blood. The results of the trial, along with the scandal at USAID, made for a painful period. In a manner similar to the loss of enthusiasm for eradication 20 years earlier, there was much pessimism regarding the feasibility of ever developing a malaria vaccine.

Diggs' mandate at USAID was to reorganize and reorder the priorities of the MVDP as well as to trim the program to fit the reduced budget. Continuing support of the University of Maryland Center for Vaccine Development was seen as not cost-effective since clinical trials were not envisioned for the next several years. Similarly, the development of a field trial capability in Papua New Guinea was not going to be used for vaccine trials in the near future. USAID funding for both was terminated. Fortunately, the Center for Vaccine Development had multiple streams of income and survived without too much difficulty. The Papua New Guinea operation was stressed more severely but muddled through, and the facility built by USAID was put to good use. A similar fate was met by the USAID-funded malaria laboratory at the Biomedical Research Institute in Rockville, MD, as well as most of the rest of the old portfolio. Under the "healing process," all investigator-initiated research projects were subject to periodic external peer review, and none survived more than a few years.

The first real strategic effort in which Diggs was involved at USAID had to do with blood stage antigens. By 1990, WRAIR was devoting essentially all of its effort to trying to develop CSP-based vaccines. This focus was appropriate given the prevalent notion that preblood stage vaccines would be more appropriate for military personnel in that a blood stage infection and the attendant pathology could be prevented. On the other hand, blood stage vaccines would be expected to moderate, but not necessarily prevent, clinical disease.

With this in mind, Diggs felt that the MVDP should focus on blood stage vaccines. At that time, USAID was still funding the University of Hawaii group that had provided the first evidence of protection of New World monkeys by immunization with native MSP-1 in Freund's adjuvant.

In the mid-1990s, expression of recombinant protein in *E. coli* was generally in disrepute. Hermann Bujard in Heidelberg, however, was actively working with *E. coli* expression systems, and Jeffrey Lyon at WRAIR elected to take a sabbatical in Bujard's laboratory. Lyon came back with much of the technology under his belt. At that time, FASEB had a "help wanted" service, and Diggs placed an advertisement for a molecular biologist with experience in expressing foreign proteins in *E. coli*. Evelina Angov, who had just finished a postdoctoral stint at NIH, satisfied the criteria and was hired through an interagency agreement from USAID to WRAIR. She began work with Lyon, and within a year they had obtained encouraging results in the expression of MSP-1 in *E. coli*. At that point, they indicated that they needed considerable help to move from where they were to the next level, and as a consequence a robust process development unit at WRAIR was initiated.

At the time it was already known that MSP-1 was polymorphic (see p. 177); however, the question of cross-reactivity among the various forms remained open. Some studies suggested that a close match would be needed, while others indicated that cross-protection could be expected. Diggs opined that if a single-component vaccine in a field situation where many MSP-1 forms were present was not tried, he would never know whether it would work. He decided to push on. At the onset, it was decided to make the 3D7 version of the molecule since the researchers wanted to see whether it might show some protection under conditions of homologous challenge in the standard sporozoite challenge model, which uses the 3D7 strain of *P. falciparum*. The problem with this approach is that these experimentally induced human infections must be terminated as soon as there are parasites detectable in the blood, which might not allow time for a blood stage vaccine to take effect. There were doubts that protection would be achieved, but, again, the study was necessary.

At this point there was no vaccine—only an antigen. When the characterization of the antigen, including responses of animals, was presented to GSK (with which WRAIR had been working for years on CSP-based vaccines), the GSK workers were impressed and agreed to collaborate toward clinical evaluation of a vaccine formulated with the WRAIR MSP-1 and their proprietary adjuvant, AS02A, which by this time had been used with RTS,S to produce a vaccine that showed efficacy in human trials with the sporozoite challenge model.

An initial Phase 1 (safety) study identified no safety problems and showed that the vaccine was immunogenic. A Phase 2a trial with four groups that received either (i) MSP-1/AS02a, (ii) RTS,S/AS02a, (iii) a co-formulation of the two vaccines, or (iv) one vaccine in one arm and the other in the opposite arm was performed. There was no protection with the MSP-1 vaccine; however, there was also no detectable interference with RTS,S/AS02a-mediated protection.

At about this time (1999), the MVI at PATH came into being and its investigators were looking around for "advanced" malaria vaccines to test. They entered into an agreement with WRAIR to field test MSP-1. Thus began a 5-year odyssey involving WRAIR, GSK, MVI, USAID, and the Kenya Medical Research Institute, culminating in a negative result to be published very soon. At present Diggs is still the Senior Technical Advisor to the USAID MVDP, and he continues to work toward the development of malaria vaccines to protect children and other vulnerable people in areas of endemic infection.

My pathway to malaria vaccines was quite different from those of Carter Diggs, Mary Stevenson, and Denise Doolan. I was born in a small private hospital on the East Side of Manhattan in 1933. My parents were immigrants from Russia who met and married after coming to America. Our family was distinctly at the lower end of the middle class. I attended the local public schools, and after graduating from James Monroe High School I enrolled at the City College of New York (CCNY). CCNY was famous for its academic rigor and a student body composed of high achievers. Although I intended to teach high school biology, as the time of graduation approached, several biology professors encouraged me to go on to graduate school. Through personal contact it was arranged that W. C. Allee (who, because of his age, had been forced to retire from the University of Chicago and was now heading the Zoology Department at the University of Florida, Gainesville) accept me as a graduate student; Allee in turn arranged a research assistantship for me with the protozoologist James B. Lackey at the Department of Sanitary Engineering.

My first independent research project as a microbe hunter involved a survey of oligotrich ciliates (protozoans) in the Gulf of Mexico. Just as I was becoming proficient with ciliate taxonomy and behavior, I received a letter from the Selective Service Board—I was drafted into the U.S. Army. This Korean War interruption of my graduate studies turned out to be an opportunity to learn some parasitology and to travel the world. I was sent to Fort Sam Houston in Texas and then to Valley Forge Army Hospital in Pennsylvania to be trained as a medical technologist. I was shipped overseas, where, through good fortune, I was assigned to medical laboratories in Salzburg, Austria, and Darmstadt and Heidelberg, Germany. In Germany and Austria I worked in clinical laboratories doing hematology, parasitology, microbiology, phlebotomy, and blood chemistry. Upon my discharge from the U.S. Army (where, by an act of Congress, I received the exalted rank of Private First Class) I decided not to return to Florida to continue my graduate studies; instead, I began to teach science and math in a junior high school in Yonkers, NY. At the same time I took a few graduate courses at CCNY and assisted in the teaching of the evening session introductory biology laboratories at CCNY.

During the summer of 1957, thanks to a CCNY Biology Club Scholarship, I took the Invertebrate Zoology course at the Marine Biological Laboratory (MBL) in Woods Hole, MA. At the MBL I was able to spend all day, every day, for 10 weeks studying live invertebrates. The field trips were extraordinary, especially for a South Bronx slum kid, and the faculty was exceptional. At the end of that summer I was invited to be the laboratory assistant for the course. I readily accepted, and for the next two summers I shared the work with another young parasitologist, Frank Friedl, from the University of Minnesota. The experiences in the U.S. Army and the MBL crystallized my interests in combining protozoology with clinical disease. I decided to return to graduate school and scoured university catalogs for those offering the greatest number of courses in parasitology, and I eventually settled on Northwestern University (or, more accurately, they settled on me). I was awarded an Abbott Laboratory Fellowship with Robert Hull (a student of the eminent protozoologist R. R. Kudo). Hull (who had little training in either parasitology or biochemistry) had an NIH grant and, together with his graduate student (Father Truong, a Vietnamese priest), had begun a project on malaria immunity with *P. lophurae* and chickens.

I elected to do some biochemistry despite Hull's laboratory having minimal instrumentation: an ancient Warburg manometer, an assortment of low-speed centrifuges, a Klett colorimeter, a DU spectrophotometer, and a paper electrophoresis apparatus but no possibility of working with radioisotopes. Hull, who did no bench work, appeared not to be very knowledgeable about malaria, so no specific malaria project was assigned to me. It soon became apparent that if I wanted to study the biochemistry of malaria I would have to do it on my own by learning more about the parasite and then would have to find mentors who could teach me biochemistry.

During my time as a graduate student, *P. lophurae* was maintained in chickens by intravenous inoculation. The method is rather simple; however, only a small amount of blood (5 to 10 ml) can be obtained from chicks (relative to ducklings), and because of age immunity the animals would not provide sufficient numbers of parasites when they were older than 5 to 6 weeks and weighed more than 300 gs. As a result, most of my time during the first year involved passing the infection by blood inoculation every 3 to 4 days, measuring blood volumes, determining whether there were statistically significant differences between blood smears, and trying to block natural immunity by using carbon ink. By taking courses in biochemistry (with Lazlo Lorand) and biophysics (with Irving Klotz), I became more familiar with biochemical techniques and approaches.

My fascination with malaria pigment began when I peered through a microscope to examine a blood film taken from a chicken infected with *P. lophurae* and saw the parasites filled with refractile golden-brown granules of hemozoin. As a consequence, one of my doctoral thesis projects became a biochemical characterization of *P. lophurae* hemozoin, and another project involved studying changes in the serum proteins during infection and recovery (441, 445). The winter before I completed my Ph.D. (1960) I spent the Christmas holidays in New York City visiting with my parents. During that time I dropped in to see Frank Friedl, with whom I had shared the Invertebrate Zoology laboratory assistant duties during the summers of 1958 and 1959. Frank had finished his Ph.D. and was now a postdoctoral fellow at the Rockefeller University, working in the laboratory of the distinguished helminthologist Norman R. Stoll. Frank suggested that I stop in and see William Trager on the third floor of Theobald Smith Hall, just across from Stoll's laboratories. Trager was the "dean" of malaria researchers, and although I had no appointment, he graciously listened to my account of what I was doing and heard my future plans. This accidental encounter affected my scientific career for the next 45 years, when Trager offered me an NIH postdoctoral fellowship at the Rockefeller University beginning in the fall of 1960.

At the time, Trager's laboratory was concerned with nutritional studies of *P. lophurae*. He maintained the infection by blood passage in ducklings, in which (unlike in chickens) the developmental cycles were synchronous and the degree of infection was high (as many as 90% of the red cells were infected in 4 days), and the larger body size of the duckling (1 to 2 kg) allowed for the possibility of obtaining up to 100 ml of blood from each duckling. The biochemical equipment consisted of a brand-new Beckman DU spectrophotometer (placed in a broom closet!), which Trager was very anxious to see used, an assortment of power supplies for electrophoresis, refrigerated centrifuges, fraction collectors, water baths, and so on. When I arrived at Rockefeller, my research interests were vague. This was to change quickly, largely as a result of Trager's having attended a conference in Florida where the other speakers were Julius Marmur (Brandeis) and Ernest Bueding (Johns Hopkins). As early as 1955, Bueding (1910 to 1986) and his coworkers had shown, using immunologic methods, that the "Einheit" in biochemistry did not hold: there was enzyme heterogeneity in schistosomes as well as in the tissues of the same animal. Of course, in this he was at odds with some of the most influential biochemists of the time (64). During the meeting Bueding asked Trager if he thought that malaria parasites had enzyme heterogeneity. Returning from the meeting, Trager "volunteered" me to provide Marmur, who was carrying out a survey of the nucleotide base composition of DNA from a variety of protozoans by using the "melting" properties of DNA, with a large quantity of erythrocyte-free *P. lophurae*. He also suggested that I not concern myself with *P. lophurae* DNA but instead concentrate on the possible heterogeneity of lactate dehydrogenase (LDH) in *P. lophurae* and its duckling host. With the help of Elliot Vessel and another Trager postdoctoral investigator, Philip D'Alesandro, I was soon able to separate *P. lophurae* LDH from that of the red blood cell by starch gel and starch block electrophoresis (439), and to Trager's delight the Beckman DU was fully utilized! When Marmur's colleagues at Brandeis, Nathan Kaplan and Allan Wilson, heard of the availability of gram quantities of relatively pure malaria parasites, they too wanted some for their own work. The Kaplan laboratory was actively synthesizing analogs of NAD and had found that the ratio of enzymatic activity with NAD and its various analogs (especially the acetylpyridine NAD) was a very sensitive measure of the differences of various dehydrogenases (especially LDH) in different species and in different organs. I was invited to Brandeis to present a seminar and bring a fresh preparation of packed frozen *P. lophurae*. During that visit to Brandeis, it was discovered that *P. lophurae* LDH had an exceptionally high affinity for the acetylpyridine analog of NAD (436). This proved to be the case for other species of *Plasmodium*, and today LDH activity with acetylpyridine NAD serves as the basis for the diagnostic tests OptiMal and Malstat. It has also been used to quantitatively measure the adherence of *P. falciparum*-infected red cells in vitro (380) and sequestration in comatose children (431).

Although I was quite happy at Rockefeller—truly a researcher's paradise—I still longed to see whether I could carry out an independent research program, and I also wanted an opportunity to teach. In 1962 I was offered the position of Assistant Professor of Biology at the University of California at Riverside (UCR), where one of my former Professors from CCNY, Herman Spieth, had moved in 1954 to head up the Department of Biology. At the time I interviewed at UCR, Spieth was the Chancellor, and it was upon his recommendation that I was eventually invited to join the UCR faculty. During my early years at UCR I continued using *P. lophurae* grown in ducklings because it allowed us to obtain large quantities of erythrocyte-"free" parasites each week very inexpensively. At seminars, in discussing our research, I had a standard line that went: a gram of parasites that costs only a buck a duck.

I was convinced (as were Trager, Taliaferro, and Russell) that the avian malaria model was an appropriate one for human malaria since *P. falciparum* was unavailable, having not yet been cultured continuously in the laboratory. Although I did characterize the antigens of *P. lophurae* in 1963 and reported on this work at a Sadun-sponsored meeting at WRAIR (International Panel Workshop on Cultivation of Plasmodia and Immunology of Malaria, 27 to 28 August 1963), I subsequently abandoned work on malarial antigens and immunity for a decade, since biochemical work, particularly isoenzymes, the structure of malarial DNA and precursors for its synthesis, the sources of amino acids for parasite protein synthesis, the chemical characterization of hemozoin, and membrane transport of glucose, amino acids and purines, occupied most of my efforts.

The transport studies with *P. lophurae* led, almost naturally, to a characterization of the transport interface—the surface membranes of the parasitized red cell and the parasite itself. In 1974, work on membrane characterization was undertaken in collaboration with a postdoctoral fellow from Japan, Yuzo Takahashi. We demonstrated by electron microscopy that there were distinct antigenic alterations in the membranes of erythrocytes infected with *P. lophurae*. We next moved on to membrane isolation and fractionation. Some of the basic techniques for the isolation and characterization of membranes were acquired during a summer spent at the MBL at Woods Hole, MA, when Leonard Warren (Wistar Institute) and I shared a laboratory and worked on both duckling and dogfish red cell membranes. In 1976, after Trager and Jensen reported on their successful continuous in vitro cultivation of *P. falciparum*, the focus of our research on red blood cell membranes moved from the malaria of birds to that of humans. One of the most interesting aspects (at least to me) of the *P. falciparum*-infected red cell was the dramatic change in the external surface: the appearance of pimples or knobs (see p. 217). I became especially interested in determining the biochemical composition of the knob and in turn how the molecular constituents contributed to the antigenic changes of the *P. falciparum*-infected red cell. In effect, biochemistry led to immunology, and that in turn led to an interest in vaccines.

Prior to 1984, our laboratory (as well as others) attempted to find parasite proteins inserted into the red cell membrane of cells infected with *P. falciparum*. Indeed, our inability to find convincing evidence for the insertion of parasite proteins in *P. lophurae*- and *P. falciparum*-infected red cells led us to hypothesize (438) that in addition to parasite proteins the surface of the red cell could be modified through a reorganization of the red cell's own membrane proteins. The approach we used to determine the validity of the hypothesis involved immunological studies similar to those employed by others, but rather than using immune sera from monkeys or humans, we prepared monoclonal antibodies against live *P. falciparum*-infected red cells. Some of these monoclonal antibodies not only bound to the external surface of the infected red cell but also were able to prevent adhesion (559). A variety of biochemical analyses convinced us that the monoclonal antibodies did not recognize a parasite protein but instead recognized a red blood cell membrane protein, the anion transporter named band 3 (because it was the third band to appear after electrophoresis in a gel), albeit in a modified form. This would lead to a decade-long search for its role in immunity.

Band 3 is the most abundant membrane protein in the human red blood cell, being present at a million copies per cell. It is a membrane-spanning protein with eight exterior loops and a highly acidic domain that serves as a center of interaction with cytoplasmic proteins such as ankyrin, glycolytic enzymes such as glyceraldehyde-3-phosphate dehydrogenase and aldolase, hemoglobin, and hemichrome. Using the amino acid sequence of band 3 protein, peptides corresponding to particular regions of the protein were synthesized, and antibodies to these were made in mice and rabbits by using FCA. To our delight (and surprise), the synthetic peptides containing the amino acid sequence HPLQKTY (histidine, proline, leucine, glutamine, lysine, threonine, and tyrosine, corresponding to band 3 residues 546 to 553) blocked adhesion of infected red cells in vitro, as did antibodies made against the synthetic peptides (112, 113). When the synthetic peptide was infused into *P. falciparum*-infected *Aotus* or *Saimiri* monkeys, there was evidence for an antisequestering effect (112). Serum from humans living in areas where malaria was endemic had higher reactivity to the peptide than did serum from individuals in areas where infection was not endemic (226), yet such individuals were healthy and showed no signs of being anemic. We named the peptide "pfalhesin" and went on to show that its receptor contained a peptide sequence (RGD—arginine, glycine, and aspartic acid) found in thrombospondin (156) and fibronectin (157), surface molecules of the endothelial cells lining the post-capillary venules, the place where sequestration occurs. After this work, we used the anion transport inhibitor DIDS to block adhesion (557). In human red cells DIDS binding occurs in a region close to that containing HPLQKTY, namely, YETFSKLIKIFQDH (tyrosine, glutamic acid, threonine, phenylalanine, serine, lysine, leucine, isoleucine, lysine, isoleucine, phenylalanine, glutamine, aspartic acid, and histidine, corresponding to amino acid residues 534 to 547 of band 3), and synthetic peptides with this sequence (called the DIDS-binding region [DBR]), as well as antibodies prepared in chickens and rabbits against this peptide sequence, blocked adhesion (557). The DBR peptide specifically mediated binding to CD36. Sera from healthy individuals from a region where malaria was endemic—so-called immune sera—were shown to contain antibodies to the DBR peptide (558). We proposed that the DBR and pfalhesin were unexposed in the uninfected red cell but that upon infection when band 3 clusters, these buried (cryptic) peptide sequences "pop up" to become exposed on the knobby surface of the *P. falciparum*-infected red cell. In another study we were able to show that oxidative stress also resulted in band 3 clustering and exposure of DBR.

Taken together, our studies showed that membrane modifications of *P. falciparum*-infected red cells need not involve parasite proteins; however, a metabolically active parasite was essential to promote the membrane alterations (i.e., conformational change and exposure of a cryptic amino acid sequence of band 3 protein) through what we think is an oxidative mechanism (443). The DBR and pfalhesin are constituents of the knob-associated "glue" that contributes to sequestration.

However, DBR and pfalhesin may do more than contribute to knob-mediated sequestration. Abnormal hemoglobins such as hemoglobin C are associated with protection against severe malaria (168). Although PfEMP1 is abnormally distributed on these cells (168), and this may contribute to their reduced adherence, a more likely explanation is that in these cells there is increased deposition of oxidized hemoglobin (hemichrome) with parasite development, and as a result band 3 clusters, pfalhesin, and the DBR antigen are exposed, leading to opsonization by autologous IgG and complement. Membrane alterations akin to those observed in *P. falciparum*-

infected red cells also occur in senescent red cells (444, 558). Since these changes also occur in sickle, thalassemic, and glucose-6-phosphate dehydrogenase-deficient red cells, it has been suggested that when parasites develop in these cells, deposition of hemichromes and oxidative aggregation of band 3 protein occur. When this occurs in ring-stage *P. falciparum*-infected red cells with these genetic abnormalities, their removal is promoted, leading to a reduced parasite burden (19, 366).

Our biochemical studies of red blood cell antigens thus uncovered an immunological mechanism for the reduction in the number of blood parasites in individuals who are less susceptible to falciparum malaria. Changes in band 3 protein also lower the numbers of sequestered infected red cells in the deep tissues, thus reducing the likelihood of the development of the severe condition, cerebral malaria. In addition, exposure of this band 3-related antigen not only may be responsible for removal of aged red blood cells but also may contribute to strain-transcending immunity (355, 473).

We have come full circle. Our interest in the immunological basis of premunition was sparked by observations made on chickens resistant to reinfection (see p. 284–285). The demonstration of premunition in human malaria (473), as well as our serendipitous discovery of a strain-independent immunity when using monoclonal antibodies, led to an understanding of how a red cell membrane protein could possibly play a role in strain-transcending immunity in both human and rodent malarias. This somewhat peripatetic research pathway led to my becoming an immunologist, of sorts.

Whether a vaccine based on the normally hidden amino acid sequences of the band 3 protein would protect or cause an autoimmune reaction and result in anemia is not known. However, based on findings with sera from individuals living in an area where malaria is endemic who are healthy and nonanemic (226), this seems unlikely. Further, it remains to be determined whether such band 3-related peptides or a vaccine produced against peptides such as these could act in an antisequestering capacity.

13

Why the Vaccine
Remains Elusive

Twenty-five years ago, Sydney Cohen, who along with Ian McGregor discovered that infection and malaria-related symptoms could be cleared from infants by injections of immunoglobulin from African adults living in a region where malaria is endemic, made a prediction: "An experimental vaccine . . . may be only a decade away . . . and if it is very effective malaria will be eradicated like smallpox" (539). Cohen's euphoric forecast remains unrealized. In the countdown to actualization, the quest for a malaria vaccine—part mirage and part speculation—has been marked by numerous disappointments; however, there are many who remain confident there will be a malaria vaccine but grudgingly acknowledge that the time frame will be very different from the estimates made in the 1980s. When the vaccine arrives, the achievement will be due to the tireless efforts by a band of investigators whose lexicon never included the word "failure."

Malaria vaccines, it is assumed, will assist in transforming the nations in the developing world from a state of poverty to one rich in health. Tourists and warriors will be protected, and children in countries where malaria is endemic will be spared from unnecessary death by childhood vaccinations in much the same way as those in the developed world are spared from other infections today. Vaccines now exist for an array of bacterial diseases such as anthrax, diphtheria, tetanus, typhoid fever, whooping cough (pertussis), pneumococcal pneumonia, and meningococcal disease, as well as for the viral diseases hepatitis A, hepatitis B, human papillomavirus infection, influenza, polio, measles, rubella (German measles), shingles, smallpox, chicken pox, and yellow fever (see Table 1, p. 89). The availability of these vaccines has led the CDC to recommend the following childhood immunizations: hepatitis A and B, polio, rotavirus, influenza,

MMR (measles, mumps, rubella), TdaP (diphtheria, tetanus, pertussis), DtaP (diphtheria, tetanus, pertussis), varicella, pneumococcal pneumonia, and meningitis. When such a program of immunizations is implemented, by the age of 2 children in the United States will have received 28 doses of vaccines and 77% of kindergartners who have completed their vaccinations will be protected. There is, however, no vaccine for malaria, a disease that kills 1 million people, 80% of whom are African children under the age of 5 years.

Sensing this, the Bill and Melinda Gates Foundation, which has already donated $1 billion to a variety of malaria programs, has called for the world to launch a new campaign to eradicate the disease (10, 192, 236). They argue that new scientific advances such as new antimalarial drugs, including the four fixed-dose artemisinin combination therapy, as well as the eight vaccine candidates, three of which are already in clinical trials, including the much-vaunted RTS,S (see p. 255–261), along with growing financial and political support from the Global Fund To Fight AIDS, Tuberculosis and Malaria, the U.S. President's Malaria Initiative and the World Bank's Booster Program for Malaria Control for Africa as well as increased funding from other sources including USAID, the U.S. Army, the U.S. Navy, NIH, MRC, the Wellcome Trust, the Swiss Agency for Development and Cooperation, and the Netherlands Ministry for Development Cooperation, make the goal of eradication realistic. Indeed, recent progress in controlling malaria in some parts of the world has come from four interventions: sleeping under insecticide-impregnated bed nets (see p. 214), spraying houses with insecticides, giving preventive drug treatments to pregnant women, and providing timely treatment of the sick with effective antimalarial drugs. Realistically, however, eradication today may be the same impossible goal faced by the WHO in the 1950s when it set out to rid the world of malaria by vector control and drug treatment. However, even achieving something less than eradication will be a decades-long project that will require new tools, including an effective vaccine, to lift the burden of this disease from many. A strong health care system will also be necessary and "unless well thought out and executed it could divert scarce resources, disrupt other important initiatives and should it flounder it could undermine support for global health initiatives in general" (10).

What are the impediments to producing a protective malaria vaccine, and why has it taken so long to bring so few candidate vaccines to clinical trials and none to market? Why have vaccines succeeded with other microbial diseases but failed with malaria? The successful vaccines for infectious diseases fall into two categories: bacteria or their toxic products and viruses. Genetically speaking, these are both relatively simple pathogens. Influenza virus has 8 genes, mumps virus has 9, measles virus has 10, and smallpox virus has 200 to 400; the protein antigens used for vaccine development, i.e., to induce antibodies, is relatively small: two for influenza virus and perhaps several dozen for smallpox virus. For smallpox, measles, yellow fever, and the Sabin polio vaccines, success was achieved by attenuating the live virus, whereas with the Salk polio vaccine, the virulent virus inactivated by formalin retains its potency to induce an immune response. The successful hepatitis B virus vaccine is an example of a subunit or recombinant vaccine. In these cases vaccine developers were able to induce antibodies to neutralize key viral proteins. Unfortunately, a similar kind of vaccine will not affect the malaria parasite in the same way as it affects the smallpox virus, which remains the only pathogen that humans have ever successfully removed from the face of the Earth (see p. 77–80). Indeed, antibody responses are unlikely to afford any protection against the sequestered liver and red blood cell stages and may even have a limited effect on the infectious merozoites as they move between cells. Bacteria are somewhat more complex than viruses, having on average 1,000 genes, and the vaccines against tetanus, diphtheria, and pertussis target not the entire microbe but often its single toxic protein. Again, the action is mainly through the induction of neutralizing antibodies.

Considering the number of bacteria and viruses that cause human diseases, the number of protective vaccines is actually quite small, but against one-celled parasites such as *Plasmodium* (or many-celled parasites such as roundworms and flatworms) the number of vaccines is zero. The reasons for this are many. First and perhaps foremost is the size and complexity of malaria parasites and these other many-celled pathogens. Only since 2002 have researchers been able to look at the complete gene content of the 14 chromosomes of malaria parasites and deduce the metabolic pathways, identify novel drug targets, and develop new potential vaccine candidates from among the 5,300 genes typically found in the lethal *P. falciparum* strains (181). Adding to the complexity is that half of these genes have neither a known function nor a counterpart in any other living creature. Malaria has four different life stages, each of which is characterized by its own particular array of genes that direct the synthesis of a mosaic of antigens. For example, one study identified 1,289 proteins, with 714 present in asexual stages, 931 present in gametocytes, and 645 present in gametes

(288); in another study, between 200 and 500 genes were found to be active in salivary-gland sporozoites (292) and 246 were active in gametocytes (567). Therefore, a protective or transmission-blocking vaccine will probably have to be directed against a unique surface component, which is likely to be stage specific. Building an effective vaccine against malaria is tantamount to requiring that the immunization be effective against several microbes all at the same time. Malaria parasites, living as they do inside cells, have evolved a series of strategies that allow them to confuse, hide from, and misdirect the responses of the immune system. Once resident in a red blood cell or liver cell, malaria parasites are unlikely to be exposed to the lethal effects of antibody, and it is only during the brief transit time while moving between host cells that the parasites are vulnerable.

According to the PATH MVI, launched in 1999 to accelerate the development of malaria vaccines and funded by a $50 million grant from the Bill and Melinda Gates Foundation, at least 40 promising antigens have been identified (134), and the WHO portfolio of vaccines undergoing clinical trials lists 31 candidates (544). Of these vaccine candidates, most are limited to single-stage antigens, although it is widely accepted that a multistage, multiantigen vaccine would be ideal. Most of the vaccines under development today were discovered before the genetic blueprint of *P. falciparum* was known, and the existing combinations being tested or proposed for testing are seemingly concocted from the range available to the organization responsible for the test. It is also clear that there are not enough resources to fully test the vaccine candidates we have on hand, let alone those yet to be discovered (544). Unresolved is how these combinations should be developed and whether a particular antigen in a combination will interfere with another. There is still no candidate vaccine able to stimulate immunological memory such that protection is long lasting (487). Often overlooked by those selecting antigens for consideration for incorporation in a vaccine is whether challenging the malaria parasite with a vaccine will result in the emergence of vaccine-resistant mutants. Should this not be monitored before deployment, as is often done with antimalaria drugs?

There appears to be a "species barrier," meaning that although it has been relatively easy to achieve some protection in laboratory mice by using subunit vaccines consisting of a variety of antigens, including DNA-vectored vaccines, adenovirus-vectored vaccines, and prime-boost approaches, these have routinely failed in clinical trials (see p. 262 and 300). One explanation may be the test system itself: the mouse is not the natural host for the rodent malarias used extensively for testing subunit vaccines or irradiated sporozoite vaccines. Some argue that since there is no accepted laboratory correlate of clinical immunity and a reliable and predictive animal model does not exist, the testing of putative vaccines must be done early on in human volunteers. The drawbacks of the human challenge model are the time and expenses required to manufacture the vaccine, test its safety, obtain regulatory approval, and then conduct the clinical trial itself. In addition, when vaccine candidates are regarded as failures after a human trial, a negative public perception sets in, and this may limit the funding (and enthusiasm) for continued trials. To avoid such pitfalls, a great deal of clinical work is often required, including the establishment of go/no-go criteria applied to tests of animals in order to limit the number of candidate vaccines brought forward. This exercise can be more than frustrating in terms of time, expense, and applicability.

Vaccine developers who subscribe to the use of monkey models suggest that malaria vaccine trials with nonhuman primates are less expensive and allow a choice of antigen before the need to produce clinical-grade material. However, monkeys are expensive, and there are ecological and ethical concerns about their use. Investigators who reject the use of monkey models argue that it is an invalid model since it uses an unnatural host challenged by an unnatural route and using unnatural dosages. Indeed, in most experimental models of liver stage immunity, very large numbers of irradiated sporozoites (50,000 to 100,000) are injected by intravenous inoculation whereas in nature mosquitoes inoculate a small number (~20 or fewer), yet we now know that sporozoites injected along with the mosquito saliva are deposited into the dermis and not directly into the blood circulation (467, 523). These sporozoites move randomly in the skin until they contact the cells lining the capillaries or lymph vessels, where they glide around and along until they enter and are carried away rapidly by the blood circulation or more slowly by the lymphatic system. This slow trickling may extend for up to 6 h and may lead to a prolonged and very different immune response from that in parasites injected subcutaneously or directly into the circulation. Indeed, it was shown that much larger numbers of irradiated sporozoites were required to elicit the same response through intravenous injection than was induced by the bite of five to nine irradiated *P. yoelii*-infected mosquitoes or by skin injection of 5,000 irradiated sporozoites (190). These findings may have a profound bearing on the deployment of a malaria vaccine.

Many of the blood stage antigens are highly variable and may be strain specific; hence, one must be able to find the invariant epitopes. Such antigens may elicit a reduction in parasite density but can be poorly immunogenic unless coupled with an appropriate adjuvant. There are, however, a limited number of adjuvants available for human use, and FCA—shown to be very effective in animal models—is not one of them. Formulation of antigens with an adjuvant is empirical, as is optimizing the dose and timing of immunizations; all require many time-consuming trials. Vaccines designed to block merozoite invasion of red blood cells (p. 230–231) through the induction of antibodies may not be effective because the antibodies cannot gain access to the "docking" machinery, or the invasion process itself is completed in a matter of minutes. A further complication is that there may be multiple pathways of invasion available to the parasite, so that blocking one simply results in the use of an alternate pathway.

There is a bias in the present-day vaccines in that they contain antigens produced by GMP and are recognized by antibody. The efficacy of these vaccines is ordinarily measured by their antiparasitic effect in human subjects in both endemic and nonendemic settings. Nevertheless, there are complications in defining clinical efficacy. What is the critical end point: a delay in the appearance of parasites in the blood? a reduction in the numbers of parasites in the blood? sterile immunity? a reduction in the rate of mosquito transmission? Where will the appropriate clinical trials be held? What are the required mosquito transmission rates for assessment of efficacy?

The limited success with protein-based vaccines—ones that usually elicit an antibody response—and the recognition that cell-mediated immunity may be critical for protection against liver and possibly blood stages have prompted investigations of DNA and vectored vaccines that induce strong cell-mediated immunity. Currently, however, none of these candidate vaccines has demonstrated more than marginal protection against naturally acquired malaria infection, and there is little potential for any of them to be licensed in the next 5 to 10 years. One of the reasons offered for this is the paucity of antigens being tested and of the many vaccines now in clinical trials; the vaccines contain less than a dozen antigens (3).

Some investigators argue that the path to a protective vaccine will be determined by a better understanding of the immune mechanisms that operate during natural infections as well as after immunization. This may be so; however, the mechanism of action has not been elucidated for many of the vaccines now on the market that provide excellent protection against a variety of infectious diseases. The highly successful smallpox vaccine is the prime example of a vaccine that has needed almost no understanding of host biology, pathogenesis, immunology, or vaccinology (396). The vaccine works, it is reasonably safe, and little else is required. Indeed, while the cowpox virus expresses hundreds of proteins, the targets and mechanisms of protective immunity have never been clearly identified. It is theorized that cell-mediated immunity is critical to the clearance of the primary smallpox infection whereas antibody responses protect against reinfection. This leads to the question: can the same strategy be used with malaria? Possibly, if the obstacles to an attenuated vaccine (see p. 264) can be overcome; however, it must be recognized that the eradication of smallpox was not solely due to a robust immune response to the virus. First, there were no animal reservoirs; second, the methods for preserving the antigenicity of the vaccine proved to be effective; and third, the vaccine was easily administered. Such unique conditions do not exist for malaria; therefore, it is unlikely that eradication can be achieved by a protective vaccine alone.

Recognizing the needs, in December 2006 a Global Research Forum, held in Bangkok, Thailand, involving 230 experts from 100 organizations from 35 countries, called for joint action to accelerate the development of highly effective malaria vaccines. A strategic plan, The Malaria Vaccine Technology Roadmap, was formulated with the stated goal of having a malaria vaccine by 2025 that would have a protective efficacy of 80% against clinical disease and would provide protection for longer than 4 years. An interim landmark was to have a licensed first-generation vaccine by 2015 with 50% protective efficacy against severe disease and death and to last longer than 1 year. It remains to be seen how many bumps in the road there will be (358).

However, even when a malaria vaccine is in hand, there may be considerable frustration involved in scaling up and implementing vaccination programs, as exemplified for vaccines for polio and measles, diseases for which eradication has not been achieved. Moreover, the economics, financing, marketing, and distribution of a pharmaceutical that targets the economically disadvantaged populations of the tropical world, where the cost of one series of immunizations might exceed the per capita health spending of the poorest nations, will represent a formidable challenge. Indeed, to reach the goal proposed by the World Health Assembly in 2005, i.e., reducing the global malaria burden by at least 50% by 2010 and 75% by 2015 (and without a vaccine!), it would cost between $3.8 billion to $4.5 billion annually between now and 2015.

Pharmaceutical firms are unlikely to invest the resources necessary to produce a malaria vaccine and satisfy their investors without some promise of a reasonable return on their investment. Absent such a commitment, philanthropic and governmental agencies will have to partner with pharmaceutical companies to support malaria vaccines. Such collaborations do exist, as is the case in financing and distributing antiretroviral therapies to patients in developing countries. However, it remains to be seen whether such cooperation can be maintained for decades while we wait for The Malaria Vaccine Technology Roadmap to become a highway that is able to bring protection against one of humankind's greatest killers.

References

1. **Abdulla, S., R. Oberholzer, O. Juma, S. Kubhoja, F. Machera, C. Membi, S. Omari, A. Urassa, H. Mshinda, A. Jumanne, N. Salim, M. Shomari, T. Aebi, D. M. Schellenberg, T. Carter, T. Villafana, M. A. Demoitié, M. C. Dubois, A. Leach, M. Lievens, J. Vekemans, J. Ghen, W. R. Ballou, and M. Tanner.** 2008. Safety and immunogenicity of RTS,S/AS02D malaria vaccine in infants. *N. Engl. J. Med.* **359:**2533–2544.

1a. **Achtman, A. H., P. C. Bull, R. Stephens, and J. Langhorne.** 2005. Longevity of the immune response and memory to blood-stage malaria infection. *Curr. Top. Microbiol. Immunol.* **297:**71–102.

2. **Acosta, C. J., C. M. Galindo, D. Schellenberg, J. J. Aponte, E. Kahigwa, H. Urassa, J. R. Schellenberg, H. Masanja, R. Hayes, A. Y. Kitua, F. Lwilla, H. Mshinda, C. Menendez, M. Tanner, and P. L. Alonso.** 1999. Evaluation of the SPf66 vaccine for malaria control when delivered through the EPI scheme in Tanzania. *Trop. Med. Int. Health* **4:**368–376.

3. **Aide, P., Q. Bassat, and P. L. Alonso.** 2007. Towards an effective malaria vaccine. *Arch. Dis. Child.* **92:**476–479.

4. **Aikawa, M., J. R. Rabbege, I. Udeinya, and L. H. Miller.** 1983. Electron microscopy of knobs in *Plasmodium falciparum*-infected erythrocytes. *J. Parasitol.* **69:**435–437.

5. **Alkin, S. S.** 2004. Monoclonal antibodies: the story of a discovery that revolutionized science and medicine. *Nat. Rev. Immunol.* **4:**153–156.

6. **Alonso, P. L., T. Smith, J. R. Schellenberg, H. Masanja, S. Mwankusye, H. Urassa, I. Bastos de Azevedo, J. Chongela, S. Kobero, C. Menendez, et al.** 1994. Randomised trial of efficacy of SPf66 vaccine against *Plasmodium falciparum* malaria in children in southern Tanzania. *Lancet* **344:**1175–1181.

7. **Altman, L. K.** 1987. *Who Goes First? The Story of Self-Experimentation in Medicine.* Random House, New York, NY.

8. **Anonymous.** 1989. *AID's Malaria Vaccine Research Activities.* General Accounting Office, Washington, DC.

9. **Anonymous.** 1975. Malaria vaccine on the horizon. *Br. Med. J.* **1:**231–232. (Editorial.)

10. **Anonymous.** 2007. Is malaria eradication possible? *Lancet* **370:**1459.

11. **Anonymous.** 1998, posting date. Miodrag Ristic Investigation File, 1963–67, 1980–92.

12. **Anonymous.** 1963. Obituary. Jules Freund 1890–1960. *J. Immunol.* **90:**331–336.

13. **Anonymous.** 1993. Retired prof convicted of fraud. Researcher to perform 300 hours of service, pay $50,000. http://www.illinimedia.com/di/archives/1993/February/22/prof.html. Illini Media Co. Champaign, IL.

14. **Anonymous.** 1993. Scientist accused of theft retains post. *Science* **260:**1415.

15. **Aponte, J. J., P. Aide, M. Renom, I. Mandomando, Q. Bassat, J. Sacarlal, M. N. Manaca, S. Lafuente, A. Barbosa, A. Leach, M. Lievens, J. Vekemans, B. Sigauque, M. C. Dubois, M. A. Demoitie, M. Sillman, B. Savarese, J. G. McNeil, E. Macete, W. R. Ballou, J. Cohen, and P. L. Alonso.** 2007. Safety of the RTS,S/AS02D candidate malaria vaccine in infants living in a highly endemic area of Mozambique: a double blind randomised controlled phase I/IIb trial. *Lancet* **370:**1543–1551.

16. **Arakawa, T., A. Komesu, H. Otsuki, J. Sattabongkot, R. Udomsangpetch, Y. Matsumoto, N. Tsuji, Y. Wu, M. Torii, and T. Tsuboi.** 2005. Nasal immunization with a malaria transmission-blocking vaccine candidate, Pfs25, induces complete protective immunity in mice against field isolates of *Plasmodium falciparum*. *Infect. Immun.* **73:**7375–7380.

17. **Ardeshir, F., J. E. Flint, and R. T. Reese.** 1985. Expression of *Plasmodium falciparum* surface antigens in *Escherichia coli*. *Proc. Natl. Acad. Sci. USA* **82:**2518–2522.

18. **Ardeshir, F., J. E. Flint, S. J. Richman, and R. T. Reese.** 1987. A 75 kd merozoite surface protein of *Plasmodium falciparum* which is related to the 70 kd heat-shock proteins. *EMBO J.* **6:**493–499.

19. **Arese, P., K. Ayi, A. Skorokhod, and F. Turrini.** 2006. Removal of early parasite forms from circulation as a mechanism of resistance against malaria in widespread red blood cell mutations, p. 25–53. *In* K. R. Dronamraju and P. Arese (ed.), *Malaria: Genetic and Evolutionary Aspects.* Springer, New York, NY.

20. **Aucouturier, J., S. Ascarateil, and L. Dupuis.** 2006. The use of oil adjuvants in therapeutic vaccines. *Vaccine* **24S2:**S2/44–S42/45.

21. **Avril, M., B. Gamain, C. Lepolard, N. Viaud, A. Scherf, and J. Gysin.** 2006. Characterization of anti-var2CSA-PfEMP1 cytoadhesion inhibitory mouse monoclonal antibodies. *Microbes Infect. Inst. Pasteur* **8:**2863–2871.

22. **Babon, J. J., W. D. Morgan, G. Kelly, J. F. Eccleston, J. Feeney, and A. A. Holder.** 2007. Structural studies on *Plasmodium vivax* merozoite surface protein-1. *Mol. Biochem. Parasitol.* **153:**31–40.

23. **Ballou, W. R.** 2007. Obstacles to the development of a safe and effective attenuated pre-erythrocytic stage malaria vaccine. *Microbes Infect. Inst. Pasteur* **9:**761–766.

24. **Ballou, W. R., and C. P. Cahill.** 2007. Two decades of commitment to malaria vaccine development: GlaxoSmithKline Biologicals. *Am. J. Trop. Med. Hyg.* **77:**289–295.

25. **Ballou, W. R., S. L. Hoffman, J. A. Sherwood, M. R. Hollingdale, F. A. Neva, W. T. Hockmeyer, D. M. Gordon, I. Schneider, R. A. Wirtz, J. F. Young, et al.** 1987. Safety and efficacy of a recombinant DNA *Plasmodium falciparum* sporozoite vaccine. *Lancet* **329:**1277–1281.

26. **Bannister, L., A. Margos, and J. Hopkins.** 2005. Making a home for *Plasmodium* post-genomics: ultrastructural organization of the blood stages, p. 24–49. *In* I. W. Sherman (ed.), *Molecular Approaches to Malaria.* ASM Press, Washington, DC.

27. **Bannister, L., and G. H. Mitchell.** 2003. The ins, outs and roundabouts of malaria. *Trends Parasitol.* **19:**209–213.

28. **Bannister, L. H., G. A. Butcher, E. D. Dennis, and G. H. Mitchell.** 1975. Structure and invasive behaviour of *Plasmodium knowlesi* merozoites in vitro. *Parasitology* **71:**483–491.

29. **Barcia, J. J.** 2007. The Giemsa stain: its history and applications. *Int. J. Surg. Pathol.* **15:**292–296.

30. **Bardes, C. L.** 2008. *Pale Faces. The Masks of Anemia.* Bellevue, New York, NY.

31. **Barnwell, J. W., R. J. Howard, H. G. Coon, and L. H. Miller.** 1983. Splenic requirement for antigenic variation and expression of the variant antigen on the erythrocyte membrane in cloned *Plasmodium knowlesi* malaria. *Infect. Immun.* **40:**985–994.

32. **Barnwell, J. W., R. J. Howard, and L. H. Miller.** 1982. Altered expression of *Plasmodium knowlesi* variant antigen on the erythrocyte membrane in splenectomized rhesus monkeys. *J. Immunol.* **128:**224–226.

33. **Barnwell, J. W., M. E. Nichols, and P. Rubinstein.** 1989. In vitro

evaluation of the role of the Duffy blood group in erythrocyte invasion by *Plasmodium vivax*. *J. Exp. Med.* **169:**1795–1802.

34. **Baruch, D. I.** 1999. Adhesive receptors on malaria-parasitized red cells. *Baillieres Best Pract. Res. Clin. Haematol.* **12:**747–761.

35. **Baruch, D. I., B. Gamain, and L. H. Miller.** 2003. DNA immunization with the cysteine-rich interdomain region 1 of the *Plasmodium falciparum* variant antigen elicits limited cross-reactive antibody responses. *Infect. Immun.* **71:**4536–4543.

36. **Baruch, D. I., B. L. Pasloske, H. B. Singh, X. Bi, X. C. Ma, M. Feldman, T. F. Taraschi, and R. J. Howard.** 1995. Cloning the *P. falciparum* gene encoding PfEMP1, a malarial variant antigen and adherence receptor on the surface of parasitized human erythrocytes. *Cell* **82:**77–87.

37. **Baumler, E.** 1984. *Paul Ehrlich, Scientist for Life.* Holmes and Meier, New York, NY.

38. **Beaudoin, R. L.** 1977. Should cultivated exoerythrocytic parasites be considered as a source of antigen for a malaria vaccine? *Bull. W. H. O.* **55:**373–376.

39. **Beaudoin, R. L., C. P. Strome, F. Mitchell, and T. A. Tubergen.** 1977. *Plasmodium berghei*: immunization of mice against the ANKA strain using the unaltered sporozoite as an antigen. *Exp. Parasitol.* **42:**1–5.

40. **Beaudoin, R. L., C. P. Strome, T. A. Tubergen, and F. Mitchell.** 1976. *Plasmodium berghei berghei*: irradiated sporozoites of the ANKA strain as immunizing antigens in mice. *Exp. Parasitol.* **39:**438–443.

40a. **Bejon, P., J. Lusingu, A. Olotu, A. Leach, M. Lievens, J. Vekemans, S. Mshamu, T. Lang, J. Gould, M. C. Dubois, M. A. Demoitié, J. F. Stallaert, P. Vansadia, T. Carter, P. Njunguna, K. O. Awuondo, A. Malabeja, O. Abdul, S. Gesase, N. Mturi, C. J. Drakeley, B. Savarese, T. Villafana, W. R. Ballou, J. Cohen, G. M. Riley, M. M. Lemnge, K. Marsh, and L. von Seidlein.** 2008. Efficacy of RTS,S/AS01E vaccine against malaria in children 5 to 17 months of age. *N. Engl. J. Med.* **359:**2521–2532.

41. **Belnoue, E., F. T. Costa, T. Frankenberg, A. M. Vigario, T. Voza, N. Leroy, M. M. Rodrigues, I. Landau, G. Snounou, and L. Renia.** 2004. Protective T cell immunity against malaria liver stage after vaccination with live sporozoites under chloroquine treatment. *J. Immunol.* **172:**2487–2495.

42. **Biagini, G. A., P. M. O'Neill, P. G. Bray, and S. A. Ward.** 2005. Current drug development portfolio for antimalarial therapies. *Curr. Opin. Pharmacol.* **5:**1–6.

43. **Biello, D.** 5 March 2008, posting date. Lending a helping arm: volunteers risk malaria to test vaccine. *Sci. Am. Online*

44. **Bignami, A., and G. Bastianelli.** 1890. Observazioni suile febbri malariche estive-autunnali. *Rif. Medi.* **232:**1334–1335.

45. **Biswas, P.** 1999. Patterns of parasitemia, antibodies, complement and circulating immune complexes in drug-suppressed simian *Plasmodium knowlesi* malaria. *Indian J. Malariol.* **36:**33–41.

46. **Bitragunta, S., M. V. Murhekar, Y. J. Hutin, P. P. Penumur, and M. D. Gupte.** 2008. Persistence of diphtheria, Hyderabad, India, 2003–2006. *Emerg. Infect. Dis.* **14:**1144–1146.

47. **Blackman, M. J., T. J. Scott-Finnigan, S. Shai, and A. A. Holder.** 1994. Antibodies inhibit the protease-mediated processing of a malaria merozoite surface protein. *J. Exp. Med.* **180:**389–393.

48. **Boffey, P. M.** 1984. Malaria vaccine is near, U.S. health officials say. *New York Times.* New York, NY.

49. **Boffey, P. M.** 1988. U.S. Malaria Program, roiled by harassment dispute, faces new security. *New York Times.* New York, NY.

50. **Bojang, K. A., S. K. Obaro, U. D'Alessandro, S. Bennett, P. Langerock, G. A. Targett, and B. M. Greenwood.** 1998. An efficacy trial of the malaria vaccine SPf66 in Gambian infants—second year of follow-up. *Vaccine* **16:**62–67.

51. **Boutlis, C. S., E. M. Riley, N. M. Anstey, and J. B. de Souza.** 2005. Glycosylphosphatidylinositols in malaria pathogenesis and immunity: potential for therapeutic inhibition and vaccination. *Curr. Top. Microbiol. Immunol.* **297:**145–185.

52. **Brattig, N. W., K. Kowalsky, X. Liu, G. D. Burchard, F. Kamena, and P. H. Seeberger.** 2008. *Plasmodium falciparum* glycosylphosphatidylinositol toxin interacts with the membrane of non-parasitized red blood cells: a putative mechanism contributing to malaria anemia. *Microbes Infect. Inst. Pasteur* **10:**885–891.

53. **Bray, R. S.** 1976. Vaccination against *Plasmodium falciparum*: a negative result. *Trans. R. Soc. Trop. Med. Hyg.* **70:**258.

54. **Brock, T.** 1999. *Robert Koch, a Life in Medicine.* ASM Press, Washington, DC.

55. **Brooke, J.** 1993. Colombian physician challenges malaria on the home front. *New York Times*, New York, NY.

56. **Brown, E. M.** 2000. Why Wagner-Jauregg won the Nobel prize for discovering malaria therapy for general paresis of the insane. *Hist. Psychiatry* **11:**371–382.

57. **Brown, H.** 2007. Obituary. Sir Ian Alexaner McGregor. *Lancet* **369:**134.
58. **Brown, K. N.** 1973. Antibody-induced variation in malaria parasites. *Nature* **242:**49–50.
59. **Brown, K. N.** 1976. Resistance to malaria, p. 268–295. *In* S. Cohen and E. Sadun (ed.), *Immunology of Parasitic Infections.* Blackwell, Oxford, United Kingdom.
60. **Brown, K. N., and I. N. Brown.** 1965. Immunity to malaria: antigenic variation in chronic infections of *Plasmodium knowlesi. Nature* **208:** 1286–1288.
61. **Brown, K. N., I. N. Brown, and L. A. Hills.** 1970. Immunity to malaria. I. Protection against *Plasmodium knowlesi* shown by monkeys sensitized with drug-suppressed infections or by dead parasites in Freund's adjuvant. *Exp. Parasitol.* **28:**304–317.
62. **Brown, K. N., and J. Williamson.** 1962. Antigens of brucei trypanosomes. *Nature* **194:**1253–1255.
63. **Brown, P.** 1991. Trials and tribulations of a malaria vaccine: one of the most hopeful candidates for a malaria vaccine emerged from an area where the disease is endemic. Its results have raised expectations and scientific controversy. *New Sci.* **129:**18–20.
64. **Bueding, E.** 1962. Comparative biochemistry of parasitic helminths. *Comp. Biochem. Physiol.* **4:**343–351.
65. **Butcher, G. A., and S. Cohen.** 1972. Antigenic variation and protective immunity in *Plasmodium knowlesi* malaria. *Immunology* **23:**503–521.
66. **Butcher, G. A., S. Cohen, and P. C. Garnham.** 1970. Passive immunity in *Plasmodium knowlesi* malaria. *Trans. R. Soc. Trop. Med. Hyg.* **64:**850–856.
67. **Butcher, G. A., G. H. Mitchell, and S. Cohen.** 1973. Mechanism of host specificity in malarial infection. *Nature* **244:**40–41. (Letter.)
68. **Bynum, J., and C. Overy.** 1998. *The Beast in the Mosquito: the Correspondence of Ronald Ross and Patrick Manson.* Rodopi, Amsterdam, The Netherlands.
69. **Camus, D., and T. J. Hadley.** 1985. A *Plasmodium falciparum* antigen that binds to host erythrocytes and merozoites. *Science* **230:**553–556.
70. **Carter, R.** 2002. Spatial simulation of malaria transmission and its control by malaria transmission blocking vaccination. *Int. J. Parasitol.* **32:**1617–1624.
71. **Carter, R.** 2001. Transmission blocking malaria vaccines. *Vaccine* **19:**2309–2314.

72. **Carter, R., and D. H. Chen.** 1976. Malaria transmission blocked by immunisation with gametes of the malaria parasite. *Nature* **263**:57–60.

73. **Carter, R., R. W. Gwadz, and I. Green.** 1979. *Plasmodium gallinaceum*: transmission-blocking immunity in chickens. II. The effect of antigamete antibodies in vitro and in vivo and their elaboration during infection. *Exp. Parasitol.* **47**:194–208.

74. **Carter, R., R. W. Gwadz, and F. M. McAuliffe.** 1979. *Plasmodium gallinaceum*: transmission-blocking immunity in chickens. I. Comparative immunogenicity of gametocyte- and gamete-containing preparations. *Exp. Parasitol.* **47**:185–193.

75. **Carter, R., and D. C. Kaushal.** 1984. Characterization of antigens on mosquito midgut stages of *Plasmodium gallinaceum*. III. Changes in zygote surface proteins during transformation to mature ookinete. *Mol. Biochem. Parasitol.* **13**:235–241.

76. **Cassier, M.** 2005. Appropriatrion and commercialization of the Pasteur anthrax vaccine. *Stud. Hist. Philos. Biol. Biomed. Sci.* **36**:722–742.

77. **Cates, L., S. James, G. Jennings, W. Hockmeyer, and R. Ballou.** 2003. External evaluation of the USAID Malaria Vaccine Development Project. U.S. Agency for International Development, Washington, DC.

78. **Cavasini, C. E., L. C. Mattos, A. A. Couto, C. R. Bonini-Domingos, S. H. Valencia, W. C. Neiras, R. T. Alves, A. R. Rossit, L. Castilho, and R. L. Machado.** 2007. *Plasmodium vivax* infection among Duffy antigen-negative individuals from the Brazilian Amazon region: an exception? *Trans. R. Soc. Trop. Med. Hyg.* **101**:1042–1044.

79. **Chakravarty, S., I. A. Cockburn, S. Kuk, M. G. Overstreet, J. B. Sacci, and F. Zavala.** 2007. CD8[+] T lymphocytes protective against malaria liver stages are primed in skin-draining lymph nodes. *Nat. Med.* **13**:1035–1041.

80. **Chang, K. H., and M. M. Stevenson.** 2004. Malarial anaemia: mechanisms and implications of insufficient erythropoiesis during blood-stage malaria. *Int. J. Parasitol.* **34**:1501–1516.

81. **Chatterjee, S., J. L. Perignon, E. Van Marck, and P. Druilhe.** 2006. How reliable are models for malaria vaccine development? Lessons from irradiated sporozoite immunizations. *J. Postgrad. Med.* **52**:321–324.

82. **Chen, Q., A. Heddini, A. Barragan, V. Fernandez, S. F. Pearce, and M. Wahlgren.** 2000. The semiconserved head structure of *Plasmodium falciparum* erythrocyte membrane protein 1 mediates binding to multiple independent host receptors. *J. Exp. Med.* **192**:1–10.

83. **Chen, Q., F. Pettersson, A. M. Vogt, B. Schmidt, S. Ahuja, P. Lilje-strom, and M. Wahlgren.** 2004. Immunization with PfEMP1-DBL1al-pha generates antibodies that disrupt rosettes and protect against the sequestration of *Plasmodium falciparum*-infected erythrocytes. *Vaccine* **22:**2701–2712.

84. **Chernin, E.** 1984. The malariatherapy of neurosyphilis. *J. Parasitol.* **70:**611–617.

85. **Chitnis, C. E.** 2001. Molecular insights into receptors used by malaria parasites for erythrocyte invasion. *Curr. Opin. Hematol.* **8:**85–91.

86. **Clark, I. A., A. C. Allison, and F. E. Cox,** 1976. Protection of mice against *Babesia* and *Plasmodium. Nature* **259:**309–311.

87. **Clark, I. A., E. J. Wills, J. E. Richmond, and A. C. Allison.** 1977. Suppression of babesiosis in BCG–infected mice and its correlation with tumor inhibition. *Infect. Immun.* **17:**430–438.

88. **Clyde, D. F.** 1990. Immunity to falciparum and vivax malaria induced by irradiated sporozoites: a review of the University of Maryland studies 1971–75. *Bull. W. H. O.* **68**(Suppl. 1)**:**9–12.

89. **Clyde, D. F.** 1975. Immunization of man against falciparum and vivax malaria by use of attenuated sporozoites. *Am. J. Trop. Med. Hyg.* **24:**397–401.

90. **Clyde, D. F., H. Most, and J. Vanderberg.** 1973. Immunization of man against sporozoite-induced falciparum malaria. *Am. J. Med. Sci.* **266:**169–177.

91. **Coatney, G. R., W. E. Collins, M. Warren, and P. G. Contacos.** 1971. *The Primate Malarias.* U.S. Government Printing Office, Bethesda, MD.

92. **Cochrane, A. H., R. Nussenzweig, and E. Nardin.** 1980. Immunization against sporozoites, p. 163–202. *In* J. Kreier (ed.), *Malaria,* vol. 3. Academic Press, San Diego, CA.

93. **Cockburn, I. A., M. J. Mackinnon, A. O'Donnell, S. J. Allen, J. M. Moulds, M. Baisor, M. Bockarie, J. C. Reeder, and J. A. Rowe.** 2004. A human complement receptor 1 polymorphism that reduces *Plasmodium falciparum* rosetting confers protection against severe malaria. *Proc. Natl. Acad. Sci. USA* **101:**272–277.

94. **Coggeshall, L. T.** 1938. *Plasmodium lophurae,* a new species of malaria pathogenic for the domestic fowl. *Am. J. Hyg.* **27:**615–618.

95. **Coggeshall, L. T., and M. Eaton.** 1938. Complement fixation reaction in monkey malaria. *J. Exp. Med.* **66:**177–190.

96. **Coggeshall, L. T., and H. Kumm.** 1937. Demonstration of passive immunity in experimental monkey malaria. *J. Exp. Med.* **66:**177–190.

97. **Coggeshall, L. T., and H. Kumm.** 1938. Effect of repeated superinfection upon the patency of immune serum of monkeys harboring chronic infections of *Plasmodium knowlesi. J. Exp. Med.* **68:**17–27.

98. **Cohen, S.** 1963. Gamma-globulin metabolism. *Br. Med. J.* **19:**202–206.

99. **Cohen, S., and G. A. Butcher.** 1970. Properties of protective malarial antibody. *Nature* **225:**732–734.

100. **Cohen, S., G. A. Butcher, and R. B. Crandall.** 1969. Action of malarial antibody in vitro. *Nature* **223:**368–371.

101. **Cohen, S., I. McGregor, and S. Carrington.** 1961. Gamma-globulin and acquired immunity to human malaria. *Nature* **192:**733–737.

102. **Collins, W. E.** 2002. Nonhuman primate models. I. Nonhuman primate host-parasite combinations. *Methods Mol. Med.* **72:**77–84.

103. **Collins, W. E.** 2002. Nonhuman primate models. II. Infection of *Saimiri* and *Aotus* monkeys with *Plasmodium vivax. Methods Mol. Med.* **72:**85–92.

103a. **Collins, W. E., and J. W. Barnwell.** 2008. A helpful beginning for malaria vaccines. *N. Engl. J. Med.* **359:**2599–2601.

104. **Collins, W. E., and G. M. Jeffery.** 1999. A retrospective examination of secondary sporozoite- and trophozoite-induced infections with *Plasmodium falciparum*: development of parasitologic and clinical immunity following secondary infection. *Am. J. Trop. Med. Hyg.* **61:**20–35.

105. **Collins, W. E., and G. M. Jeffery.** 1999. A retrospective examination of sporozoite- and trophozoite-induced infections with *Plasmodium falciparum* in patients previously infected with heterologous species of *Plasmodium*: effect on development of parasitologic and clinical immunity. *Am. J. Trop. Med. Hyg.* **61:**36–43.

106. **Collins, W. E., and G. M. Jeffery.** 1999. A retrospective examination of the patterns of recrudescence in patients infected with *Plasmodium falciparum. Am. J. Trop. Med. Hyg.* **61:**44–48.

107. **Collins, W. E., and G. M. Jeffery.** 2007. *Plasmodium malariae*: parasite and disease. *Clin. Microbiol. Rev.* **20:**579–592.

108. **Collins, W. E., G. M. Jeffery, and J. M. Roberts.** 2004. A retrospective examination of reinfection of humans with *Plasmodium vivax. Am. J. Trop. Med. Hyg.* **70:**642–644.

109. **Collins, W. E., and G. M. Jeffrey.** 2005. *Plasmodium ovale*: parasite and disease. *Clin. Microbiol. Rev.* **18:**570–581.

110. **Collins, W. E., J. C. Skinner, M. Pappaioanou, J. R. Broderson, N. S. Ma, V. Filipski, P. S. Stanfill, and L. Rogers.** 1988. Infection of Peruvian

Aotus nancymai monkeys with different strains of *Plasmodium falciparum*, *P. vivax*, and *P. malariae*. *J. Parasitol.* **74**:392–398.

111. **Cowman, A. F., and B. S. Crabb.** 2006. Invasion of red blood cells by malaria parasites. *Cell* **124**:755–766.

112. **Crandall, I., W. E. Collins, J. Gysin, and I. W. Sherman.** 1993. Synthetic peptides based on motifs present in human band 3 protein inhibit cytoadherence/sequestration of the malaria parasite *Plasmodium falciparum*. *Proc. Natl. Acad. Sci. USA* **90**:4703–4707.

113. **Crandall, I., and I. W. Sherman.** 1994. Antibodies to synthetic peptides based on band 3 motifs react specifically with *Plasmodium falciparum* (human malaria)-infected erythrocytes and block cytoadherence. *Parasitology* **108**:389–396.

114. **Dahlback, M., T. S. Rask, P. H. Andersen, M. A. Nielsen, N. T. Ndam, M. Resende, L. Turner, P. Deloron, L. Hviid, O. Lund, A. G. Pedersen, T. G. Theander, and A. Salanti.** 2006. Epitope mapping and topographic analysis of VAR2CSA DBL3X involved in *P. falciparum* placental sequestration. *PLoS Pathog.* **2**:e124.

115. **Dame, J. B., J. L. Williams, T. F. McCutchan, J. L. Weber, R. A. Wirtz, W. T. Hockmeyer, W. L. Maloy, J. D. Haynes, I. Schneider, D. Roberts, et al.** 1984. Structure of the gene encoding the immunodominant surface antigen on the sporozoite of the human malaria parasite *Plasmodium falciparum*. *Science* **225**:593–599.

116. **Daubersies, P., B. Ollomo, J. P. Sauzet, K. Brahimi, B. L. Perlaza, W. Eling, H. Moukana, P. Rouquet, C. de Taisne, and P. Druilhe.** 2008. Genetic immunisation by liver stage antigen 3 protects chimpanzees against malaria despite low immune responses. *PLoS ONE* **3**:e2659.

117. **Daubersies, P., A. W. Thomas, P. Millet, K. Brahimi, J. A. Langermans, B. Ollomo, L. BenMohamed, B. Slierendregt, W. Eling, A. Van Belkum, G. Dubreuil, J. F. Meis, C. Guerin-Marchand, S. Cayphas, J. Cohen, H. Gras-Masse, and P. Druilhe.** 2000. Protection against *Plasmodium falciparum* malaria in chimpanzees by immunization with the conserved pre-erythrocytic liver-stage antigen 3. *Nat. Med.* **6**:1258–1263.

118. **Deane, L., J. Neto, and I. Silveira.** 1966. Experimental infection of a splenectomized squirrel monkey *Saimiri sciureus*. *Trans. R. Soc. Trop. Med. Hyg.* **60**:811–812.

119. **Deans, A. M., S. Nery, D. J. Conway, O. Kai, K. Marsh, and J. A. Rowe.** 2007. Invasion pathways and malaria severity in Kenyan *Plasmodium falciparum* clinical isolates. *Infect. Immun.* **75**:3014–3020.

120. **Deans, J.** 1984. Protective antigens of blood stage *Plasmodium knowlesi* parasites. *Philos. Trans. R. Soc. Lond. Ser. B* **307**:159–169.

121. **Deans, J. A., T. Alderson, A. W. Thomas, G. H. Mitchell, E. S. Lennox, and S. Cohen.** 1982. Rat monoclonal antibodies which inhibit the in vitro multiplication of *Plasmodium knowlesi*. *Clin. Exp. Immunol.* **49**:297–309.

122. **Deans, J. A., A. M. Knight, W. C. Jean, A. P. Waters, S. Cohen, and G. H. Mitchell.** 1988. Vaccination trials in rhesus monkeys with a minor, invariant, *Plasmodium knowlesi* 66 kD merozoite antigen. *Parasite Immunol.* **10**:535–552.

123. **Debre, P.** 1998. *Louis Pasteur*. Johns Hopkins Press, Baltimore, MD.

124. **de Kruif, P.** 1926. *Microbe Hunters*. Harcourt Brace, San Diego, CA.

125. **Dennis, E. D., G. H. Mitchell, G. A. Butcher, and S. Cohen.** 1975. In vitro isolation of *Plasmodium knowlesi* merozoites using polycarbonate sieves. *Parasitology* **71**:475–481.

126. **Desowitz, R.** 2002. *Federal Bodysnatchers and the New Guinea Virus*. Norton, New York, NY.

127. **Desowitz, R.** 1991. *The Malaria Capers. More Tales of Parasites and People, Research and Reality*. Norton, New York, NY.

128. **Diggs, C. L.** 1980. Prospects for development of vaccines against *Plasmodium falciparum* infection, p. 299–315. *In* J. Kreier (ed.), *Malaria*, vol. 3. Academic Press, Inc., San Diego, CA.

129. **Diggs, C. L., F. Hines, and B. T. Wellde.** 1995. *Plasmodium falciparum*: passive immunization of *Aotus lemurinus griseimembra* with immune serum. *Exp. Parasitol.* **80**:291–296.

130. **Dobell, C.** 1960. *Antony van Leeuwenhoek and His "Little Animals"; Being Some Account of the Father of Protozoology and Bacteriology and His Multifarious Discoveries in These Disciplines*. Dover, New York, NY.

131. **Dobson, M. J.** 1997. *Contours of Death and Disease in Early Modern England*. Cambridge University Press, Cambridge, United Kingdom.

132. **Doolan, D., J. Dobano, and J. K. Baird.** 2009. Acquired immunity to malaria. *Clin. Microbiol. Rev.* **22**:13–36.

133. **Doolan, D., and S. L. Hoffman.** 2002. Nucleic acid vaccines against malaria. *Chem. Immunol.* **80**:308–321.

134. **Doolan, D., and A. Stewart.** 2007. Status of malaria R & D in 2007. *Expert Rev. Vaccines* **6**:903–905.

135. **Doolan, D. L., R. C. Hedstrom, W. O. Rogers, Y. Charoenvit, M. Rogers, P. de la Vega, and S. L. Hoffman.** 1996. Identification and characterization of the protective hepatocyte erythrocyte protein 17

kDa gene of *Plasmodium yoelii,* homolog of *Plasmodium falciparum* exported protein 1. *J. Biol. Chem.* **271:**17861–17868.

136. **Doolan, D. L., and S. L. Hoffman.** 1997. Pre-erythrocytic-stage immune effector mechanisms in *Plasmodium* spp. infections. *Philos. Trans. R. Soc. Lond. Ser. B* **352:**1361–1367.

137. **Doolan, D. L., and S. L. Hoffman.** 2000. The complexity of protective immunity against liver-stage malaria. *J. Immunol.* **165:**1453–1462.

138. **Doolan, D. L., and N. Martinez-Alier.** 2006. Immune response to pre-erythrocytic stages of malaria parasites. *Curr. Mol. Med.* **6:**169–185.

139. **Doolan, D. L., Y. Mu, B. Unal, S. Sundaresh, S. Hirst, C. Valdez, A. Randall, D. Molina, X. Liang, D. A. Freilich, J. A. Oloo, P. L. Blair, J. C. Aguiar, P. Baldi, D. H. Davies, and P. L. Felgner.** 2008. Profiling humoral immune responses to *P. falciparum* infection with protein microarrays. *Proteomics* **8:**4680–4694.

140. **Doolan, D. L., M. Sedegah, R. C. Hedstrom, P. Hobart, Y. Charoenvit, and S. L. Hoffman.** 1996. Circumventing genetic restriction of protection against malaria with multigene DNA immunization: $CD8^+$ cell-, interferon gamma-, and nitric oxide-dependent immunity. *J. Exp. Med.* **183:**1739–1746.

141. **Doolan, D. L., S. Southwood, D. A. Freilich, J. Sidney, N. L. Graber, L. Shatney, L. Bebris, L. Florens, C. Dobano, A. A. Witney, E. Appella, S. L. Hoffman, J. R. Yates III, D. J. Carucci, and A. Sette.** 2003. Identification of *Plasmodium falciparum* antigens by antigenic analysis of genomic and proteomic data. *Proc. Natl. Acad. Sci. USA* **100:**9952–9957.

142. **Douradinha, B., M. R. van Dijk, R. Ataide, G. J. van Gemert, J. Thompson, J. F. Franetich, D. Mazier, A. J. Luty, R. Sauerwein, C. J. Janse, A. P. Waters, and M. M. Mota.** 2007. Genetically attenuated P36p-deficient *Plasmodium berghei* sporozoites confer long-lasting and partial cross-species protection. *Int. J. Parasitol.* **37:**1511–1519.

143. **Draper, S. J., A. C. Moore, A. L. Goodman, C. A. Long, A. A. Holder, S. C. Gilbert, F. Hill, and A. V. Hill.** 2008. Effective induction of high-titer antibodies by viral vector vaccines. *Nat. Med.* **14:**819–821.

144. **Druilhe, P., and J. W. Barnwell.** 2007. Pre-erythrocytic stage malaria vaccines: time for a change in path. *Curr. Opin. Microbiol.* **10:**371–378.

145. **Druilhe, P., and C. Marchand.** 1989. From sporozoite to liver stages: the saga of the irradiated sporozoite vaccine, p. 39–50. *In* K. P. W.

McAdams (ed.), *Frontiers of Infectious Diseases. New Strategies in Parasitology.* Churchill Livingstone, London, United Kingdom.

146. **Druilhe, P., L. Renia, and D. Fidock.** 1998. Immunity to liver stages, p. 513–544. *In* I. W. Sherman (ed.), *Malaria. Parasite Biology, Pathogenesis, and Protection.* ASM Press, Washington, DC.

147. **Dubos, R.** 1976. *Louis Pasteur, Free Lance of Science.* Scribners Sons, New York, NY.

148. **Duffy, M. F., A. Caragounis, R. Noviyanti, H. M. Kyriacou, E. K. Choong, K. Boysen, J. Healer, J. A. Rowe, M. E. Molyneux, G. V. Brown, and S. J. Rogerson.** 2006. Transcribed *var* genes associated with placental malaria in Malawian women. *Infect. Immun.* **74:**4875–4883.

149. **Duffy, M. F., A. G. Maier, T. J. Byrne, A. J. Marty, S. R. Elliott, M. T. O'Neill, P. D. Payne, S. J. Rogerson, A. F. Cowman, B. S. Crabb, and G. V. Brown.** 2006. VAR2CSA is the principal ligand for chondroitin sulfate A in two allogeneic isolates of *Plasmodium falciparum. Mol. Biochem. Parasitol.* **148:**117–124.

150. **Duffy, P. E.** 2007. *Plasmodium* in the placenta: parasites, parity, protection, prevention and possibly preeclampsia. *Parasitology* **134:** 1877–1881.

151. **Duffy, P. E., and M. Fried.** 2005. Malaria in the pregnant woman. *Curr. Top. Microbiol. Immunol.* **295:**169–200.

152. **Duraisingh, M. T., A. G. Maier, T. Triglia, and A. F. Cowman.** 2003. Erythrocyte-binding antigen 175 mediates invasion in *Plasmodium falciparum* utilizing sialic acid-dependent and -independent pathways. *Proc. Natl. Acad. Sci. USA* **100:**4796–4801.

153. **Eaton, M. D.** 1938. The agglutination of *Plasmodium knowlesi* by immune serum. *J. Exp. Med.* **67:**857.

154. **Eaton, M. D.** 1939. The soluble malarial antigen in the serum of monkeys infected with *Plasmodium knowlesi. J. Exp. Med.* **69:**517–532.

155. **Eaton, M. D., and L. T. Coggeshall.** 1939. Production in monkeys of complement fixing antibodies without active immunity by injection of killed *Plasmodium knowlesi. J. Exp. Med.* **70:**141–146.

156. **Eda, S., J. Lawler, and I. W. Sherman.** 1999. *Plasmodium falciparum*-infected erythrocyte adhesion to the type 3 repeat domain of thrombospondin-1 is mediated by a modified band 3 protein. *Mol. Biochem. Parasitol.* **100:**195–205.

157. **Eda, S., and I. W. Sherman.** 2004. *Plasmodium falciparum*-infected erythrocytes bind to the RGD motif of fibronectin via the band 3-related adhesin. *Exp. Parasitol.* **107:**157–162.

158. **Eisen, H. N.** 2001. Michael Heidelberger 1888–1991. *Biogra. Mem. Natl. Acad. Sci.* **80**:122–140.

159. **Ellis, J., L. S. Ozaki, R. W. Gwadz, A. H. Cochrane, V. Nussenzweig, R. S. Nussenzweig, and G. N. Godson.** 1983. Cloning and expression in *E. coli* of the malarial sporozoite surface antigen gene from *Plasmodium knowlesi*. *Nature* **302**:536–538.

160. **Enea, V., D. Arnot, E. C. Schmidt, A. Cochrane, R. Gwadz, and R. S. Nussenzweig.** 1984. Circumsporozoite gene of *Plasmodium cynomolgi* (Gombak): cDNA cloning and expression of the repetitive circumsporozoite epitope. *Proc. Natl. Acad. Sci. USA* **81**:7520–7524.

161. **Epstein, J. E.** 2007. What will a partly protective malaria vaccine mean to mothers in Africa? *Lancet* **370**:1523–1524.

162. **Epstein, J. E., B. Giersing, G. Mullen, V. Moorthy, and T. L. Richie.** 2007. Malaria vaccines: are we getting closer? *Curr. Opin. Mol. Ther.* **9**:12–24.

163. **Epstein, J. E., S. Rao, F. Williams, D. Freilich, T. Luke, M. Sedegah, P. de la Vega, J. Sacci, T. L. Richie, and S. L. Hoffman.** 2007. Safety and clinical outcome of experimental challenge of human volunteers with *Plasmodium falciparum*-infected mosquitoes: an update. *J. Infect. Dis.* **196**:145–154.

164. **Espinal, C., E. Moreno, J. Umana, J. Ramirez, and M. Montilla.** 1984. Susceptibility of different populations of Colombian *Aotus* monkeys to the FCB-1 strain of *Plasmodium falciparum*. *Am. J. Trop. Med. Hyg.* **33**:777–782.

165. **Eugui, E. M., and A. C. Allison.** 1979. Malaria infections in different strains of mice and their correlation with natural killer activity. *Bull. W. H. O.* **57** (Suppl. 1):231–238.

166. **Eyles, D.** 1952. Studies on *Plasmodium gallinaceum*. I. Factors associated with the malaria infection in the vertebrate host which influence the degree of infection in the mosquito. *Am. J. Hyg.* **55**:386–391.

167. **Eyles, D.** 1952. Studies on *Plasmodium gallinaceum*. II. Factors in the blood of the vertebrate host influencing mosquito infection. *Am. J. Hyg.* **55**:276–290.

168. **Fairhurst, R. M., H. Fujioka, K. Hayton, K. F. Collins, and T. E. Wellems.** 2003. Aberrant development of *Plasmodium falciparum* in hemoglobin CC red cells: implications for the malaria protective effect of the homozygous state. *Blood* **101**:3309–3315.

169. **Fleischer, B.** 2004. 100 years ago: Giemsa's solution of staining plasmodia. *Trop. Med. Inter. Health* **9**:755–756.

170. **Fortin, A., M. M. Stevenson, and P. Gros.** 2002. Susceptibility to malaria as a complex trait: big pressure from a tiny creature. *Hum. Mol. Genet.* **11:**2469–2478.

171. **Franke-Fayard, B., C. J. Janse, M. Cunha-Rodrigues, J. Ramesar, P. Buscher, I. Que, C. Lowik, P. J. Voshol, M. A. den Boer, S. G. van Duinen, M. Febbraio, M. M. Mota, and A. P. Waters.** 2005. Murine malaria parasite sequestration: CD36 is the major receptor, but cerebral pathology is unlinked to sequestration. *Proc. Natl. Acad. Sci. USA* **102:**11468–11473.

172. **Freeman, R. R., A. J. Trejdosiewicz, and G. A. Cross.** 1980. Protective monoclonal antibodies recognising stage-specific merozoite antigens of a rodent malaria parasite. *Nature* **284:**366–368.

173. **Freund, J., K. J. Thomson, H. E. Sommer, A. W. Walter, and E. L. Schenkein.** 1945. Immunization of Rhesus monkeys against malarial infection (*P. knowlesi*) with killed parasites and adjuvants. *Science* **102:**202–204.

174. **Freund, J., K. Thomson, H. Sommer, A. Walter, and T. Pisani.** 1948. Immunization of monkeys against malaria by means of killed parasites with adjuvants. *Am. J. Trop. Med.* **28:**1–22.

175. **Freund, J. S., H. Sommer, and A. Walter.** 1945. Immunization against malaria: vaccination of ducks with killed parasites incorporated with adjuvants. *Science* **102:**200–202.

176. **Frevert, U., and E. Nardin.** 2008. Cellular effector mechanisms against *Plasmodium* liver stages. *Cell. Microbiol.* **10:**1956–1967.

177. **Fried, M., K. K. Hixson, L. Anderson, Y. Ogata, T. K. Mutabingwa, and P. E. Duffy.** 2007. The distinct proteome of placental malaria parasites. *Mol. Biochem. Parasitol.* **155:**57–65.

178. **Fried, M., F. Nosten, A. Brockman, B. J. Brabin, and P. E. Duffy.** 1998. Maternal antibodies block malaria. *Nature* **395:**851–852.

179. **Galinski, M. R., A. R. Dluzewski, and J. W. Barnwell.** 2005. A mechanistic approach to merozoite invasion of red blood cells: merozoite biogenesis, rupture, and invasion of erythrocytes, p. 113–168. *In* I. W. Sherman (ed.), *Molecular Approaches to Malaria.* ASM Press, Washington, DC.

180. **Gamage-Mendis, A. C., J. Rajakaruna, R. Carter, and K. N. Mendis.** 1992. Transmission blocking immunity to human *Plasmodium vivax* malaria in an endemic population in Kataragama, Sri Lanka. *Parasite Immunol.* **14:**385–396.

181. **Gardner, M. J., N. Hall, E. Fung, O. White, M. Berriman, R. W. Hyman, J. M. Carlton, A. Pain, K. E. Nelson, S. Bowman, I. T. Paulsen,**

K. James, J. A. Eisen, K. Rutherford, S. L. Salzberg, A. Craig, S. Kyes, M. S. Chan, V. Nene, S. J. Shallom, B. Suh, J. Peterson, S. Angiuoli, M. Pertea, J. Allen, J. Selengut, D. Haft, M. W. Mather, A. B. Vaidya, D. M. Martin, A. H. Fairlamb, M. J. Fraunholz, D. S. Roos, S. A. Ralph, G. I. McFadden, L. M. Cummings, G. M. Subramanian, C. Mungall, J. C. Venter, D. J. Carucci, S. L. Hoffman, C. Newbold, R. W. Davis, C. M. Fraser, and B. Barrell. 2002. Genome sequence of the human malaria parasite *Plasmodium falciparum*. *Nature* **419**:498–511.

182. Garnham, P. C. C. 1947. Exo-erythrocytic schizogony in *Plasmodium kochi*: a preliminary note. *Trans. R. Soc. Trop. Med. Hyg.* **40**:719–722.

183. Gaur, D., J. R. Storry, M. E. Reid, J. W. Barnwell, and L. H. Miller. 2003. *Plasmodium falciparum* is able to invade erythrocytes through a trypsin-resistant pathway independent of glycophorin B. *Infect. Immun.* **71**:6742–6746.

184. Geiman, Q. M., and M. J. Meagher. 1967. Susceptibility of a New World monkey to *Plasmodium falciparum* from man. *Nature* **215**:437–439.

185. Geison, G. 1990. Pasteur, Roux and rabies: scientific versus clinical mentalities. *J. Hist. Med. Allied Sci.* **45**:341–365.

186. Geison, G. 1978. Pasteur's work on rabies: re-examining the ethical issues. *Hastings Center Rep.* **8**:26–33.

187. Geison, G. 1995. *The Private Science of Louis Pasteur*. Princeton University Press, Princeton, NJ.

188. Good, M. F. 2005. Vaccine-induced immunity to malaria parasites and the need for novel strategies. *Trends Parasitol.* **21**:29–34.

189. Good, M. F., and D. L. Doolan. 1999. Immune effector mechanisms in malaria. *Curr. Opin. Immunol.* **11**:412–419.

190. Good, M. F., and D. L. Doolan. 2007. Malaria's journey through the lymph node. *Nat. Med.* **13**:1023–1024.

191. Gordon, D. M., T. W. McGovern, U. Krzych, J. C. Cohen, I. Schneider, R. LaChance, D. G. Heppner, G. Yuan, M. Hollingdale, M. Slaoui, et al. 1995. Safety, immunogenicity, and efficacy of a recombinantly produced *Plasmodium falciparum* circumsporozoite protein-hepatitis B surface antigen subunit vaccine. *J. Infect. Dis.* **171**:1576–1585.

192. Grabowsky, M. 2008. The billion dollar malaria moment. *Nature* **451**:1051–1052.

193. Gratepanche, S., B. Gamain, J. D. Smith, B. A. Robinson, A. Saul,

and L. H. Miller. 2003. Induction of crossreactive antibodies against the *Plasmodium falciparum* variant protein. *Proc. Natl. Acad. Sci. USA* **100**:13007–13012.

194. **Graves, P., H. Gelband, and P. Garner.** 1998. The SPf66 malaria vaccine: what is the evidence for efficacy? *Parasitol. Today* **14**:218–220.

195. **Graves, P. M.** 1998. Comparison of the cost-effectiveness of vaccines and insecticide impregnation of mosquito nets for the prevention of malaria. *Ann. Trop. Med. Parasitol.* **92**:399–410.

196. **Graves, P. M., R. Carter, T. R. Burkot, I. A. Quakyi, and N. Kumar.** 1988. Antibodies to *Plasmodium falciparum* gamete surface antigens in Papua New Guinea sera. *Parasite Immunol.* **10**:209–218.

197. **Greenwood, B. M., K. Bojang, C. Whitty, and G. Targett.** 2005. Malaria. *Lancet* **365**:1487–1498.

198. **Grotendorst, C. A., N. Kumar, R. Carter, and D. C. Kaushal.** 1984. A surface protein expressed during the transformation of zygotes of *Plasmodium gallinaceum* is a target of transmission-blocking antibodies. *Infect. Immun.* **45**:775–777.

199. **Gruenberg, J., D. R. Allred, and I. W. Sherman.** 1983. Scanning electron microscope-analysis of the protrusions (knobs) present on the surface of *Plasmodium falciparum*-infected erythrocytes. *J. Cell Biol.* **97**:795–802.

200. **Grun, J., and W. Weidanz.** 1983. Antibody-independent immunity to reinfection malaria in B cell deficient mice. *Infect. Immun.* **41**:1197–1204.

201. **Gruner, A. C., M. Mauduit, R. Tewari, J. F. Romero, N. Depinay, M. Kayibanda, E. Lallemand, J. M. Chavatte, A. Crisanti, P. Sinnis, D. Mazier, G. Corradin, G. Snounou, and L. Renia.** 2007. Sterile protection against malaria is independent of immune responses to the circumsporozoite protein. *PLoS ONE* **2**:e1371.

202. **Guerin-Marchand, C., P. Druilhe, B. Galey, A. Londono, J. Patarapotikul, R. L. Beaudoin, C. Dubeaux, A. Tartar, O. Mercereau-Puijalon, and G. Langsley.** 1987. A liver-stage-specific antigen of *Plasmodium falciparum* characterized by gene cloning. *Nature* **329**:164–167.

203. **Guillemin, J.** 2002. Choosing scientific patrimony: Sir Ronald Ross, Alphonse Laveran and the mosquito hypothesis for malaria. *J. Hist. Med.* **57**:385–409.

204. **Gwadz, R. W.** 1976. Successful immunization against the sexual stages of *Plasmodium gallinaceum*. *Science* **193**:1150–1151.

205. **Gwadz, R. W., and L. C. Koontz.** 1984. *Plasmodium knowlesi*: persistence of transmission blocking immunity in monkeys immunized with gamete antigens. *Infect. Immun.* **44:**137–140.

206. **Gysin, J.** 1998. Animal models: primates, p. 419–441. *In* I. W. Sherman (ed.), *Malaria. Parasite Biology, Pathognesis and Protection.* ASM Press, Washington, DC.

207. **Gysin, J., M. Hommel, and L. P. da Silva.** 1980. Experimental infection of the squirrel monkey (*Saimiri sciureus*) with *Plasmodium falciparum*. *J. Parasitol.* **66:**1003–1009.

208. **Hadley, T. J., and L. H. Miller.** 1988. Invasion of erythrocytes by malaria parasites: erythrocyte ligands and parasite receptors. *Prog. Allergy* **41:**49–71.

209. **Hafalla, J. C., A. Morrot, G. Sano, G. Milon, J. J. Lafaille, and F. Zavala.** 2003. Early self-regulatory mechanisms control the magnitude of CD8$^+$ T cell responses against liver stages of murine malaria. *J. Immunol.* **171:**964–970.

210. **Hafalla, J. C., G. Sano, L. H. Carvalho, A. Morrot, and F. Zavala.** 2002. Short-term antigen presentation and single clonal burst limit the magnitude of the CD8(+) T cell responses to malaria liver stages. *Proc. Natl. Acad. Sci. USA* **99:**11819–11824.

211. **Harris, P. K., S. Yeoh, A. R. Dluzewski, R. A. O'Donnell, C. Withers-Martinez, F. Hackett, L. H. Bannister, G. H. Mitchell, and M. J. Blackman.** 2005. Molecular identification of a malaria merozoite surface sheddase. *PLoS Pathog.* **1:**241–251.

212. **Harrison, G.** 1978. *Mosquitoes, Malaria and Man: a History of Hostilities since 1880.* Dutton, New York, NY.

213. **Haynes, J. D., C. L. Diggs, F. A. Hines, and R. E. Desjardins.** 1976. Culture of human malaria parasites *Plasmodium falciparum. Nature* **263:**767–769.

214. **Heidelberger, M., W. A. Coates, and M. Mayer.** 1946. Studies in human malaria. II. Attempts to influence relapsing vivax malaria by treatment of patients with vaccine (*P. vivax*). *J. Immunol.* **52:**101–107.

215. **Heidelberger, M., M. Mayer, and C. Demarest.** 1946. Studies in human malaria. I. The preparation of vaccines and suspensions containing plasmodia. *J. Immunol.* **52:**325–330.

216. **Heidelberger, M., C. Prout, J. Hindle, and A. Rose.** 1946. Studies in human malaria. III. An attempt at vaccination of paretics against blood-borne infection with *Plasmodium vivax. J. Immunol.* **52:**109–118.

216a. Heidelberger, M., M. Mayer, A. Alving, B. Craige, R. Jones, T. Pullman, and M. Whorton. 1946. Studies in human malaria. IV. An attempt at vaccination of volunteers against mosquito-borne infection with *Pl. vivax. J. Immunol.* **52:**113–118.

217. Heppner, D. G., D. M. Gordon, M. Gross, B. Wellde, W. Leitner, U. Krzych, I. Schneider, R. A. Wirtz, R. L. Richards, A. Trofa, T. Hall, J. C. Sadoff, P. Boerger, C. R. Alving, D. R. Sylvester, T. G. Porter, and W. R. Ballou. 1996. Safety, immunogenicity, and efficacy of *Plasmodium falciparum* repeatless circumsporozoite protein vaccine encapsulated in liposomes. *J. Infect. Dis.* **174:**361–366.

218. Heppner, D. G., Jr., K. E. Kester, C. F. Ockenhouse, N. Tornieporth, O. Ofori, J. A. Lyon, V. A. Stewart, P. Dubois, D. E. Lanar, U. Krzych, P. Moris, E. Angov, J. F. Cummings, A. Leach, B. T. Hall, S. Dutta, R. Schwenk, C. Hillier, A. Barbosa, L. A. Ware, L. Nair, C. A. Darko, M. R. Withers, B. Ogutu, M. E. Polhemus, M. Fukuda, S. Pichyangkul, M. Gettyacamin, C. Diggs, L. Soisson, J. Milman, M. C. Dubois, N. Garcon, K. Tucker, J. Wittes, C. V. Plowe, M. A. Thera, O. K. Duombo, M. G. Pau, J. Goudsmit, W. R. Ballou, and J. Cohen. 2005. Towards an RTS,S-based, multi-stage, multi-antigen vaccine against falciparum malaria: progress at the Walter Reed Army Institute of Research. *Vaccine* **23:**2243–2250.

219. Herrera, S., B. L. Perlaza, A. Bonelo, and M. Arevalo-Herrera. 2002. Aotus monkeys: their great value for anti-malaria vaccines and drug testing. *Int. J. Parasitol.* **32:**1625–1635.

220. Herrington, D., J. Davis, E. Nardin, M. Beier, J. Cortese, H. Eddy, G. Losonsky, M. Hollingdale, M. Sztein, M. Levine, et al. 1991. Successful immunization of humans with irradiated malaria sporozoites: humoral and cellular responses of the protected individuals. *Am. J. Trop. Med. Hyg.* **45:**539–547.

221. Herrington, D. A., E. H. Nardin, G. Losonsky, I. C. Bathurst, P. J. Barr, M. R. Hollingdale, R. Edelman, and M. M. Levine. 1991. Safety and immunogenicity of a recombinant sporozoite malaria vaccine against *Plasmodium vivax. Am. J. Trop. Med. Hyg.* **45:**695–701.

222. Hillier, C. J., L. A. Ware, A. Barbosa, E. Angov, J. A. Lyon, D. G. Heppner, and D. E. Lanar. 2005. Process development and analysis of liver-stage antigen 1, a preerythrocyte-stage protein-based vaccine for *Plasmodium falciparum. Infect. Immun.* **73:**2109–2115.

223. Hoffman, S. L., E. D. Franke, M. R. Hollingdale, and P. Druilhe. 1996. Attacking the infected hepatocyte, p. 35–75. *In* S. Hoffman

(ed.), *Malaria Vaccine Development: a Multi-Immune Response Approach.* ASM Press, Washington, DC.

224. **Hoffman, S. L., L. M. Goh, T. C. Luke, I. Schneider, T. P. Le, D. L. Doolan, J. Sacci, P. de la Vega, M. Dowler, C. Paul, D. M. Gordon, J. A. Stoute, L. W. Church, M. Sedegah, D. G. Heppner, W. R. Ballou, and T. L. Richie.** 2002. Protection of humans against malaria by immunization with radiation-attenuated *Plasmodium falciparum* sporozoites. *J. Infect. Dis.* **185:**1155–1164.

225. **Hoffman, S. L., M. Sedegah, and R. C. Hedstrom.** 1994. Protection against malaria by immunization with a *Plasmodium yoelii* circumsporozoite protein nucleic acid vaccine. *Vaccine* **12:**1529–1533.

226. **Hogh, B., E. Petersen, I. Crandall, A. Gottschau, and I. W. Sherman.** 1994. Immune responses to band 3 neoantigens on *Plasmodium falciparum*-infected erythrocytes in subjects living in an area of intense malaria transmission are associated with low parasite density and high hematocrit value. *Infect. Immun.* **62:**4362–4366.

227. **Holbrook, T.** 1980. Immunizatioin against exo-erythrocytic stages of malaria parasites, p. 203–230. *In* J. Kreier (ed.), *Malaria*, vol. 3. Academic Press, San Diego, CA.

228. **Holder, A. A., and R. R. Freeman.** 1982. Biosynthesis and processing of a *Plasmodium falciparum* schizont antigen recognized by immune serum and a monoclonal antibody. *J. Exp. Med.* **156:**1528–1538.

229. **Holder, A. A., and R. R. Freeman.** 1981. Immunization against blood-stage rodent malaria using purified parasite antigens. *Nature* **294:**361–364.

230. **Holder, A. A., R. R. Freeman, and S. C. Nicholls.** 1988. Immunization against *Plasmodium falciparum* with recombinant polypeptides produced in *Escherichia coli. Parasite Immunol.* **10:**607–617.

231. **Holder, A. A., J. A. Guevara Patino, C. Uthaipibull, S. E. Syed, I. T. Ling, T. Scott-Finnigan, and M. J. Blackman.** 1999. Merozoite surface protein 1, immune evasion, and vaccines against asexual blood stage malaria. *Parasitologia* **41:**409–414.

232. **Hollingdale, M. R., and U. Krzych.** 2002. Immune responses to liver-stage parasites: implications for vaccine development. *Chem. Immunol.* **80:**97–124.

233. **Hollingdale, M. R., J. L. Leef, M. McCullough, and R. L. Beaudoin.** 1981. In vitro cultivation of the exoerythrocytic stage of *Plasmodium berghei* from sporozoites. *Science* **213:**1021–1022.

234. **Hollingdale, M. R., P. Leland, and A. L. Schwartz.** 1983. In vitro cul-

tivation of the exoerythrocytic stage of *Plasmodium berghei* in a hepatoma cell line. *Am. J. Trop. Med. Hyg.* **32**:682–684.

235. **Holmes, O. W.** 1909–1914. *The Contagiousness of Puerperal Fever.* P. F. Collier, New York, NY.

236. **Hopkin, M.** 2008. Malaria: the big push. *Nature* **451**:1047–1049.

237. **Howard, R. F., and R. T. Reese.** 1990. *Plasmodium falciparum*: heterooligomeric complexes of rhoptry polypeptides. *Exp. Parasitol.* **71**: 330–342.

238. **Howard, R. F., H. A. Stanley, and R. T. Reese.** 1988. Characterization of a high-molecular-weight phosphoprotein synthesized by the human malarial parasite *Plasmodium falciparum*. *Gene* **64**:65–75.

239. **Howard, R. F., A. Varki, and R. T. Reese.** 1984. Merozoite proteins synthesized in *P. falciparum* schizonts. *Prog. Clin. Biol. Res.* **155**:45–61.

240. **Howard, R. J.** 1984. Antigenic variation of bloodstage malaria parasites. *Philos. Trans. R. Soc. Lond. Ser. B* **307**:141–158.

241. **Howard, R. J., J. W. Barnwell, and V. Kao.** 1983. Antigenic variation of *Plasmodium knowlesi* malaria: identification of the variant antigen on infected erythrocytes. *Proc. Natl. Acad. Sci. USA* **80**:4129–4133.

242. **Howard, R. J., J. W. Barnwell, V. Kao, W. A. Daniel, and S. B. Aley.** 1982. Radioiodination of new protein antigens on the surface of *Plasmodium knowlesi* schizont-infected erythrocytes. *Mol. Biochem. Parasitol.* **6**:343–367.

243. **Hsu, S., H. F. Hsu, and L. F. Burmeister.** 1981. *Schistosoma mansoni*: vaccination of mice with highly X-irradiated cercariae. *Exp. Parasitol.* **52**:91–104.

244. **Huff, C. G.** 1964. Cultivation of the exoerythrocytic stages of malarial parasites. *Am. J. Trop. Med. Hyg.* **13**:171–177.

245. **Huff, C. G.** 1969. Exoerythrocytic stages of avian and reptilian malarial parasites. *Exp. Parasitol.* **24**:383–421.

246. **Huff, C. G.** 1947. Life cycle of malarial parasites. *Annu. Rev. Microbiol.* **1**:43–60.

247. **Huff, C. G.** 1968. Recent experimental research on avian malaria. *Adv. Parasitol.* **6**:293–311.

248. **Huff, C. G., and W. Bloom.** 1935. A malarial parasite infecting all blood and blood-forming cells of birds. *J. Infect. Dis.* **57**:315–336.

249. **Huff, C. G., and F. Coulston.** 1944. The development of *Plasmodium gallinaceum* from sporozoite to erythrocytic trophozoite. *J. Infect. Dis.* **75**:231–249.

250. **Huff, C. G., D. F. Marchbank, and T. Shiroshi.** 1958. Changes in in-

fectiousness of malarial gametes. II. Analysis of the possible causative factors. *Exp. Parasitol.* **7**:399–417.

251. **Hui, G. S., and W. A. Siddiqui.** 1987. Serum from Pf195 protected *Aotus* monkeys inhibit *Plasmodium falciparum* growth in vitro. *Exp. Parasitol.* **64**:519–522.

252. **Hui, G. S., L. Q. Tam, S. P. Chang, S. E. Case, C. Hashiro, W. A. Siddiqui, T. Shiba, S. Kusumoto, and S. Kotani.** 1991. Synthetic low-toxicity muramyl dipeptide and monophosphoryl lipid A replace Freund complete adjuvant in inducing growth-inhibitory antibodies to the *Plasmodium falciparum* major merozoite surface protein, gp195. *Infect. Immun.* **59**:1585–1591.

253. **Humphreys, M.** 2003. Whose body? Which disease? Studying malaria while treating neurosyphilis, p. 53–74. *In* J. M. E. Goodman, A. and L. Marks (ed.), *Useful Bodies. Humans in the Service of Medical Science in the Twentieth Century.* Johns Hopkins University Press, Baltimore, MD.

254. **Hviid, L., and A. Salanti.** 2007. VAR2CSA and protective immunity against pregnancy-associated *Plasmodium falciparum* malaria. *Parasitology* **134**:1871–1876.

255. **Ing, R., M. Segura, N. Thawani, M. Tam, and M. M. Stevenson.** 2006. Interaction of mouse dendritic cells and malaria-infected erythrocytes: uptake, maturation, and antigen presentation. *J. Immunol.* **176**:441–450.

256. **Isikoff, M.** 1989. Mismanagement in malaria program. *Washington Post*, Washington, DC.

257. **Jacobs, H.** 1943. Immunization against malaria. *Am. J. Trop. Med.* **23**:597–606.

258. **James, S. P., W. D. Nicol, and P. G. Shute** 1932. A study of induced malignant tertian malaria. *Proc. R. Soc. Med.* **25**:1153–1186.

259. **Janeway, C. A., and R. Medzhitov.** 2002. Innate immune reccognition. *Annu. Rev. Immunol.* **20**:197–216.

260. **Janssen, C. S., R. S. Phillips, C. M. Turner, and M. P. Barrett.** 2004. *Plasmodium* interspersed repeats: the major multigene superfamily of malaria parasites. *Nucleic Acids Res.* **32**:5712–5720.

261. **Jensen, J. B.** 2005. Reflections on the continuous cultivation of *Plasmodium falciparum*. *J. Parasitol.* **91**:487–491.

262. **Jobe, O., J. Lumsden, A. K. Mueller, J. Williams, H. Silva-Rivera, S. H. Kappe, R. J. Schwenk, K. Matuschewski, and U. Krzych.** 2007. Genetically attenuated *Plasmodium berghei* liver stages induce sterile

protracted protection that is mediated by major histocompatibility complex Class I-dependent interferon-gamma-producing CD8$^+$ T cells. *J. Infect. Dis.* **196:**599–607.

263. **John, C. C., A. J. Tande, A. M. Moormann, P. O. Sumba, D. E. Lanar, X. M. Min, and J. W. Kazura.** 2008. Antibodies to pre-erythrocytic *Plasmodium falciparum* antigens and risk of clinical malaria in Kenyan children. *J. Infect. Dis.* **197:**519–526.

264. **Kan, S. C., K. M. Yamaga, K. J. Kramer, S. E. Case, and W. A. Siddiqui.** 1984. *Plasmodium falciparum*: protein antigens identified by analysis of serum samples from vaccinated *Aotus* monkeys. *Infect. Immun.* **43:**276–282.

265. **Kapuscinsk, R.** 2002. *Shadow of the Sun.* Vantage, New York, NY.

266. **Kaslow, D. C.** 1999. Development of a transmission-blocking vaccine: malaria, mosquitoes and medicine, p. 175–182. *In* L. Paoletti and P. McInnes (ed.), *Vaccines: from Concept to Clinic.* CRC Press, Boca Raton, FL.

267. **Kaslow, D. C.** 2002. Transmission blocking vaccines. *Chem. Immunol.* **80:**287–307.

268. **Kaslow, D. C.** 1996. Transmission-blocking vaccines, p. 181–227. *In* S. L. Hoffman (ed.), *Malaria Vaccine Development.* ASM Press, Washington, DC.

269. **Kean, B., K. E. Mott, and A. J. Russell.** 1978. *Tropical Medicine and Parasitology: Classic Investigations,* vol. 1, p. 23–34. Cornell University Press, Ithaca, NY.

270. **Kester, K. E., J. F. Cummings, C. F. Ockenhouse, R. Nielsen, B. T. Hall, D. M. Gordon, R. J. Schwenk, U. Krzych, C. A. Holland, G. Richmond, M. G. Dowler, J. Williams, R. A. Wirtz, N. Tornieporth, L. Vigneron, M. Delchambre, M. A. Demoitie, W. R. Ballou, J. Cohen, and D. G. Heppner, Jr.** 2008. Phase 2a trial of 0, 1, and 3 month and 0, 7, and 28 day immunization schedules of malaria vaccine RTS,S/AS02 in malaria-naive adults at the Walter Reed Army Institute of Research. *Vaccine* **26:**2191–2202.

271. **Kilejian, A.** 1974. A unique histidine-rich polypeptide from the malarial parasite *Plasmodium lophurae. J. Biol. Chem.* **249:**4650–4655.

272. **Kilejian, A.** 1975. Circular mitochondrial DNA from the avian malarial parasite *Plasmodium lophurae. Biochim. Biophys. Acta* **390:**276–284.

273. **Kilejian, A.** 1976. Does a histidine-rich protein from *Plasmodium lophurae* have a function in merozoite invasion? *J. Protozool.* **23:**272–276.

274. **Kilejian, A.** 1978. Histidine-rich protein as a model malaria vaccine. *Science* **201**:922–924.

275. **King, C. L., P. Michon, A. R. Shakri, A. Marcotty, D. Stanisic, P. A. Zimmerman, J. L. Cole-Tobian, I. Mueller, and C. E. Chitnis.** 2008. Naturally acquired Duffy-binding protein-specific binding inhibitory antibodies confer protection from blood-stage *Plasmodium vivax* infection. *Proc. Natl. Acad. Sci. USA* **105**:8363–8368.

276. **Kremer, M., and R. Glennerster.** 2004. *Strong Medicine. Creating Incentives for Phamaceutical Research on Neglected Diseases.* Princeton University Press, Princeton, NJ.

277. **Krotoski, W. A.** 1989. The hypnozoite and malarial relapse. *Prog. Clin. Parasitol.* **1**:1–19.

278. **Krotoski, W. A., D. M. Krotoski, P. C. Garnham, R. S. Bray, R. Killick-Kendrick, C. C. Draper, G. A. Targett, and M. W. Guy.** 1980. Relapses in primate malaria: discovery of two populations of exoerythrocytic stages. Preliminary note. *Br. Med. J.* **280**:153–154.

279. **Kubler-Kielb, J., F. Majadly, Y. Wu, D. L. Narum, C. Guo, L. H. Miller, J. Shiloach, J. B. Robbins, and R. Schneerson.** 2007. Longlasting and transmission-blocking activity of antibodies to *Plasmodium falciparum* elicited in mice by protein conjugates of Pfs25. *Proc. Natl. Acad. Sci. USA* **104**:293–298.

280. **Kumar, K. A., G. Sano, S. Boscardin, R. S. Nussenzweig, M. C. Nussenzweig, F. Zavala, and V. Nussenzweig.** 2006. The circumsporozoite protein is an immunodominant protective antigen in irradiated sporozoites. *Nature* **444**:937–940.

281. **Kumar, N., and R. Carter.** 1985. Biosynthesis of two stage-specific membrane proteins during transformation of *Plasmodium gallinaceum* zygotes into ookinetes. *Mol. Biochem. Parasitol.* **14**:127–139.

282. **Kyes, S. A., S. M. Kraemer, and J. D. Smith.** 2007. Antigenic variation in *Plasmodium falciparum*: gene organization and regulation of the *var* multigene family. *Eukaryot. Cell* **6**:1511–1520.

283. **Labaied, M., A. Harupa, R. F. Dumpit, I. Coppens, S. A. Mikolajczak, and S. H. Kappe.** 2007. *Plasmodium yoelii* sporozoites with simultaneous deletion of P52 and P36 are completely attenuated and confer sterile immunity against infection. *Infect. Immun.* **75**:3758–3768.

284. **Lal, A. A., V. F. de la Cruz, G. H. Campbell, P. M. Procell, W. E. Collins, and T. F. McCutchan.** 1988. Structure of the circumsporozoite gene of *Plasmodium malariae*. *Mol. Biochem. Parasitol.* **30**:291–294.

285. **Landau, I., and P. Gautret.** 1998. Animal mdoels: rodents, p. 401–417. *In* I. W. Sherman (ed.), *Malaria. Parasite Biology, Pathogenesis, and Protection.* ASM Press, Washington, DC.

286. **Langhorne, J., F. M. Ndungu, A. M. Sponaas, and K. Marsh.** 2008. Immunity to malaria: more questions than answers. *Nat. Immunol.* **9:**725–732.

287. **Langreth, S. G., and R. T. Reese.** 1979. Antigenicity of the infected-erythrocyte and merozoite surfaces in Falciparum malaria. *J. Exp. Med.* **150:**1241–1254.

288. **Lasonder, E., Y. Ishihama, J. S. Andersen, A. M. Vermunt, A. Pain, R. W. Sauerwein, W. M. Eling, N. Hall, A. P. Waters, H. G. Stunnenberg, and M. Mann.** 2002. Analysis of the *Plasmodium falciparum* proteome by high-accuracy mass spectrometry. *Nature* **419:**537–542.

289. **Leahy, M.** 2006. Breaking the cycle. *Washington Post,* Washington, DC.

290. **Leech, J. H., J. W. Barnwell, L. H. Miller, and R. J. Howard.** 1984. Identification of a strain-specific malarial antigen exposed on the surface of *Plasmodium falciparum*-infected erythrocytes. *J. Exp. Med.* **159:**1567–1575.

291. **Lengler, C.** 2002. Insecticide treated bed nets and curtains for preventing malaria. *Cochrane Database Syst. Reviews* **2:**1–38.

292. **Le Roch, K. G., and E. A. Winzeler.** 2005. The transcriptome of the malaria parasite, p. 68–84. *In* I. W. Sherman (ed.), *Molecular Approaches to Malaria.* ASM Press, Washington, DC.

293. **Li, S., E. Locke, J. Bruder, D. Clarke, D. L. Doolan, M. J. Havenga, A. V. Hill, P. Liljestrom, T. P. Monath, H. Y. Naim, C. Ockenhouse, D. C. Tang, K. R. Van Kampen, J. F. Viret, F. Zavala, and F. Dubovsky.** 2007. Viral vectors for malaria vaccine development. *Vaccine* **25:**2567–2574.

294. **Lindsay, S. W., and M. E. Gibson.** 1988. Bednets revisited—old idea, new angle. *Parasitol. Today* **4:**270–272.

295. **Lobo, C. A., R. Dhar, and N. Kumar.** 1999. Immunization of mice with DNA-based Pfs25 elicits potent malaria transmission-blocking antibodies. *Infect. Immun.* **67:**1688–1693.

296. **Locher, C. P., M. Paidhungat, R. G. Whalen, and J. Punnonen.** 2005. DNA shuffling and screening strategies for improving vaccine efficacy. *DNA Cell Biol.* **24:**256–263.

297. **Longenecker, B., R. Breitenbach, and J. Farmer.** 1967. Plasma protein changes in normal. thymectomized and bursectomized chickens during a *Plasmodium lophurae* infection. *Exp. Parasitol.* **21:**292–309.

298. **Longenecker, B., R. Breitenbach, and J. Farmer.** 1966. The role of the bursa Fabricius, spleen, and thymus in the control of a *Plasmodium lophurae* infection in the chicken. *J. Immunol.* **97:**594–599.

299. **Longenecker, B., R. Breitenbach, L. Congdon, and J. Farmer.** 1969. Natural and acquired antibodies to *Plasmodium lophurae* in intact and bursaless chickens. 1. Agglutination and passive immunity studies. *J. Parasitol.* **55:**418–425.

300. **Luginbuhl, A., M. Nikolic, H. P. Beck, M. Wahlgren, and H. U. Lutz.** 2007. Complement factor D, albumin, and immunoglobulin G anti-band 3 protein antibodies mimic serum in promoting rosetting of malaria-infected red blood cells. *Infect. Immun.* **75:**1771–1777.

301. **Luke, T. C., and S. L. Hoffman.** 2003. Rationale and plans for developing a non-replicating, metabolically active, radiation-attenuated *Plasmodium falciparum* sporozoite vaccine. *J. Exp. Biol.* **206:**3803–3808.

302. **MacCallum, W. G.** 1898. On the hematozoan infections of birds. *J. Exp. Med.* **3:**117–135.

303. **MacDonald, G.** 1973. *Dynamics of Tropical Disease: a Selection of Papers with a Biographical Introduction and Bibliography.* Oxford University Press, London, United Kingdom.

304. **MacDonald, G.** 1957. *The Epidemiology and Control of Malaria.* Oxford University Press, London, United Kingdom.

305. **Mackaness, G. B.** 1964. The immunological basis of acquired cellular resistance. *J. Exp. Med.* **120:**105–120.

306. **Mackinnon, M. J., P. R. Walker, and J. A. Rowe.** 2002. *Plasmodium chabaudi*: rosetting in a rodent malaria model. *Exp. Parasitol.* **101:** 121–128.

307. **Magistrado, P., A. Salanti, N. G. Tuikue Ndam, S. B. Mwakalinga, M. Resende, M. Dahlback, L. Hviid, J. Lusingu, T. G. Theander, and M. A. Nielsen.** 2008. VAR2CSA expression on the surface of placenta-derived *Plasmodium falciparum*-infected erythrocytes. *J. Infect. Dis.* **198:**1071–1074.

308. **Maier, A. G., M. Rug, M. T. O'Neill, M. Brown, S. Chakravorty, T. Szestak, J. Chesson, Y. Wu, K. Hughes, R. L. Coppel, C. Newbold, J. G. Beeson, A. Craig, B. S. Crabb, and A. F. Cowman.** 2008. Exported proteins required for virulence and rigidity of *Plasmodium falciparum*-infected human erythrocytes. *Cell* **134:**48–61.

309. **Malkin, E. M., D. J. Diemert, J. H. McArthur, J. R. Perreault, A. P. Miles, B. K. Giersing, G. E. Mullen, A. Orcutt, O. Muratova, M. Awkal, H. Zhou, J. Wang, A. Stowers, C. A. Long, S. Mahanty, L. H.**

Miller, A. Saul, and A. P. Durbin. 2005. Phase 1 clinical trial of apical membrane antigen 1: an asexual blood-stage vaccine for *Plasmodium falciparum* malaria. *Infect. Immun.* **73:**3677–3685.

310. **Manson-Bahr, P.** 1962. *Patrick Manson.* Thomas Nelson and Sons, London, United Kingdom.

311. **Manson-Bahr, P.** 1963. The story of malaria: the drama and actors. *Int. Rev. Trop. Med.* **2:**329–390.

312. **Manwell, R., and F. Goldstein.** 1940. Passive immunity in avian malaria. *Am. J. Hyg.* **30**(C):409–421.

313. **Markiewski, M., and J. Lambros.** 2007. The role of complement in inflammatory diseases. From behind the scenes into the spotlight. *Am. J. Pathol.* **171:**715–727.

314. **Marquardt, M.** 1951. *Paul Ehrlich.* Schuman, New York, NY.

315. **Marshall, E.** 1988. Crisis in AID malaria network. *Science* **241:**521–523.

316. **Marshall, E.** 1989. Malaria reseracher indicted. *Science* **245:**1326.

317. **Marshall, E.** 1996. Serious setback for Patarroyo. *Science* **273:**1652.

318. **Maurice, J.** 1993. Controversial vaccine shows promise. *Science* **259:**1689–1690.

319. **Mazumdar, P.** 2003. History of immunology, p. 23–46. *In* W. Paul (ed.), *Fundamentals of Immunology.* Lippincott-Raven, Philadelphia, PA.

320. **McAllister, B.** 1989. Ex-malaria research aide indicted. *Washington Post*, Washington, DC.

321. **McCutchan, T. F., A. A. Lal, V. F. de la Cruz, L. H. Miller, W. L. Maloy, Y. Charoenvit, R. L. Beaudoin, P. Guerry, R. Wistar, Jr., S. L. Hoffman, et al.** 1985. Sequence of the immunodominant epitope for the surface protein on sporozoites of *Plasmodium vivax*. *Science* **230:**1381–1383.

322. **McDonald, V., M. Hannon, L. Tanigoshi, and I. W. Sherman.** 1981. *Plasmodium lophurae*: immunization of Pekin ducklings with different antigen preparations. *Exp. Parasitol.* **51:**195–203.

323. **McGregor, I.** 1993. Towards a vaccine against malaria. *Br. J. Biomed. Sci.* **50:**35–42.

324. **McGregor, I. A., H. M. Gilles, J. H. Walters, A. H. Davies, and F. A. Pearson.** 1956. Effects of heavy and repeated malarial infections on Gambian infants and children: effects of erythrocytic parasitization. *Br. Med. J.* **2:**686–692.

325. **McNeil, J. G.** 2007. The soul of a new vaccine. *New York Times*, New York, NY.

326. **Mikolajczak, S. A., A. S. Aly, and S. H. Kappe.** 2007. Preerythrocytic malaria vaccine development. *Curr. Opin. Infect. Dis.* **20:**461–466.

326a.**Mikolajczak, S. A., H. Silva-Rivera, X. Peng, A. S. Tarun, N. Camargo, V. Jacobs-Lorena, T. M. Daly, L. W. Bergman, P. de la Vega, J. Williams, A. S. Aly, and S. H. Kappe.** 2008. Distinct malaria parasite sporozoites reveal transcriptional changes that cause differential tissue infection competence in the mosquito vector and mammalian host. *Mol. Cell. Biol.* **28:**6196–6207

327. **Miller, L. H.** 1969. Distribution of mature trophozoites and schizonts of *Plasmodium falciparum* in the organs of *Aotus trivirgatus,* the night monkey. *Am. J. Trop. Med. Hyg.* **18:**860–865.

328. **Miller, L. H.** 1977. Hypothesis on the mechanism of erythrocyte invasion by malaria merozoites. *Bull. W. H. O.* **55:**157–162.

329. **Miller, L. H., J. D. Haynes, F. M. McAuliffe, T. Shiroishi, J. R. Durocher, and M. H. McGinniss.** 1977. Evidence for differences in erythrocyte surface receptors for the malarial parasites, *Plasmodium falciparum* and *Plasmodium knowlesi. J. Exp. Med.* **146:**277–281.

330. **Miller, L. H., S. J. Mason, D. F. Clyde, and M. H. McGinniss.** 1976. The resistance factor to *Plasmodium vivax* in blacks. The Duffy-blood-group genotype, FyFy. *N. Engl. J. Med.* **295:**302–304.

331. **Miller, L. H., S. J. Mason, J. A. Dvorak, M. H. McGinniss, and I. K. Rothman.** 1975. Erythrocyte receptors for (*Plasmodium knowlesi*) malaria: Duffy blood group determinants. *Science* **189:**561–563.

332. **Milstein, C.** 1980. Monoclonal antibodies. *Sci. Am.* **243:**66–74.

333. **Milstein, C.** 1999. The hybridoma revolution: an offshoot of basic research. *Bioessays* **21:**966–973.

334. **Mims, C.** 2000. *The War within Us. Everyman's Guide to Infection.* Academic Press, Inc., San Diego, CA.

335. **Min-Oo, G., A. Fortin, G. Pitari, M. Tam, M. M. Stevenson, and P. Gros.** 2007. Complex genetic control of susceptibility to malaria: positional cloning of the Char9 locus. *J. Exp. Med.* **204:**511–524.

336. **Mitchell, G. H., G. A. Butcher, and S. Cohen.** 1975. Merozoite vaccination against *Plasmodium knowlesi* malaria. *Immunology* **29:**397–407.

337. **Mitchell, G. H., W. H. Richards, G. A. Butcher, and S. Cohen.** 1977. Merozoite vaccination of douroucouli monkeys against falciparum malaria. *Lancet* **i:**1335–1338.

338. **Mitchell, G. H., W. H. Richards, A. Voller, F. M. Dietrich, and P. Dukor.** 1979. Nor-MDP, saponin, corynebacteria, and pertussis or-

ganisms as immunological adjuvants in experimental malaria vaccination of macaques. *Bull. W. H. O.* **57**(Suppl. 1):189–197.

339. **Mitchell, G. H., A. W. Thomas, G. Margos, A. R. Dluzewski, and L. H. Bannister.** 2004. Apical membrane antigen 1, a major malaria vaccine candidate, mediates the close attachment of invasive merozoites to host red blood cells. *Infect. Immun.* **72**:154–158.

340. **Miura, K., D. B. Keister, O. V. Muratova, J. Sattabongkot, C. A. Long, and A. Saul.** 2007. Transmission-blocking activity induced by malaria vaccine candidates Pfs25/Pvs25 is a direct and predictable function of antibody titer. *Malaria J.* **6**:107.

341. **Muehlenbachs, A., M. Fried, J. Lachowitzer, T. K. Mutabingwa, and P. E. Duffy.** 2007. Genome-wide expression analysis of placental malaria reveals features of lymphoid neogenesis during chronic infection. *J. Immunol.* **179**:557–565.

342. **Mueller, A. K., M. Labaied, S. H. Kappe, and K. Matuschewski.** 2005. Genetically modified *Plasmodium* parasites as a protective experimental malaria vaccine. *Nature* **433**:164–167.

343. **Muller, O., K. Cham, S. Jaffar, and B. Greenwood.** 1997. The Gambian National Impregnated Bednet Programme: evaluation of the 1994 cost recovery trial. *Social Sci. Med.* **44**:1903–1909.

344. **Mulligan, H. W., P. F. Russell, and B. N. Mohan.** 1941. Active immunization of fowls against *Plasmodium gallinaceum* by injections of killed homologous sporozoites. *J. Malaria Inst. India* **4**:25–34.

345. **Nardin, E., R. Nussenzweig, I. McGregor, and J. Bryan.** 1979. Antisporozoite antibodies. Their frequent occurrence in individuals living in an area of hyperendemic malaria. *Science* **206**:597–599.

346. **Ndungu, F., B. C. Urban, K. Marsh, and J. Langhorne.** 2005. Regulation of immune responses by *Plasmodium*-infected red blood cells. *Parasite Immunol.* **27**:373–378.

347. **Nijhout, M. M., and R. Carter.** 1978. Gamete development in malaria parasites: bicarbonate-dependent stimulation by pH in vitro. *Parasitology* **76**:39–53.

348. **Nuland, S. B.** 2003. *The Doctors' Plague: Germs, Childbed Fever and the Strange Case of Ignac Semmelweiss.* Norton, New York, NY.

349. **Nunes, M. C., and A. Scherf.** 2007. *Plasmodium falciparum* during pregnancy: a puzzling parasite tissue adhesion tropism. *Parasitology* **134**:1863–1869.

350. **Nussenzweig, R., and V. Nussenzweig.** 1984. Development of sporozoite vaccines. *Philos. Trans. R. Soc. Lond. Ser. B* **307**: 117–128.

351. **Nussenzweig, R. S., J. Vanderberg, H. Most, and C. Orton.** 1967. Protective immunity produced by the injection of X-irradiated sporozoites of *Plasmodium berghei*. *Nature* **216**:160–162.

352. **Nussenzweig, R. S., J. Vanderberg, G. L. Spitalny, C. I. Rivera, C. Orton, and H. Most.** 1972. Sporozoite-induced immunity in mammalian malaria. A review. *Am. J. Trop. Med. Hyg.* **21**:722–728.

353. **Nussenzweig, V., and R. Nussenzweig.** 1989. Rationale for the development of an engineered sporozoite malaria vaccine. *Adv. Immunol.* **45**:283–334.

354. **Ockenhouse, C. F., P. F. Sun, D. E. Lanar, B. T. Wellde, B. T. Hall, K. Kester, J. A. Stoute, A. Magill, U. Krzych, L. Farley, R. A. Wirtz, J. C. Sadoff, D. C. Kaslow, S. Kumar, L. W. Church, J. M. Crutcher, B. Wizel, S. Hoffman, A. Lalvani, A. V. Hill, J. A. Tine, K. P. Guito, C. de Taisne, R. Anders, W. R. Ballou, et al.** 1998. Phase I/IIa safety, immunogenicity, and efficacy trial of NYVAC-Pf7, a pox-vectored, multiantigen, multistage vaccine candidate for *Plasmodium falciparum* malaria. *J. Infect. Dis.* **177**:1664–1673.

355. **O'Dea, K. P., P. G. McKean, and K. N. Brown.** 2002. *Plasmodium chabaudi chabaudi* AS: modification of acute infection in CBA/Ca mice as a result of pre-treatment with erythrocyte band 3 in adjuvant. *Exp. Parasitol.* **102**:66–71.

356. **Offit, P. A.** 2007. *Vaccinated: One Man's Quest To Defeat the World's Deadliest Diseases.* Smithsonian Books, Washington, DC.

357. **Okie, S.** 1989. 2 indicted in theft of $130,000 from AID grant. *Washington Post*, Washington, DC.

358. **Okie, S.** 2008. A new attack on malaria. *N. Engl. J. Med.* **358**:2425–2428.

359. **Oransky, I.** 2003. David Francis Clyde. *Lancet* **361**:439.

360. **Orjih, A. U., A. H. Cochrane, and R. S. Nussenzweig.** 1982. Comparative studies on the immunogenicity of infective and attenuated sporozoites of *Plasmodium berghei*. *Trans. R. Soc. Trop. Med. Hyg.* **76**: 57–61.

361. **Outchkourov, N. S., W. Roeffen, A. Kaan, J. Jansen, A. Luty, D. Schuiffel, G. J. van Gemert, M. van de Vegte-Bolmer, R. W. Sauerwein, and H. G. Stunnenberg.** 2008. Correctly folded Pfs48/45 protein of *Plasmodium falciparum* elicits malaria transmission-blocking immunity in mice. *Proc. Natl. Acad. Sci. USA* **105**:4301–4305.

362. **Overstreet, M. G., I. A. Cockburn, and F. Zavala.** 2008. Protective CD8$^+$ T cells against *Plasmodium* liver stages: immunobiology of an 'unnatural' immune response. *Immunol. Rev.* **225**:272–283.

363. **Owuor, B. O., C. O. Odhiambo, W. O. Otieno, C. Adhiambo, D. W. Makawiti, and J. A. Stoute.** 2008. Reduced immune complex binding capacity and increased complement susceptibility of red cells from children with severe malaria associated anemia. *Mol. Med.* **14:** 89–97.

364. **Palmer, K. L., G. S. Hui, W. A. Siddiqui, and E. L. Palmer.** 1982. A large-scale in vitro production system for *Plasmodium falciparum. J. Parasitol.* **68:**1180–1183.

365. **Pandey, K. C., S. Singh, P. Pattnaik, C. R. Pillai, U. Pillai, A. Lynn, S. K. Jain, and C. E. Chitnis.** 2002. Bacterially expressed and refolded receptor binding domain of *Plasmodium falciparum* EBA-175 elicits invasion inhibitory antibodies. *Mol. Biochem. Parasitol.* **123:**23–33.

366. **Pantaleo, A., G. Giribaldi, F. Mannu, P. Arese, and F. Turrini.** 2008. Naturally occurring anti-band 3 antibodies and red blood cell removal under physiological and pathological conditions. *Autoimmun. Rev.* **7:**457–462.

367. **Pasteur, V.-R.** 1966. *Louis Pasteur: A Great Life in Brief.* Knopf, New York, NY.

368. **Patarroyo, M. E., R. Amador, P. Clavijo, A. Moreno, F. Guzman, P. Romero, R. Tascon, A. Franco, L. A. Murillo, G. Ponton, et al.** 1988. A synthetic vaccine protects humans against challenge with asexual blood stages of *Plasmodium falciparum* malaria. *Nature* **332:**158–161.

369. **Patarroyo, M. E., P. Romero, M. L. Torres, P. Clavijo, A. Moreno, A. Martinez, R. Rodriguez, F. Guzman, and E. Cabezas.** 1987. Induction of protective immunity against experimental infection with malaria using synthetic peptides. *Nature* **328:**629–632.

370. **Perlaza, B. L., C. Zapata, A. Z. Valencia, S. Hurtado, G. Quintero, J. P. Sauzet, K. Brahimi, C. Blanc, M. Arevalo-Herrera, P. Druilhe, and S. Herrera.** 2003. Immunogenicity and protective efficacy of *Plasmodium falciparum* liver-stage Ag-3 in *Aotus lemurinus griseimembra* monkeys. *Eur. J. Immunol.* **33:**1321–1327.

371. **Permin, A., and J. Juhl.** 2002. The development of *Plasmodium gallinaceum* infections in chickens following single infections with three different dose levels. *Vet. Parasitol.* **105:**1–10.

372. **Petrovsky, N., and J. C. Auilar.** 2004. Vaccine adjuvants: current state and future trends. *Immunol. Cell Biol.* **82:**488–496.

373. **Pinzon-Charry, A., and M. F. Good.** 2008. Malaria vaccines: the case for a whole-organism approach. *Expert Opin. Biol. Ther.* **8:**441–448.

374. **Playfair, J., and G. Bancroft.** 2004. *Infection Immunity.* Oxford University Press, Oxford, United Kingdom.

375. **Plotkin, S., and S. Plotkin.** 2004. A short history of vaccination, p. 1–30. *In* S. Plotkin and W. A. Orenstein (ed.), *Vaccines,* 4th ed. The W. B. Saunders Co., Philadelphia, PA.

376. **Pombo, D. J., G. Lawrence, C. Hirunpetcharat, C. Rzepczyk, M. Bryden, N. Cloonan, K. Anderson, Y. Mahakunkijcharoen, L. B. Martin, D. Wilson, S. Elliott, S. Elliott, D. P. Eisen, J. B. Weinberg, A. Saul, and M. F. Good.** 2002. Immunity to malaria after administration of ultra-low doses of red cells infected with *Plasmodium falciparum. Lancet* **360:**610–617.

377. **Porter, R.** 1963. Chemical structure of gamma-globulin and antibodies. *Br. Med. J.* **19:**197–201.

378. **Potocnjak, P., N. Yoshida, R. S. Nussenzweig, and V. Nussenzweig.** 1980. Monovalent fragments (Fab) of monoclonal antibodies to a sporozoite surface antigen (Pb44) protect mice against malarial infection. *J. Exp. Med.* **151:**1504–1513.

379. **Pradel, G.** 2007. Proteins of the malaria parasite sexual stages: expression, function and potential for transmission blocking strategies. *Parasitology* **134:**1911–1929.

380. **Prudhomme, J., and I. W. Sherman.** 1999. A high capacity in vitro assay for measuring the cytoadherence of *Plasmodium falciparum* infected erythrocytes. *J. Immunol. Methods* **229:**169–176.

381. **Putnam, F.** 1975. *The Plasma Proteins. Structure, Function and Genetic Control,* vol. 1. Academic Press, San Diego, CA.

382. **Radke, M., and E. H. Sadun.** 1963. Resistance produced in mice by exposure to irradiated *Schistosoma mansoni* cercariae. *Exp. Parasitol.* **13:**134–142.

383. **Ranawaka, M. B., Y. D. Munesinghe, D. M. de Silva, R. Carter, and K. N. Mendis.** 1988. Boosting of transmission-blocking immunity during natural *Plasmodium vivax* infections in humans depends upon frequent reinfection. *Infect. Immun.* **56:**1820–1824.

384. **Rank, R., and W. Weidanz.** 1976. Non-sterilizing immunity in avian malaria: an antibody-independent phenomenon. *Proc. Soc. Exp. Biol. Med.* **151:**257–259.

385. **Rasti, N., F. Namusoke, A. Chene, Q. Chen, T. Staalsoe, M. Q. Klinkert, F. Mirembe, F. Kironde, and M. Wahlgren.** 2006. Nonimmune immunoglobulin binding and multiple adhesion characterize

Plasmodium falciparum-infected erythrocytes of placental origin. *Proc. Natl. Acad. Sci. USA* **103**:13795–13800.

386. **Reese, R. T., and M. R. Motyl.** 1979. Inhibition of the in vitro growth of *Plasmodium falciparum*. I. The effects of immune serum and purified immunoglobulin from owl monkeys. *J. Immunol.* **123**:1894–1899.

387. **Reese, R. T., W. Trager, J. B. Jensen, D. A. Miller, and R. Tantravahi.** 1978. Immunization against malaria with antigen from *Plasmodium falciparum* cultivated in vitro. *Proc. Natl. Acad. Sci. USA* **75**:5665–5668.

388. **Remarque, E., B. Faber, C. Kocken, and A. Thomas.** 2007. Apical membrane antigen 1: a malaria vaccine candidate. *Trends Parasitol.* **24**:74–84.

389. **Remarque, E. J., B. W. Faber, C. H. Kocken, and A. W. Thomas.** 2008. A diversity-covering approach to immunization with *Plasmodium falciparum* apical membrane antigen 1 induces broader allelic recognition and growth inhibition responses in rabbits. *Infect. Immun.* **76**:2660–2670.

390. **Renia, L., A. C. Gruner, M. Mauduit, and G. Snounou.** 2006. Vaccination against malaria with live parasites. *Expert Rev. Vaccines* **5**:473–481.

391. **Reyes-Sandoval, A., J. T. Harty, and S. M. Todryk.** 2007. Viral vector vaccines make memory T cells against malaria. *Immunology* **121:** 158–165.

392. **Reyes-Sandoval, A., S. Sridhar, T. Berthoud, A. C. Moore, J. T. Harty, S. C. Gilbert, G. Gao, H. C. Ertl, J. C. Wilson, and A. V. Hill.** 2008. Single-dose immunogenicity and protective efficacy of simian adenoviral vectors against *Plasmodium berghei*. *Eur. J. Immunol.* **38:** 732–741.

393. **Richards, W. H.** 1977. Vaccination against sporozoite challenge. A review. *Trans. R. Soc. Trop. Med. Hyg.* **71**:279–280.

394. **Richards, W. H. G.** 1966. Active immunization of chicks against *Plasmodium gallinaceum* by inactivated homologous sporozoites and erythrocytic parasites. *Nature* **212**:1492–1494.

395. **Richards, W. H. G., and V. Latter.** 1977. Malarial immunity in *Plasmodium gallinaceum*. *Parasitology* **75**:xxxix.

396. **Richie, T.** 2006. High road, low road? Choices and challenges on the pathway to a malaria vaccine. *Parasitology* **133**(Suppl.):S113–S144.

397. **Richman, S. J., and R. T. Reese.** 1988. Immunologic modeling of a 75-kDa malarial protein with carrier-free synthetic peptides. *Proc. Natl. Acad. Sci. USA* **85**:1662–1666.

398. **Rieckmann, K. H.** 1990. Human immunization with attenuated sporozoites. *Bull. W. H. O.* **68**(Suppl. 1):13–16.

399. **Roestenberg, M., M. McCall, T. E. Mollnes, M. van Deuren, T. Sprong, I. Klasen, C. C. Hermsen, R. W. Sauerwein, and A. van der Ven.** 2007. Complement activation in experimental malaria. *Trans. R. Soc. Trop. Med. Hyg.* **101**:643–649.

400. **Rogers, W. O., W. R. Weiss, A. Kumar, J. C. Aguiar, J. A. Tine, R. Gwadz, J. G. Harre, K. Gowda, D. Rathore, S. Kumar, and S. L. Hoffman.** 2002. Protection of rhesus macaques against lethal *Plasmodium knowlesi* malaria by a heterologous DNA priming and poxvirus boosting immunization regimen. *Infect. Immun.* **70**:4329–4335.

401. **Rogerson, S. J., L. Hviid, P. E. Duffy, R. F. Leke, and D. W. Taylor.** 2007. Malaria in pregnancy: pathogenesis and immunity. *Lancet Infect. Dis.* **7**:105–117.

402. **Rosenthal, P. J.** 2001. *Antimalarial Chemotherapy.* Humana Press, Totowa, NJ.

403. **Ross, R.** 1923. *Memoirs: with a Full Account of the Great Malaria Problem and Its Solution.* Murray, London, United Kingdom.

404. **Ross, R.** 1910. *The Prevention of Malaria.* Dutton, New York, NY.

405. **Rowe, J. A.** 2005. Rosetting, p. 416–426. *In* I. W. Sherman (ed.), *Molecular Approaches to Malaria.* ASM Press, Washington, DC.

406. **Rowe, J. A., I. G. Handel, M. A. Thera, A. M. Deans, K. E. Lyke, A. Kone, D. A. Diallo, A. Raza, O. Kai, K. Marsh, C. V. Plowe, O. K. Doumbo, and J. M. Moulds.** 2007. Blood group O protects against severe *Plasmodium falciparum* malaria through the mechanism of reduced rosetting. *Proc. Natl. Acad. Sci. USA* **104**:17471–17476.

407. **Rudzinska, M. A.** 1969. The fine structure of malaria parasites. *Int. Rev. Cytol.* **25**:161–199.

408. **Russell, P., H. W. Mulligan, and B. N. Mohan.** 1942. Specific agglutinogenic properties. *J. Malaria Inst. India* **4**:311–319.

409. **Russell, P. F.** 1950. International preventive medicine. *Sci. Monthly* **71**:393–400.

410. **Russell, P. F.** 1955. *Man's Mastery of Malaria.* Oxford University Press, London, United Kingdom.

411. **Russell, P. F., H. W. Mulligan, and B. D. Mohan.** 1941. Specific agglutinogenic properties of inactivated sporozoites of *P. gallinaceum. J. Malaria Inst. India* **4**:15–24.

412. **Ryan, J. R., J. A. Stoute, J. Amon, R. F. Dunton, R. Mtalib, J. Koros, B. Owour, S. Luckhart, R. A. Wirtz, J. W. Barnwell, and R. Rosen-**

berg. 2006. Evidence for transmission of *Plasmodium vivax* among a Duffy antigen negative population in Western Kenya. *Am. J. Trop. Med. Hyg.* **75**:575–581.

413. **Sadun, E., L. Norman, and M. M. Brooke.** 1957. The production of antibodies in rabbits infected with irradiated *Trichinella spiralis. Am. J. Trop. Med. Hyg.* **6**:271–279.

414. **Sadun, E. H.** 1966. Introduction to the International Panel Workshop on Biological Research in Malaria. *Mil. Med.* **131**(Suppl.):847–852.

415. **Sadun, E. H., B. T. Wellde, and R. L. Hickman.** 1969. Resistance produced in owl monkeys (*Aotus trivirgatus*) by inoculation with irradiated *Plasmodium falciparum. Mil. Med.* **134**:1165–1175.

416. **Salanti, A., M. Dahlback, L. Turner, M. A. Nielsen, L. Barfod, P. Magistrado, A. T. Jensen, T. Lavstsen, M. F. Ofori, K. Marsh, L. Hviid, and T. G. Theander.** 2004. Evidence for the involvement of VAR2CSA in pregnancy-associated malaria. *J. Exp. Med.* **200**:1197–1203.

417. **Salvatore, D., A. N. Hodder, W. Zeng, L. E. Brown, R. F. Anders, and D. C. Jackson.** 2002. Identification of antigenically active tryptic fragments of apical membrane antigen-1 (AMA1) of *Plasmodium chabaudi* malaria: strategies for assembly of immunologically active peptides. *Vaccine* **20**:3477–3484.

418. **Sauerwein, R.** 2007. Transmission blocking vaccines: the bonus of effective malaria control. *Microbes Infect.* **9**:792–795.

419. **Saul, A.** 2007. Mosquito stage, transmission blocking vaccines for malaria. *Curr. Opin. Infect. Dis.* **20**:476–481.

420. **Saul, A., G. Lawrence, A. Allworth, S. Elliott, K. Anderson, C. Rzepczyk, L. B. Martin, D. Taylor, D. P. Eisen, D. O. Irving, D. Pye, P. E. Crewther, A. N. Hodder, V. J. Murphy, and R. F. Anders.** 2005. A human phase 1 vaccine clinical trial of the *Plasmodium falciparum* malaria vaccine candidate apical membrane antigen 1 in Montanide ISA720 adjuvant. *Vaccine* **23**:3076–3083.

421. **Schenkel, G., E. Cabrera, M. Barr, and P. Silverman,** 1975. A new adjuvant for use in vaccination against malaria. *J. Parasitol.* **61**:549–550.

422. **Scherf, A., J. J. Lopez-Rubio, and L. Riviere.** 2008. Antigenic variation in *Plasmodium falciparum. Annu. Rev. Microbiol.* **62**:445–470.

423. **Schofield, L.** 2007. Rational approaches to developing an anti-disease vaccine against malaria. *Microbes Infect.* **9**:784–791.

424. **Schofield, L., and F. Hackett.** 1993. Signal transduction in host cells by a glycosylphosphatidylinositol toxin of malaria parasites. *J. Exp. Med.* **177**:145–153.

425. **Schofield, L., M. C. Hewitt, K. Evans, M. A. Siomos, and P. H. Seeberger.** 2002. Synthetic GPI as a candidate anti-toxic vaccine in a model of malaria. *Nature* **418:**785–789.

426. **Sedegah, M., R. Hedstrom, P. Hobart, and S. L. Hoffman.** 1994. Protection against malaria by immunization with plasmid DNA encoding circumsporozoite protein. *Proc. Natl. Acad. Sci. USA* **91:**9866–9870.

427. **Sergent, E.** 1963. Latent infection and premunition. Some definitions of microbiology and immunology, p. 39–47. *In* P. C. C. Garnham, A. Pierce, and I. Roit (ed.), *Immunity to Protozoa.* Blackwell, London, United Kingdom.

428. **Sergent, E., and E. Sergent.** 1956. History of the concept of 'relative immunity' or 'premunition' correlated to latent infection. *Indian J. Malariol.* **10:**53–80.

429. **Sergent, E., and E. Sergent.** 1910. Sur l' immunité dans a paludisme des oiseaux. *C. R. Acad. Sci.* **151:**407–409.

430. **Serghides, L., S. N. Patel, K. Ayi, and K. C. Kain.** 2006. Placental chondroitin sulfate A-binding malarial isolates evade innate phagocytic clearance. *J. Infect. Dis.* **194:**133–139.

431. **Seydel, K. B., D. A. Milner, Jr., S. B. Kamiza, M. E. Molyneux, and T. E. Taylor.** 2006. The distribution and intensity of parasite sequestration in comatose Malawian children. *J. Infect. Dis.* **194:**208–215.

432. **Sharma, P., S. J. Richman, and R. T. Reese.** 1992. Antibody responses stimulated in rabbits, guinea-pigs and mice by recombinant and synthetic portions of a 75 kDa malarial merozoite protein. *Vaccine* **10:**540–546.

433. **Sherman, I. W.** 1998. A brief history of malaria and discovery of the parasite's life cycle, p. 3–10. *In* I. W. Sherman (ed.), *Malaria. Parasite Biology, Pathogenesis, and Protection.* ASM Press, Washington, DC.

434. **Sherman, I. W.** 1964. Antigens of *Plasmodium lophurae. J. Protozool.* **11:**409–417.

435. Reference deleted.

436. **Sherman, I. W.** 1966. Heterogeneity of lactic dehydrogenase in avian malaria demonstrated by the use of coenzyme analogs, p. 73. *In* A. Corradetti (ed.), *Proceedings of the First International Congress of Parasitology.* Pergamon, Oxford, United Kingdom.

437. **Sherman, I. W.** 1966. In vitro studies of factors affecting penetration of duck erythrocytes by avian malaria (*Plasmodium lophurae*). *J. Parasitol.* **52:**17–22.

438. **Sherman, I. W.** 1985. Membrane structure and function of malaria parasites and the infected erythrocyte. *Parasitology* **91**:609–645.

439. **Sherman, I. W.** 1961. Molecular heterogeneity of lactic dehydrogenase in avian malaria (*Plasmodium lophurae*). *J. Exp. Med.* **114**:1049–1062.

440. **Sherman, I. W.** 1981. *Plasmodium lophuare*: protective immunogenicity of the histidine-rich protein. *Exp. Parasitol.* **52**:292–295.

441. **Sherman, I. W.** 1960. Serum alterations in avian malaria. *J. Protozool.* **7**:171–176.

442. **Sherman, I. W.** 2006. *The Power of Plagues.* ASM Press, Washington, DC.

443. **Sherman, I. W., S. Eda, and E. Winograd.** 2003. Cytoadherence and sequestration in *Plasmodium falciparum*: defining the ties that bind. *Microbes Infect.* **5**:897–909.

444. **Sherman, I. W., S. Eda, and E. Winograd.** 2004. Erythrocyte aging and malaria. *Cell. Mol. Biol.* **50**:159–169.

445. **Sherman, I. W., and R. W. Hull.** 1960. The pigment (hemozoin) and proteins of the avian malaria parasite *Plasmodium lophurae*. *J. Protozool.* **7**:409–416.

446. **Sherman, I. W., L. Mole, V. McDonald, and L. Tanigoshi.** 1983. Failure to protect ducklings against malaria by vaccination with histidine-rich protein. *Trans. R. Soc. Trop. Med. Hyg.* **77**:87–90.

447. **Sherman, I. W., and E. Winograd.** 1990. Antigens on the *Plasmodium falciparum* infected erythrocyte surface are not parasite derived. *Parasitol. Today* **6**:317–320.

448. **Shortt, H. E., N. H. Fairley, G. Covell, P. G. Shute, and P. C. Garnham.** 1951. The pre-erythrocytic stage of *Plasmodium falciparum*. *Trans. R. Soc. Trop. Med. Hyg.* **44**:405–419.

449. **Shortt, H. E., and P. C. C. Garnham.** 1948. Pre–erythrocytic stage in mamalian malaria parasites. *Nature* **161**:126.

450. **Shott, J. P., S. M. McGrath, M. G. Pau, J. H. Custers, O. Ophorst, M. A. Demoitie, M. C. Dubois, J. Komisar, M. Cobb, K. E. Kester, P. Dubois, J. Cohen, J. Goudsmit, D. G. Heppner, and V. A. Stewart.** 2008. Adenovirus 5 and 35 vectors expressing *Plasmodium falciparum* circumsporozoite surface protein elicit potent antigen-specific cellular IFN-gamma and antibody responses in mice. *Vaccine* **26**:2818–2823.

451. **Shuler, A. V.** 1985. *Malaria. Meeting the Global Challenge.* Oelshlager, Gunn & Hain, Boston, MA.

452. **Shute, P. G.** 1958. Thirty years of malaria therapy. *J. Trop. Med.* **61**:57–61.

453. **Shute, P. G., and G. Covell.** 1967. Malariatherapy's contribution to malaria research. *Protozoology* **11**:33–40.

454. **Siddiqui, W. A.** 1977. An effective immunization of experimental monkeys against a human malaria parasite, *Plasmodium falciparum. Science* **197**:388–389.

455. **Siddiqui, W. A.** 1980. Immunization against asexual blood-inhabiting stages of plasmodia, p. 231–262. *In* J. Kreier (ed.), *Malaria*, vol. 3. Academic Press, San Diego, CA.

456. **Siddiqui, W. A., K. Kramer, and S. M. Richard-Crum.** 1978. In vitro cultivation and partial purification of *Plasmodium falciparum* antigen suitable for vaccination studies in *Aotus* monkeys. *J. Parasitol.* **64**:168–169.

457. **Siddiqui, W. A., L. Q. Tam, K. J. Kramer, G. S. Hui, S. E. Case, K. M. Yamaga, S. P. Chang, E. B. Chan, and S. C. Kan.** 1987. Merozoite surface coat precursor protein completely protects *Aotus* monkeys against *Plasmodium falciparum* malaria. *Proc. Natl. Acad. Sci. USA* **84**: 3014–3018.

458. **Siddiqui, W. A., D. W. Taylor, S. C. Kan, K. Kramer, and S. Richmond-Crum.** 1978. Partial protection of *Plasmodium falciparum*-vaccinated *Aotus* trivirgatus against a challenge of a heterologous strain. *Am. J. Trop. Med. Hyg.* **27**:1277–1278.

459. **Siddiqui, W. A., D. W. Taylor, S. C. Kan, K. Kramer, S. M. Richmond-Crum, S. Kotani, T. Shiba, and S. Kusumoto.** 1978. Vaccination of experimental monkeys against *Plasmodium falciparum*: a possible safe adjuvant. *Science* **201**:1237–1239.

460. **Silverman, P.** 1975. Malaria vaccines. *Br. Med. J.* **2**:335. (Letter.)

461. **Silverstein, A.** 1999. The history of immunology, p. 19–35. *In* W. Paul (ed.), *Fundamentals of Immunology.* Lippincott-Raven, Philadelphia.

462. **Simpson, G. L., R. H. Schenkel, and P. H. Silverman.** 1974. Vaccination of rhesus monkeys against malaria by use of sucrose density gradient fractions of *Plasmodium knowlesi* antigens. *Nature* **247**:304–305.

463. **Singh, B., L. K. Sung, A. Matusop, A. Radhakrishnan, S. S. Shamsul, J. Cox-Singh, A. Thomas, and D. J. Conway.** 2004. A large focus of naturally acquired *Plasmodium knowlesi* infections in human beings. *Lancet* **363**:1017–1024.

464. **Singh, J., A. Ray, and C. Nair.** 1953. Isolation of a new strain of *Plasmodium knowlesi. Nature* **172**:122.

465. **Singh, S., K. Miura, H. Zhou, O. Muratova, B. Keegan, A. Miles, L. B. Martin, A. J. Saul, L. H. Miller, and C. A. Long.** 2006. Immunity to recombinant *Plasmodium falciparum* merozoite surface protein

1 (MSP1): protection in *Aotus nancymai* monkeys strongly correlates with anti-MSP1 antibody titer and in vitro parasite-inhibitory activity. *Infect. Immun.* **74:**4573–4580.

466. **Sinnis, P., and V. Nussenzweig.** 1996. Preventing sporozoite invasion of hepatocytes, p. 15–33. *In* S. Hoffman (ed.), *Malaria Vaccine Development: a Multi-Immune Response Approach.* ASM Press, Washington, DC.

467. **Sinnis, P., and F. Zavala.** 2008. The skin stage of malaria infection: biology and relevance to the malaria vaccine effort. *Future Microbiol.* **3:**275–278.

468. **Slater, L.** 2005. Malarial birds: modeling infectious human disease in animals. *Bull. Hist. Med.* **79:**261–294.

469. **Smith, D., and L. Sanford.** 1985. Laveran's germ. The reception and use of a medical discovery. *Am. J. Trop. Med. Hyg.* **34:**2–20.

470. **Smith, E. A., and J. A. Erickson.** 1980. The Agency of International Development Program for malaria vaccine research and development, p. 331–334. *In* J. Kreier (ed.), *Malaria,* vol. 3. Academic Press, San Diego, CA.

471. **Smith, J. D., C. E. Chitnis, A. G. Craig, D. J. Roberts, D. E. Hudson-Taylor, D. S. Peterson, R. Pinches, C. I. Newbold, and L. H. Miller.** 1995. Switches in expression of *Plasmodium falciparum var* genes correlate with changes in antigenic and cytoadherent phenotypes of infected erythrocytes. *Cell* **82:**101–110.

472. **Smith, K. A.** 2005. Wanted, an anthrax vaccine: dead or alive? *Med. Immunol.* **4:**5.

473. **Smith, T., I. Felger, M. Tanner, and H. P. Beck.** 1999. Premunition in *Plasmodium falciparum* infection: insights from the epidemiology of multiple infections. *Trans. R. Soc. Trop. Med. Hyg.* **93**(Suppl. 1):59–64.

474. **Sridhar, S., A. Reyes Sandoval, S. J. Draper, A. C. Moore, S. C. Gilbert, G. P. Gao, J. M. Wilson, and A. V. Hill.** 2008. Single-dose protection against *Plasmodium berghei* by a simian adenovirus vector using a human cytomegalovirus promoter containing intron A. *J. Virol.* **82:**3822–3833.

475. **Stanley, H. A., S. G. Langreth, and R. T. Reese.** 1989. *Plasmodium falciparum* antigens associated with membrane structures in the host erythrocyte cytoplasm. *Mol. Biochem. Parasitol.* **36:**139–149.

476. **Stanley, H. A., S. G. Langreth, R. T. Reese, and W. Trager.** 1982. *Plasmodium falciparum* merozoites: isolation by density gradient centrifugation using Percoll and antigenic analysis. *J. Parasitol.* **68:**1059–1067.

477. **Stanley, H. A., J. T. Mayes, N. R. Cooper, and R. T. Reese.** 1984. Complement activation by the surface of *Plasmodium falciparum* infected erythrocytes. *Mol. Immunol.* **21**:145–150.

478. **Stanley, H. A., and R. T. Reese.** 1984. In vitro inhibition of intracellular growth of *Plasmodium falciparum* by immune sera. *Am. J. Trop. Med. Hyg.* **33**:12–16.

479. **Stanley, H. A., and R. T. Reese.** 1985. Monkey-derived monoclonal antibodies against *Plasmodium falciparum*. *Proc. Natl. Acad. Sci. USA* **82**:6272–6275.

480. **Stevenson, M. M., J. J. Lyanga, and E. Skamene.** 1982. Murine malaria: genetic control of resistance to *Plasmodium chabaudi*. *Infect. Immun.* **38**:80–88.

481. **Stevenson, M. M., and E. M. Riley.** 2004. Innate immunity to malaria. *Nat. Rev. Immunol.* **4**:169–180.

482. **Stewart, V. A., S. M. McGrath, P. M. Dubois, M. G. Pau, P. Mettens, J. Shott, M. Cobb, J. R. Burge, D. Larson, L. A. Ware, M. A. Demoitie, G. J. Weverling, B. Bayat, J. H. Custers, M. C. Dubois, J. Cohen, J. Goudsmit, and D. G. Heppner, Jr.** 2007. Priming with an adenovirus 35-circumsporozoite protein (CS) vaccine followed by RTS,S/AS01B boosting significantly improves immunogenicity to *Plasmodium falciparum* CS compared to that with either malaria vaccine alone. *Infect. Immun.* **75**:2283–2290.

483. **Stoute, J. A., J. Gombe, M. R. Withers, J. Siangla, D. McKinney, M. Onyango, J. F. Cummings, J. Milman, K. Tucker, L. Soisson, V. A. Stewart, J. A. Lyon, E. Angov, A. Leach, J. Cohen, K. E. Kester, C. F. Ockenhouse, C. A. Holland, C. L. Diggs, J. Wittes, and D. G. Heppner, Jr.** 2007. Phase 1 randomized double-blind safety and immunogenicity trial of *Plasmodium falciparum* malaria merozoite surface protein FMP1 vaccine, adjuvanted with AS02A, in adults in western Kenya. *Vaccine* **25**:176–184.

484. **Stoute, J. A., M. Slaoui, D. G. Heppner, P. Momin, K. E. Kester, P. Desmons, B. T. Wellde, N. Garcon, U. Krzych, and M. Marchand for the RTS,S Malaria Vaccine Evaluation Group.** 1997. A preliminary evaluation of a recombinant circumsporozoite protein vaccine against *Plasmodium falciparum* malaria. *N. Engl. J. Med.* **336**:86–91.

485. **Stowers, A., and R. Carter.** 2001. Current developments in malaria transmission-blocking vaccines. *Expert Opin. Biol. Ther.* **1**:619–628.

486. **Stowers, A. W., V. Cioce, R. L. Shimp, M. Lawson, G. Hui, O. Muratova, D. C. Kaslow, R. Robinson, C. A. Long, and L. H. Miller.**

2001. Efficacy of two alternate vaccines based on *Plasmodium falciparum* merozoite surface protein 1 in an *Aotus* challenge trial. *Infect. Immun.* **69**:1536–1546.

487. **Struik, S. S., and E. M. Riley.** 2004. Does malaria suffer from lack of memory? *Immunol. Rev.* **201**:268–290.

488. **Su, X. Z., V. M. Heatwole, S. P. Wertheimer, F. Guinet, J. A. Herrfeldt, D. S. Peterson, J. A. Ravetch, and T. E. Wellems.** 1995. The large diverse gene family *var* encodes proteins involved in cytoadherence and antigenic variation of *Plasmodium falciparum*-infected erythrocytes. *Cell* **82**:89–100.

489. **Su, Z., M. Segura, K. Morgan, J. C. Loredo-Osti, and M. M. Stevenson.** 2005. Impairment of protective immunity to blood-stage malaria by concurrent nematode infection. *Infect. Immun.* **73**:3531–3539.

490. **Su, Z., M. Segura, and M. M. Stevenson.** 2006. Reduced protective efficacy of a blood-stage malaria vaccine by concurrent nematode infection. *Infect. Immun.* **74**:2138–2144.

491. **Taliaferro, W. H.** 1929. *The Immunology of Parasitic Infections.* The Century Company, New York, NY.

492. **Taliaferro, W. H.** 1948. The inhibition of reproduction of parasites by immune factors. *Bacteriol. Rev.* **12**:1–17.

493. **Taliaferro, W. H.** 1968. The lure of the unknown. *Annu. Rev. Microbiol.* **22**:1–15.

494. **Taliaferro, W. H., and L. G. Taliaferro.** 1945. Immunological relationships of *Plasmodium gallinaceum* and *Plasmodium lophurae*. *J. Infect. Dis.* **77**:224–248.

495. **Taliaferro, W. H., and L. G. Taliaferro.** 1950. Reproduction-inhibiting and parasiticidal effects on *Plasmodium gallinaceum* and *Plasmodium lophurae* during initial infection and homologous superinfection in chickens. *J. Infect. Dis.* **86**:275–294.

496. **Taliaferro, W. H., and L. G. Taliaferro.** 1934. The transmission of *Plasmodium falciparum* to the howler monkey, *Alouatta* sp. I. General nature of the infections and morphology of the parasites. *Am. J. Hyg.* **19**:318–334.

497. **Talmage, D. W.** 1983. William Hay Taliaferro. *Biogr. Mem.* **54**:375–407.

498. **Targett, G. A., and J. D. Fulton.** 1965. Immunization of rhesus monkeys against *Plasmodium knowlesi* malaria. *Exp. Parasitol.* **17**:180–193.

499. **Targett, G. A., and A. Voller.** 1965. Studies on antibody levels during vaccination of rhesus monkeys against *Plasmodium knowlesi*. *Br. Med. J.* **2**:1104–1106.

500. **Tarun, A. S., R. F. Dumpit, N. Camargo, M. Labaied, P. Liu, A. Takagi, R. Wang, and S. H. Kappe.** 2007. Protracted sterile protection with *Plasmodium yoelii* pre-erythrocytic genetically attenuated parasite malaria vaccines is independent of significant liver-stage persistence and is mediated by CD8$^+$ T cells. *J. Infect. Dis.* **196:**608–616.

501. **Taylor, D. W., and W. A. Siddiqui.** 1979. Susceptibility of owl monkeys to *Plasmodium falciparum* infection in relation to location of origin, phenotype, and karyotype. *J. Parasitol.* **65:**267–271.

502. **Taylor-Robinson, A. W.** 2003. Immunity to liver stage malaria: considerations for vaccine design. *Immunol. Res.* **27:**53–70.

503. **Thera, M. A., O. K. Doumbo, D. Coulibaly, D. A. Diallo, A. K. Kone, A. B. Guindo, K. Traore, A. Dicko, I. Sagara, M. S. Sissoko, M. Baby, M. Sissoko, I. Diarra, A. Niangaly, A. Dolo, M. Daou, S. I. Diawara, D. G. Heppner, V. A. Stewart, E. Angov, E. S. Bergmann-Leitner, D. E. Lanar, S. Dutta, L. Soisson, C. L. Diggs, A. Leach, A. Owusu, M. C. Dubois, J. Cohen, J. N. Nixon, A. Gregson, S. L. Takala, K. E. Lyke, and C. V. Plowe.** 2008. Safety and immunogenicity of an AMA-1 malaria vaccine in Malian adults: results of a phase 1 randomized controlled trial. *PLoS ONE* **3:**e1465.

504. **Thomas, A. W., J. A. Deans, G. H. Mitchell, T. Alderson, and S. Cohen.** 1984. The Fab fragments of monoclonal IgG to a merozoite surface antigen inhibit *Plasmodium knowlesi* invasion of erythrocytes. *Mol. Biochem. Parasitol.* **13:**187–199.

505. **Thompson, F. M., D. W. Porter, S. L. Okitsu, N. Westerfeld, D. Vogel, S. Todryk, I. Poulton, S. Correa, C. Hutchings, T. Berthoud, S. Dunachie, L. Andrews, J. L. Williams, R. Sinden, S. C. Gilbert, G. Pluschke, R. Zurbriggen, and A. V. Hill.** 2008. Evidence of blood stage efficacy with a virosomal malaria vaccine in a phase IIa clinical trial. *PLoS ONE* **3:**e1493.

506. **Thomson, K., J. Freund, H. Somer, and A. Walter.** 1947. Immunization of ducks against malaria by means of killed parasites with or without adjuvants. *Am. J. Trop. Med.* **27:**79–105.

507. **Thucydides.** 1998. *The Peloponnesian War.* Translated by Walter Blanco. Norton, New York, NY.

508. **Tournamille, C., Y. Colin, J. P. Cartron, and C. Le Van Kim.** 1995. Disruption of a GATA motif in the Duffy gene promoter abolishes erythroid gene expression in Duffy-negative individuals. *Nat. Genet.* **10:**224–228.

509. **Tournamille, C., A. Filipe, C. Badaut, M. M. Riottot, S. Longacre,**

J. P. Cartron, C. Le Van Kim, and Y. Colin. 2005. Fine mapping of the Duffy antigen binding site for the *Plasmodium vivax* Duffy-binding protein. *Mol. Biochem. Parasitol.* **144**:100–103.

510. Trager, W. 1969. Comments on cultivation. *Mil. Med.* **131**(Suppl. 1):1034–1035.

511. Trager, W. 1995. Malaria vaccine. *Science* **267**:1577.

512. Trager, W. 1987. The cultivation of *Plasmodium falciparum*: applications in basic and applied research. *Ann. Trop. Med. Parasitol.* **81**:511–529.

513. Trager, W., and J. Jensen. 1997. Continuous culture of *Plasmodium falciparum*: its impact on malaria research. *Int. J. Parasitol.* **27**:17.

514. Trager, W., and J. Jensen. 1980. Cultivation of erythrocytic and exoerythrocytic stages of plasmodia, p. 271–319. *In* J. P. Kreier (ed.), *Malaria*, vol. 2. Academic Press, New York, NY.

515. Trager, W., H. N. Lanners, H. A. Stanley, and S. G. Langreth. 1983. Immunization of owl monkeys to *Plasmodium falciparum* with merozoites from cultures of a knobless clone. *Parasite Immunol.* **5**:225–236.

516. Trager, W., M. A. Rudzinska, and P. C. Bradbury. 1966. The fine structure of *Plasmodium falciparum* and its host erythrocytes in natural malarial infections in man. *Bull. W. H. O.* **35**:883–885.

517. Truong, H. 1956. *An Analysis of Natural and Acquired Immunity to* Plasmodium lophurae *in chicks.* Northwestern, Evanston, IL.

518. Tsuji, M., and F. Zavala. 2003. T cells as mediators of protective immunity against liver stages of *Plasmodium*. *Trends Parasitol.* **19**:88–93.

519. Tuteja, R. 2002. DNA vaccine against malaria: a long way to go. *Crit. Rev. Biochem. Mol. Biol.* **37**:29–54.

520. Urban, B. C., R. Ing, and M. M. Stevenson. 2005. Early interactions between blood-stage *Plasmodium* parasites and the immune system. *Curr. Top. Microbiol. Immunol.* **297**:25–70.

521. Urban, B. C., and S. Todryk. 2006. Malaria pigment paralyzes dendritic cells. *J. Biol.* **5**:4.

522. Valero, M. V., L. R. Amador, C. Galindo, J. Figueroa, M. S. Bello, L. A. Murillo, A. L. Mora, G. Patarroyo, C. L. Rocha, M. Rojas, et al. 1993. Vaccination with SPf66, a chemically synthesised vaccine, against *Plasmodium falciparum* malaria in Colombia. *Lancet* **341**:705–710.

523. Vanderberg, J., A. K. Mueller, K. Heiss, K. Goetz, K. Matuschewski, M. Deckert, and D. Schluter. 2007. Assessment of antibody protection against malaria sporozoites must be done by mosquito injection of sporozoites. *Am. J. Pathol.* **171**:1405–1406; author reply, 1406.

524. **Vanderberg, J., R. Nussenzweig, and H. Most.** 1969. Protective immunity produced by the injection of X-irradiated sporozoites of *Plasmodium berghei*. V. In vitro effects of immune serum on sporozoites. *Mil. Med.* **134**:1183–1190.

525. **Vanderberg, J., R. Nussenzweig, and H. Most.** 1970. Protective immunity produced by the bite of x-irradiated mosquitoes infected with *Plasmodium berghei*. *J. Parasitol.* **56**:350–351.

526. **Vanderberg, J. P., and U. Frevert.** 2004. Intravital microscopy demonstrating antibody-mediated immobilization of *Plasmodium berghei* sporozoites injected into skin by mosquitoes. *Int. J. Parasitol.* **34**:991–996.

527. **Vanderberg, J. P., R. S. Nussenzweig, H. Most, and C. G. Orton.** 1968. Protective immunity produced by the injection of X-irradiated sporozoites of *Plasmodium berghei*. II. Effects of radiation on sporozoites. *J. Parasitol.* **54**:1175–1180.

528. **Venter, C.** 2007. *A Life Decoded.* Viking, New York, NY.

529. **Verhave, J. P.** 1975. *Immunization with Sporozoites. An Experimental Study of* Plasmodium berghei *Malaria.* Catholic University, Nijmegen, The Netherlands.

530. **Vermeulen, A. N., T. Ponnudurai, P. J. Beckers, J. P. Verhave, M. A. Smits, and J. H. Meuwissen.** 1985. Sequential expression of antigens on sexual stages of *Plasmodium falciparum* accessible to transmission-blocking antibodies in the mosquito. *J. Exp. Med.* **162**:1460–1476.

531. **Vermeulen, A. N., J. van Deursen, R. H. Brakenhoff, T. H. Lensen, T. Ponnudurai, and J. H. Meuwissen.** 1986. Characterization of *Plasmodium falciparum* sexual stage antigens and their biosynthesis in synchronised gametocyte cultures. *Mol. Biochem. Parasitol.* **20**:155–163.

532. **Viebig, N. K., E. Levin, S. Dechavanne, S. J. Rogerson, J. Gysin, J. D. Smith, A. Scherf, and B. Gamain.** 2007. Disruption of *var2csa* gene impairs placental malaria associated adhesion phenotype. *PLoS ONE* **2**:e910.

533. **Voller, A., and W. H. Richards.** 1968. An attempt to vaccinate owl monkeys (*Aotus trivirgatus*) against falciparum malaria. *Lancet* **ii**:1172–1174.

534. **Voller, A., and R. N. Rossan.** 1969. Immunological studies on simian malaria. 3. Immunity to challenge and antigenic variation in *P. knowlesi*. *Trans. R. Soc. Trop. Med. Hyg.* **63**:507–523.

535. **Voller, A., and R. N. Rossan.** 1969. Immunological studies with simian malarias. I. Antigenic variants of *Plasmodium cynomolgi bastianellii*. *Trans. R. Soc. Trop. Med. Hyg.* **63**:46–56.

536. **Voller, A., and R. N. Rossan.** 1969. Immunological studies with simian malarias. II. Heterologous immunity in the "cynomolgi" group. *Trans. R. Soc. Trop. Med. Hyg.* **63:**57–63.

537. **Wagner-Jauregg, J.** 1927. The treatment of dementia paralytica by malaria inoculation. Nobel Prize lecture. The Nobel Foundation, Stockholm, Sweden.

538. **Wahlgren, M.** 1986. Antigens and antibodies involved in humoral immunity to *Plasmodium falciparum.* Ph.D. thesis Karolinska Institute, Stockholm, Sweden.

539. **Wallis, C.** 1984. Combatting an ancient enemy. *Time,* New York, NY.

540. **Walport, M. J.** 2001. Complement. *N. Eng. J. Med.* **344:**1058–1066.

541. **Wang, J. Y., and M. H. Roehrl.** 2005. Anthrax vaccine design: strategies to achieve comprehensive protection against spore, bacillus, and toxin. *Med. Immunol.* **4:**4.

542. **Wang, R., D. L. Doolan, T. P. Le, R. C. Hedstrom, K. M. Coonan, Y. Charoenvit, T. R. Jones, P. Hobart, M. Margalith, J. Ng, W. R. Weiss, M. Sedegah, C. de Taisne, J. A. Norman, and S. L. Hoffman.** 1998. Induction of antigen-specific cytotoxic T lymphocytes in humans by a malaria DNA vaccine. *Science* **282:**476–480.

543. **Wang, R., T. L. Richie, M. F. Baraceros, N. Rahardjo, T. Gay, J. G. Banania, Y. Charoenvit, J. E. Epstein, T. Luke, D. A. Freilich, J. Norman, and S. L. Hoffman.** 2005. Boosting of DNA vaccine-elicited gamma interferon responses in humans by exposure to malaria parasites. *Infect. Immun.* **73:**2863–2872.

544. **Waters, A. P.** 2008. Genome-informed contributions to malaria therapies: feeding somewhere down the (pipe)line. *Cell Host Microbe* **3:**280–283.

545. **Waters, A. P., A. W. Thomas, J. A. Deans, G. H. Mitchell, D. E. Hudson, L. H. Miller, T. F. McCutchan, and S. Cohen.** 1990. A merozoite receptor protein from *Plasmodium knowlesi* is highly conserved and distributed throughout *Plasmodium. J. Biol. Chem.* **265:** 17974–17979.

546. **Weinstein, P. P.** 1974. In memoriam. Elvio Herbert Sadun. *J. Parasitol.* **40:**897–899.

547. **Weiss, W. R., A. Kumar, G. Jiang, J. Williams, A. Bostick, S. Conteh, D. Fryauff, J. Aguiar, M. Singh, D. T. O'Hagan, J. B. Ulmer, and T. L. Richie.** 2007. Protection of rhesus monkeys by a DNA prime/poxvirus boost malaria vaccine depends on optimal DNA priming and inclusion of blood stage antigens. *PLoS ONE* **2:**e1063.

548. **Wellde, B. T., C. L. Diggs, and S. Anderson.** 1979. Immunization of *Aotus trivirgatus* against *Plasmodium falciparum* with irradiated blood forms. *Bull. W. H. O.* **57**(Suppl. 1):153–157.

549. **Weller, T.** 1954. The cultivation of the poliomyelitis viruses in tissue culture. Nobel Prize Lecture. The Nobel Foundation, Stockholm, Sweden.

550. **White, K., U. Krzych, D. M. Gordon, T. G. Porter, R. L. Richards, C. R. Alving, C. D. Deal, M. Hollingdale, C. Silverman, D. R. Sylvester, et al.** 1993. Induction of cytolytic and antibody responses using *Plasmodium falciparum* repeatless circumsporozoite protein encapsulated in liposomes. *Vaccine* **11**:1341–1346.

551. **White, N. J.** 1998. Malaria pathophysiology, p. 371–385. *In* I. W. Sherman (ed.), *Malaria. Parasite Biology, Pathogenesis, and Protection.* ASM Press, Washington, DC.

552. **White, N. J.** 2007. *Plasmodium knowlesi*. The fifth human malaria parasite. *Clin. Infect. Dis.* **46**:172–173.

553. **White, N. J.** 1994. Tough test for malaria vaccine. *Lancet* **344**:1548–1549.

554. **Williams, G.** 1969. *The Plague Killers.* Scribners, New York, NY.

555. **Williams, R.** 2005. Avian malaria: clinical and chemical pathology of *Plasmodium gallinaceum* in the domestic fowl *Gallus gallus. Avian Pathol.* **34**:29–47.

556. **Williamson, K. C.** 2003. Pfs 230: from malaria transmission-blocking vaccine candidate toward function. *Parasite Immunol.* **25**:351–359.

557. **Winograd, E., S. Eda, and I. W. Sherman.** 2004. Chemical modifications of band 3 protein affect the adhesion of *Plasmodium falciparum*-infected erythrocytes to CD36. *Mol. Biochem. Parasitol.* **136**:243–248.

558. **Winograd, E., J. G. Prudhomme, and I. W. Sherman.** 2005. Band 3 clustering promotes the exposure of neoantigens in *Plasmodium falciparum*-infected erythrocytes. *Mol. Biochem. Parasitol.* **142**:98–105.

559. **Winograd, E., and I. W. Sherman.** 1989. Characterization of a modified red cell membrane protein expressed on erythrocytes infected with the human malaria parasite *Plasmodium falciparum*: possible role as a cytoadherent mediating protein. *J. Cell Biol.* **108**:23–30.

560. **Wipasa, J., and E. M. Riley.** 2007. The immunological challenges of malaria vaccine development. *Expert Opin. Biol. Ther.* **7**:1841–1852.

561. **Wu, Y., R. D. Ellis, D. Shaffer, E. Fontes, E. M. Malkin, S. Mahanty, M. P. Fay, D. Narum, K. Rausch, A. P. Miles, J. Aebig, A. Orcutt, O. Muratova, G. Song, L. Lambert, D. Zhu, K. Miura, C. Long, A. Saul, L. H. Miller, and A. P. Durbin.** 2008. Phase 1 trial of malaria trans-

mission blocking vaccine candidates Pfs25 and Pvs25 formulated with montanide ISA 51. *PLoS ONE* **3**:e2636.

562. **Wykes, M., and M. F. Good.** 2007. A case for whole-parasite malaria vaccines. *Int. J. Parasitol.* **37**:705–712.

563. **Wykes, M., C. Keighly, A. Pinzon-Charry, and M. F. Good.** 2007. Dendritic cell biology during malaria. *Cell. Microbiol.* **9**:300–305.

564. **Yan, M., M. H. Roehrl, E. Basar, and J. Y. Wang.** 2008. Selection and evaluation of the immunogenicity of protective antigen mutants as anthrax vaccine candidates. *Vaccine* **26**:947–955.

565. **Yazdani, S. S., A. R. Shakri, P. Pattnaik, M. M. Rizvi, and C. E. Chitnis.** 2006. Improvement in yield and purity of a recombinant malaria vaccine candidate based on the receptor-binding domain of *Plasmodium vivax* Duffy binding protein by codon optimization. *Biotechnol. Lett.* **28**:1109–1114.

566. **Yoshida, N., R. S. Nussenzweig, P. Potocnjak, V. Nussenzweig, and M. Aikawa.** 1980. Hybridoma produces protective antibodies directed against the sporozoite stage of malaria parasite. *Science* **207**: 71–73.

567. **Young, J. A., Q. L. Fivelman, P. L. Blair, P. de la Vega, K. G. Le Roch, Y. Zhou, D. J. Carucci, D. A. Baker, and E. A. Winzeler.** 2005. The *Plasmodium falciparum* sexual development transcriptome: a microarray analysis using ontology-based pattern identification. *Mol. Biochem. Parasitol.* **143**:67–79.

568. **Young, M. D., D. C. Baerg, and R. N. Rossan.** 1975. Experimental monkey hosts for human plasmodia. *Exp. Parasitol.* **38**:136–152.

569. **Young, M. D., D. C. Baerg, and R. N. Rossan.** 1976. Studies with induced malarias in *Aotus* monkeys. *Lab. Ani. Sci.* **26**:1131–1137.

570. **Young, M. D., J. Porter, and C. Johnson.** 1966. *Plasmodium vivax* transmitted from monkey to man. *Science* **153**:1006–1007.

571. **Young, M. D., and R. N. Rossan.** 1969. *Plasmodium falciparum* induced in the squirrel monkey *Saimiri sciureus. Trans. R. Soc. Trop. Med. Hyg.* **63**:686–687.

572. **Yuen, P. C.** 1993. Recovering funds. *Science* **262**:164.

573. **Zou, L., A. P. Miles, J. Wang, and A. W. Stowers.** 2003. Expression of malaria transmission-blocking vaccine antigen Pfs25 in *Pichia pastoris* for use in human clinical trials. *Vaccine* **21**:1650–1657.

Index